INTRODUCTION TO THERMODYNAMICS AND KINETIC THEORY OF MATTER

INTRODUCTION TO THERMODYNAMICS AND KINETIC THEORY OF MATTER
Second Edition

A. I. Burshstein
Weizmann Institute of Science

Wiley-VCH Verlag GmbH & Co. KGaA

All books published by Wiley-VCH are carefully produced.
Nevertheless, authors, editors, and publisher do not warrant the information
contained in these books, including this book, to be free of errors.
Readers are advised to keep in mind that statements, data, illustrations,
procedural details or other items may inadvertently be inaccurate.

Library of Congress Card No.:
Applied for

British Library Cataloging-in-Publication Data:
A catalogue record for this book is available from the British Library

Bibliographic information published by
Die Deutsche Bibliothek
Die Deutsche Bibliothek lists this publication in the Deutsche Nationalbibliografie;
detailed bibliographic data is available in the Internet at <http://dnb.ddb.de>.

First Edition
© 1996 by John Wiley & Sons, Inc.

Second Edition
© 2005 WILEY-VCH Verlag GmbH & Co. KGaA, Weinheim

All rights reserved (including those of translation into other languages).
No part of this book may be reproduced in any form – nor transmitted or translated
into machine language without written permission from the publishers.
Registered names, trademarks, etc. used in this book, even when not specifically
marked as such, are not to be considered unprotected by law.

Printed in the Federal Republic of Germany
Printed on acid-free paper

Printing   Strauss GmbH, Mörlenbach
Bookbinding   Litges & Dopf Buchbinderei GmbH, Heppenheim

ISBN-13: 978-3-527-40598-5
ISBN-10: 3-527-40598-4

Dedicated to the bright shining memory of
"Papa" Ioffe
(Academician Abram Ioffe, my first patron)

CONTENTS

PREFACE   xi

1 Gases   1

    1.1   Statistics of Molecules / 1
           Pressure. Statistical distribution. The statistical average.

    1.2   Distribution Function / 5

    1.3   Ideal Gas Equation of State / 9

    1.4   Maxwell Distribution / 12
           Summation of probabilities. Statistical weight. Mean values. Stern experiment. Structure of the flux.

    1.5   Barometric Formula / 19

    1.6   The Boltzmann Distribution / 22

    1.7   Canonical Distribution / 25

    1.8   Dielectric Properties of Gases / 27
           Generalized coordinates and momenta. Probability distribution. Electric polarization.

    1.9   Real Gas / 32
           Intermolecular interaction. Replusion. Attraction. Equation of state. Compressibility factor.

    1.10   Basic Ideas of Statistical Thermodynamics / 47
           The Gibbs ensemble. Microcanonical distribution. Gibbs' canonical distribution. Entropy. Fluctuations. Free energy. Computational scheme. Ideal gas. Identity of microparticles. Real gas. Free volume theory.

1.11 Heat Capacity of Gases / 66
Translational motion. Rotational motion. Vibrational motion. Equipartition law. Temperature anomalies.

1.12 Harmonic Oscillator Quantization / 74
Freezing out of vibrations. Energy quantization. High temperature limit. Quantum cells. Bohr's postulate.

1.13 Freezing Out of Heat Motions / 79

1.14 Gas Paramagnetism / 85
Spin $\frac{1}{2}$. General case.

1.15 Collisions / 91
Relaxation. Velocity-dependent collision rate. The mean collision rate. Experimental verification. Free path. Poissonian statistics. Velocity relaxation. Strong and weak collisions.

1.16 Transfer phenomena / 105
Local equilibrium. Heat conduction. Viscosity. Diffusion. Refinement of calculations. Heat transfer. Ultrararefied gas. Current in gases.

2 Radiation 124

2.1 Phenomenology / 124

2.2 Kirchhoff's Theorem / 126

2.3 Ultraviolet Catastrophe / 130
Heat capacity of radiation. Field quantization. Photon model.

2.4 Photon Model / 135
The basic features. Comparison of the models. Photon weight.

2.5 Photon Gas Properties / 139
Pressure. Equation of state.

2.6 Einstein Coefficients / 142
Absorption and emission. Excitation and deactivation. Rates of induced transitions. Einstein's relations. Partial rates.

2.7 Stationary Light Transformation / 147

3 Crystals 150

3.1 Phenomenology / 150
Types of crystals. Strength. Anisotropy.

3.2 Crystal Lattice Motion / 154
Interaction forces. Acoustic vibrations. Density of states.
Optical vibrations. Molecular vibrations.

3.3 Phonon Gas / 165
Model. Debye approximation. The lattice heat capacity.
Correlation of vibrations. Heat conduction of the lattice.

3.4 Equation of State / 174
Balance of forces. Internal pressure. Motional pressure.
The Mie–Grüneisen equation. Heat expansion. Knot of isotherms.

3.5 Real Crystals / 189
Point defects. Diffusion. Ion conductivity.

4 Liquids — 198

4.1 Phenomenology / 198

4.2 Equation of State / 200
Internal pressure. Motional pressure. Balance of forces.
Boiling. Melting. Orthometric line. Decomposition of the
equation into components.

4.3 Surface Phenomena / 232
Monomolecular surface. Smooth interface.

4.4 Transfer Phenomena / 242
Heat conduction. Viscosity. The free volume theory.

4.5 Brownian Motion / 251
Nonstationary diffusion. The Langevin equation.
Perrin's experiments.

5 Thermodynamics — 257

5.1 First Law / 257
Basic concepts. Caloric. Mechanical equivalent of heat. Internal
energy. Reversible and irreversible processes. Ideal gas

5.2 Second Law / 268
Carnot cycle. Entropy. Principles of thermodynamics.
Open processes. Natural variables. Gas of van der Waals.

5.3 Method of Cycles / 285
The Clapeyron–Clausius relation. Thermoelectric cycle.
Heat pumps.

5.4 Cooling of Gases / 293
The Joule–Thomson effect. Enthalpy. The sign of the effect. Inversion curve. Boyle temperature and inversion temperature.

5.5 Free Energy / 302
Gibbs–Helmholtz relation. Thermal radiation. Two-phase system. Surface tension.

5.6 Thermodynamic Potential / 312
Chemical potential. Phase equilibrium. Quasithermodynamics.

5.7 Nernst's Principle / 323

5.8 Mixtures / 326
Gibbs' paradox.

Bibliography 331

Index 333

PREFACE

This book originated in a few courses of lectures for undergraduate physicists held at Novosibirsk University and for graduate chemists there and at the Weizmann Institute of Science (Israel). As an introduction intended for unprepared readers, the book does not require any background except in general mathematics and classical mechanics. However, an educational experiment carried out at Novosibirsk University required a fundamental study of the subject at the very beginning. The course was assumed to be sufficient for third-year students to start their research work. To attain this goal one had to resort to a more detailed and picturesque presentation of the material and to inductive logic which leads the reader from observation and fact to the interpretation, and from particular conclusions to general ones. This presentation has been highly influenced by the feedback of the student audience and by the teaching experience of the author. The English edition of the book includes for the first time a chapter on thermodynamics and the original results of the author. Hopefully it will receive as much attention as its Russian predecessors.

Corpuscular models of matter had already been suggested in ancient times. Two thousand years later this outlook gave birth to a wonderful hybrid of a stochastic and a mechanical description of large ensembles of particles, *statistical mechanics*. This approach was brought to life by the desire to explain and to predict different properties of matter based on molecular structure alone. Concurrently, or somewhat earlier, *thermodynamics* was developed, which started from its own principles and introduced *entropy*, the most fundamental property of matter, which is statistical in its origin. In this book elementary statistical and kinetic theories are outlined prior to thermodynamics, from which we need to borrow a few principal statements. However, one may just as well start with the last chapter, where the basic concept of thermodynamics is outlined, and then proceed to the beginning of the book.

The present monograph is intended both for a preliminary acquaintance with these disciplines and for their advanced study. The attention paid to gases and to condensed matter is more symmetrical than in most textbooks. For instance, equations of state are usually derived for ideal and real gases, while solids and liquids are left out of consideration. Yet, the phenomenology and the basic features of different aggregate states should be studied from the common viewpoint. This helps to reveal the similarities as well as the differences between the rarefied and the condensed phases.

In the first chapter ideal and real molecular gases are considered, first within the elementary approach and then on the basis of the canonical Gibbs distribution. The classical and quantum theory of heat capacities, dielectric and magnetic properties, and transport phenomena in dense and ultradiluted gases are considered. In contrast to the classical ideal gas, the purely quantum photon gas is discussed in detail (Chapter 2). Its phonon analog is used to reduce the heat motion of the lattice to a gas-like model of point quasiparticles moving in a solid as in a container (Chapter 3). Taking account of anharmonicity in the phonon gas treatment allows the description of heat conductivity and the thermal expansion of crystals. Thereby, the behavior of gases and solids can be considered on a common footing, which contributes to the better understanding of both. On the other hand, solids and liquids may be confronted within semiquantitative free volume theory, which is an alternative to the conventional distribution function approach to the liquid phase. In Chapter 4 we use the Lennard-Jones and Devonshire equations of state to explain the remarkable peculiarities of simple liquid behavior at high densities and pressures. For the first time the quasithermodynamics of inhomogeneous media is presented (Chapter 5), which is actually the only self-consistent approach to the smooth gas–liquid interface. This approach has been successfully applied to the calculation of the density profile within the van der Waals theory of surface tension.

For a uniform quantitative description of gases, liquids, and crystals we have used the best multiparametric equations of state approved by the Russian and USA National Bureaus of Standards, which approximate experimental data within their accuracy limits. The figures and graphs that illustrate these data span a broad range of pressures, densities, and temperatures, from gases to solids, and are of interest in themselves. In particular, they show how the compressibility factor $F = p/nkT$ varies with pressure p and density n in any aggregate state and at the points of phase transitions. The "pseudoideal" states, where $F = 1$, and the points of multiple crossing of isotherms are seen on compressibility factor charts. Their origin and physical interpretation have been discussed in the literature, but here the results have been collected for the first time. These results as well as the quasithermodynamic theory of the interface are original and may be of interest not only for students, but also for scientists and engineers.

Since the book is addressed to a broad audience, we have tried to make it as self-consistent as possible. Suggested reading is composed mainly of fundamental monographs and pioneering works partly forgotten. However, a number of textbooks that may be useful for a more detailed and in-depth study of the subject are also included in the Bibliography.

I would like to take this opportunity to thank my coworker Dr. N. V. Shokhirev for a fruitful collaboration in a few original works and my secretaries in Novosibirsk (Mrs. R. I. Ratushkova and Mrs. S. Makarova) and in the Weizmann Institute of Science (Mrs. S. Newman) for technical assistance in preparing the manuscript. I am also grateful to the Abrashkevich brothers for their voluntary and comprehensive assistance with any computer problems and especially indebted to the first reader and editor of the English version, Ms. Adrienne Fairhall, for her outstanding help in shaping the language of the book.

July 1995   A. BURSHTEIN

1

GASES

1.1 STATISTICS OF MOLECULES

Historically, "gas," the gaseous phase of matter, was studied and understood before all other phases. It remains as one of a few rigorously treated systems, and therefore successful statistical interpretation of other phenomena often depends on the possibility of reducing them to a gas model. For example, a solid may be adequately represented by a mixture of electron and phonon gases. Liquids are difficult to describe qualitatively just because they cannot be modeled by a gas.

The idea that matter is composed of atoms goes back to the Greek philosophers, notably Democritus. While the existence of atoms remained just a hypothesis, it was not so easy to develop a molecular model of a gas; however, this was the only direction to take that was at all feasible. The basic features of the model are suggested by the well-known gas properties: a tendency to expand indefinitely in free space and to exert pressure on the walls of a vessel ("container"). An ensemble of infinitesimal solid spheres—point masses—would behave similarly. Having definite translational energy, the point masses would absolutely fly out, if the opportunity arose, and exert pressure on any wall met on their way. This qualitative picture could be made quantitative, if anybody could calculate the equilibrium pressure value to show that it complied with the ideal gas equation of state: $pV = RT$, where V is the volume of one mole of a gas, p is its pressure, T is the temperature, and R is the gas constant.

Pressure

If a gas is an ensemble of randomly moving point masses, then many of them collide with a wall all the time. In the ith collision the wall obtains an elementary momentum Δp_i. According to the law of conservation of momentum this is exactly equal to the change in the molecule's momentum in this ith collision against the wall. The force acting on a wall is the total momentum imparted to the wall by all collisions that happen per second:

$$F = pS = \overline{\sum_i \Delta p_i}, \qquad (1.1.1)$$

where S is the area of the wall.

This force is not constant, but varies randomly ("fluctuates") in time. Similarly, the pressure differs at from moment to moment although it fluctuates around a constant value if the gas is in an equilibrium state. The line above $\sum_i \Delta p_i$ denotes averaging, so by macroscopic pressure we imply the average over all possible values of the random variable. Since a gas consists of a great number of molecules, under equilibrium conditions any measurement, no matter when it is taken, cannot differ essentially from the mean. Noticeable deviations (fluctuations) in any large ensemble are very rare, as, for example, an accidental concentration of passengers in one airport when all other airports are empty. Besides, the sluggish response of measuring instruments smooth out most of the deviations themselves by not following very fast fluctuations of $\sum_i \Delta p_i$. Although the pointer of the instrument still "fluctuates" following slow variations of the value being measured, the obtained result is just the mean about which the readings fluctuate. Any macroscopic value is always the average of many microscopic values. With this in mind, henceforth we shall omit the sign of averaging, and the words "number of collisions" striking the wall, as well as "number of molecules" with a definite velocity, and so on will be used only in the sense of their mean values.

Though identical molecules represented by point masses lack almost any specific features, they should be distinguished by one very important parameter, namely, the velocity of motion. For example, when a molecule moves towards a wall perpendicular to the x axis, its velocity along this axis determines the magnitude of momentum imparted to the wall upon collision. If the collision is elastic: $v'_y = v_y$; $v'_z = v_z$; $v'_x = -v_x$ (the primed symbols indicate the velocity after collision, the unprimed symbols before it), then

$$\Delta p_x = mv_x - mv'_x = 2mv_x. \qquad (1.1.2)$$

To calculate the pressure one has to know the number of collisions with the wall that occur per unit time with a velocity v, which is greater than v_x but less than $v_x + dv_x$. If $dN(v_x)$ is the number of such collisions from a total number of impacts per second $N = \int dN(v_x)$, then the force acting on a wall can easily determined to be

$$F = pS = \int 2mv_x dN(v_x). \tag{1.1.3}$$

The number of collisions with such a rigidly limited velocity is infinitesimal as compared to N. This justifies the designations dN (not ΔN) and integration, instead of summation, of momenta imparted to the wall by molecules with different "attack" velocities.

Now, if we select the subensemble of molecules with some particular velocity v_x, it becomes evident that number of their collisions with a wall is

$$dN(v_x) = Sv_x dn(v_x). \tag{1.1.4}$$

In the notation accepted here, $dn(v_x)$ denotes "a small fraction of molecules" (from the total number per unit volume n) moving towards the wall with a velocity that lies between v_x and $v_x + dv_x$. The estimate (1.1.4) is similar to counting rain drops falling on the surface of the area S per second. Rain drops which are at a distance of more than v_x from the surface have no time to reach it. Only those confined in a parallelepiped with the base S and height v_x (Fig. 1.1) will succeed. Their number is equal to the volume Sv_x multiplied by the density $dn(v_x)$. As v_x is the same for all drops, it makes no difference whether the rain is slanted or straight. Therefore the molecular "rain" may be classified by only one variable—the value of v_x, because after that we can apply Eq. (1.1.4) to each group of molecules. Using it in Eq. (1.1.3) we find

$$p = \int_0^\infty 2mv_x^2 dn(v_x) = 2mn \int_0^\infty v_x^2 dW(v_x), \tag{1.1.5}$$

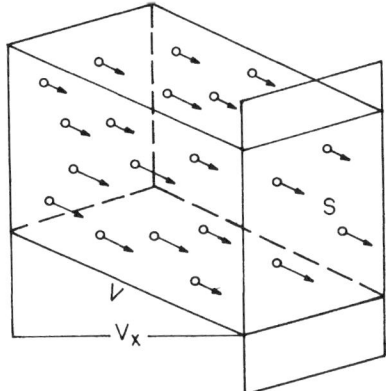

Figure 1.1 Molecules of the same velocity v, which impact with an element of area S of a wall per second. The arrows are unit vectors with direction of the velocities.

where integration is extended to all molecules moving towards the wall ($v_x > 0$). Of course, v_x cannot exceed the speed of light c. Since such rapid molecules are usually very few, they actually do not contribute to the integral so the integration may be extended to infinity, provided that temperatures are not too high.

Statistical Distribution

Equation (1.1.5) gives a microscopic, kinetic definition of pressure. The quantity involved

$$dW(v_x) = \frac{dn(v_x)}{n} \tag{1.1.6}$$

is a "probability of finding a molecule moving with the velocity v_x." It denotes the fraction of particles (from the total number per unit volume n) possessing the required property, that is, having a velocity in a small interval dv_x around the defined value. The number of molecules may be different in different intervals of the same size, but having tried all the possibilities, we count all molecules per unit volume: $\int dn(v_x) = n$. Similarly, the probability of finding a molecule with a higher or a lower velocity is not equal, but the detection of a molecule moving somewhere and somehow is an authentic event whose probability is equal to unity:

$$\int dW(v_x) = \frac{\int dn(v_x)}{n} = 1 \tag{1.1.6a}$$

This is a probability distribution normalized to 1. In this sense $dn(v_x)$ may be considered as a distribution of particles normalized by their density. The option of using either the distribution of probabilities $dW(v_x)$ or that of particles $dn(v_x)$ depends on the problem, and, sometimes, on the specific properties of a gas.

In the absence of external fields (electric, magnetic, gravitational, etc.) there is no preference for any direction of motion. Thus molecules moving towards or away from the wall are to be observed equally often, and the probability $dW(v_x)$ of detecting them remains unchanged when v_x reverses sign. Consequently,

$$\int_0^\infty v_x^2 dW(v_x) = \int_{-\infty}^0 v_x^2 dW(v_x),$$

and

$$p = 2mn \int_0^\infty v_x^2 dW(v_x) = mn \int_{-\infty}^{+\infty} v_x^2 dW(v_x) = mn\overline{v_x^2}. \quad (1.1.7)$$

The Statistical Average

The averaged value of $F(a)$ is by definition

$$\overline{F(a)} = \int F(a) dW(a).$$

In Eq. (1.1.7) this definition was applied to $F(v_x) = v_x^2$. It implies that integration is performed over the whole domain of a random variable: $-\infty \leq v_x \leq \infty$ in our case.

Since there is no preference for any direction, $\overline{v_x^2}$ does not differ from either $\overline{v_y^2}$ or $\overline{v_z^2}$. No axis has an advantage over any another. Therefore,

$$\overline{v^2} = \overline{v_x^2} + \overline{v_y^2} + \overline{v_z^2} = 3\overline{v_x^2}$$

so Eq. (1.1.7) may be written as follows

$$p = \frac{1}{3} mn\overline{v^2} = \frac{2}{3} n \frac{m\overline{v^2}}{2} = \frac{2}{3} n\bar{\epsilon}. \quad (1.1.8)$$

To continue further is impossible without ascertaining the form of the probability (1.1.6).

However, by representing n as N/V, we can note the similarity between Eq. (1.1.7) and the ideal gas equation of state $pV = RT$:

$$pV = Nm\overline{v_x^2}. \quad (1.1.9)$$

The statement of the problem follows from the above equality. Its left-hand side coincides with the corresponding side of the equation of state. To calculate the right-hand side it is necessary, first, to know the velocity distribution, and, second, to perform the averaging of v_x^2 using this distribution.

1.2 DISTRIBUTION FUNCTION

To determine the distribution function is the primary goal of statistical physics. However, from where can we get information about it? Up until the midnineteenth century an experimental study of the velocity distribution was out of the question. Even the very existence of atoms and molecules was still a hypothesis. Therefore it is surprising that only these *a priori* assumptions made by Maxwell from the basis of pure thought proved to be sufficient for the determination of this distribution.

Obviously, the number of molecules with a "definite" velocity depends on the accuracy of its determination. The smaller the interval of velocities $(v_x, v_x + dv_x)$ the lower the probability of finding molecules moving with such velocities. The probability that a molecule will have a strictly specified velocity $(dv_x = 0)$ is equal to zero. With this in mind, we can write for the isotropic space (where no direction is preferable):

$$dW(v_x) = f(v_x)dv_x, \quad dW(v_y) = f(v_y)dv_y, \quad dW(v_z) = f(v_z)dv_z, \quad (1.2.1)$$

where the distribution function $f(\xi)$ is the same for ξ either v_x, v_y, or v_z. The general relation

$$dW(a) = f(a)da \quad (1.2.2)$$

remains valid for any continuous random variable a, whether it be velocity coordinate, angular momentum, energy, etc. If a is a vector quantity, for example, velocity **v**, then to find the "probability of detecting a definite velocity" means the determination of the probability that its projections in the intervals dv_x, dv_y, and dv_z are near the given values of the components. In other words, the end of the velocity vector must be within the confines of a cube with the sides dv_x, dv_y, dv_z constructed around the point with the coordinates v_x, v_y, v_z (Fig. 1.2). The less the volume of the cube $d\mathbf{v} = dv_x dv_y dv_z$, the

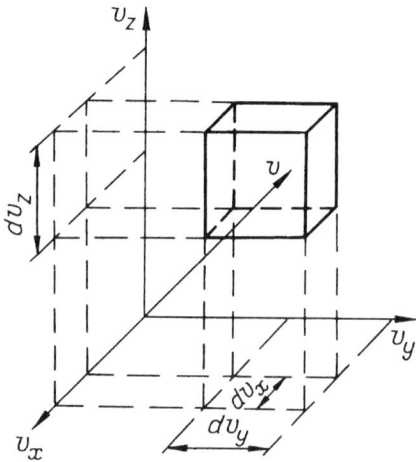

Figure 1.2 An element of velocity space.

more rarely are the molecules encountered whose velocity vector is directed to such a precisely indicated velocity space region. Thus

$$dW(\mathbf{v}) = g(\mathbf{v})dv_x dv_y dv_z = g(\mathbf{v})d\mathbf{v}. \qquad (1.2.3)$$

Since all directions are equivalent the distribution function cannot depend on spherical angles of the vector $\mathbf{v}$ (θ and φ). Hence, it must depend solely on its modulus $v = \sqrt{(v_x^2 + v_y^2 + v_z^2)}$, that is,

$$g(\mathbf{v}) = g(v). \qquad (1.2.4)$$

The identity of all distributions in Eq. (1.2.1) and the constraint imposed by Eq.(1.2.4) is all the information we can extract from the isotropy of space. Due to these limitations the choice of $f(v)$ and $g(v)$ is not absolutely arbitrary, but we need some additional idea to remove the residual uncertainty.

This idea follows from the concept of independent random motion in any direction. It seems plausible that the probability of detecting a molecule that moves along the x axis with the velocity v_x is completely independent of the projections of $\mathbf{v}$ on the other axes. If the given value v_x admits any v_y and v_z, and gives no preference to any particular value, then the above assumption is valid. This means that the three accidental events, namely, the detection of different values of three velocity components, are independent of one another. The value of one of them says nothing about any other. In this case

$$dW(\mathbf{v}) = dW(v_x) \cdot dW(v_y) \cdot dW(v_z). \qquad (1.2.5)$$

This is a mathematical formulation of the probability multiplication theorem: the probability of accidental coincidence of several independent events is equal to the product of their probabilities. For example, the probability that three cubic dice cast at a time will show 1 is $1/6 \times 1/6 \times 1/6 = 1/216$. In general, the probability of getting any three *a priori* specified numbers, for instance, 2,4,1 or 3,5,2 is the same. Accidental detection of three numbers, v_x, v_y, v_z differs from the above example only in that the probabilities of their different values are not equal, and, moreover, are still unknown. However, only the assumption that they are independent makes it possible to use the multiplication theorem (1.2.5) and this is sufficient for their determination.

Substituting distributions (1.2.1) and (1.2.3) into (1.2.5) in view of (1.2.4) yields

$$g(v) = f(v_x) \cdot f(v_y) \cdot f(v_z). \qquad (1.2.6)$$

Taking the logarithm of the above equality and differentiating it with respect to v_x, we get

$$\frac{1}{g}\frac{dg(v)}{dv}\frac{\partial v}{\partial v_x} = \frac{1}{g}\frac{dg(v)}{dv}\frac{v_x}{v} = \frac{1}{f(v_x)}\frac{df(v_x)}{dv_x}. \qquad (1.2.7)$$

Differentiation with respect to v_y and v_z gives similar results. Together with (1.2.7) they may be represented as

$$\frac{1}{g(v)}\frac{dg(v)}{dv} = \frac{1}{f(v_x)}\frac{df(v_x)}{v_x dv_x} = \frac{1}{f(v_y)}\frac{df(v_y)}{v_y dv_y} =$$
$$= \frac{1}{f(v_z)}\frac{df(v_z)}{v_z dv_z} = -2\alpha. \qquad (1.2.8)$$

Here α is a constant, since functions of different arguments may identically coincide over the whole domain only if all of them are equal. Upon integrating Eq. (1.2.8), we obtain

$$g(v) = \frac{1}{Z}\exp(-\alpha v^2), \quad f(\xi) = \frac{1}{Z_0}\exp(\alpha \xi^2), \qquad (1.2.9)$$

where $\xi = v_x, v_y, v_z$. By virtue of (1.2.6) there is a simple relationship between the integration constants Z and Z_0

$$Z = Z_0^3. \qquad (1.2.10)$$

So if the motion in different directions is statistically independent, then the distribution functions must be of the form (1.2.9) and no other. However, it should be remembered that statistical independence is just a hypothesis, and not so evident as might appear at first sight. If, for example, one of the velocity components is equal to c, the speed of light, then the other components are obviously zero, that is, strictly specified. As a result the velocity distributions of photons are qualitatively different from (1.2.9) as we shall see in the next chapter. The Maxwell hypothesis gives correct results for classical molecular gases just because at ordinary temperatures atoms and molecules cannot move with relativistic velocities. For quantum gases—bosons and fermions—the situation is different. The phonon and electron velocity distributions at low temperatures do not satisfy the functional equation (1.2.6), nor, hence, (1.2.5). Thus there is some element of luck that the idea of independent motion of molecules in different directions proved to be valid for classical (not too cold) but at the same time nonrelativistic (not very hot) gases. This is a happy thought and a great piece of luck for the theory of Maxwell, who had to proceed from more or less arbitrary premises for lack of any other.

It is also remarkable that Maxwell's hypothesis was sufficient to determine the precise shape of $f(\xi)$ and $g(v)$. The uncertainty still remaining in Eq. (1.2.9) is the unknown Z_0. This uncertainty is easily eliminated by taking into account the normalization condition

$$\int_{-\infty}^{+\infty} dW(v_x) = \frac{1}{Z_0} \int_{-\infty}^{+\infty} f(v_x) \, dv_x = 1.$$

It gives

$$Z_0 = \int_{-\infty}^{+\infty} f(\xi) d\xi = \int_{-\infty}^{+\infty} \exp\left(-\alpha \xi^2\right) d\xi = \sqrt{\frac{\pi}{\alpha}}, \quad (1.2.11)$$

which is the so-called Poisson integral. The following calculations show that it is actually equal to $\sqrt{\pi/\alpha}$:

$$Z_0^2 = \left(\int_{-\infty}^{+\infty} f(\xi) d\xi\right)^2 = \int_{-\infty}^{+\infty} d\xi \int_{-\infty}^{+\infty} d\eta \exp\left[-\alpha(\eta^2 + \xi^2)\right] =$$

$$= \int_0^{2\pi} d\varphi \int_0^{\infty} e^{-\alpha \rho^2} \rho \, d\rho = \pi \int_0^{\infty} e^{-\alpha z} dz = \frac{\pi}{\alpha}. \quad (1.2.12)$$

By substitution of (1.2.11) into (1.2.9) and (1.2.1), we obtain

$$f(v_x) = \sqrt{\frac{\alpha}{\pi}} \exp\left(-\alpha v_x^2\right), \quad dW(v_x) = \sqrt{\frac{\alpha}{\pi}} \exp\left(-\alpha v_x^2\right) dv_x. \quad (1.2.13)$$

The remaining freedom in choosing the parameter α is by no means the drawback of the theory. This degree of freedom actually exists in the system under study. The smaller is α, the more frequently molecules with high velocities, and, therefore, with high kinetic energies, are observed (Fig. 1.3). The measure of the kinetic energy of a substance is evidently its temperature. So the smaller is α, the greater is T. To make the character of this dependence clearer, it is necessary to carry out the second part of our plan: to perform the averaging of v_x^2 and use it in Eq. (1.1.9) to compare the equation obtained with the familiar gas equation of state.

1.3 IDEAL GAS EQUATION OF STATE

Let us calculate the root-mean-square velocity using Eq. (1.2.13)

$$\overline{v_x^2} = \int_{-\infty}^{\infty} v_x^2 \sqrt{\frac{\alpha}{\pi}} e^{-\alpha v_x^2} dv_x = -\sqrt{\frac{\alpha}{\pi}} \frac{d}{d\alpha} \int_{-\infty}^{\infty} e^{-\alpha v_x^2} dv_x. \quad (1.3.1)$$

This result shows that the desired average may be obtained by differentiating Z_0 as defined in Eq. (1.2.11):

$$\overline{v_x^2} = -\sqrt{\frac{\alpha}{\pi}} \cdot \frac{dZ_0}{d\alpha} = \frac{1}{2\alpha}. \quad (1.3.2)$$

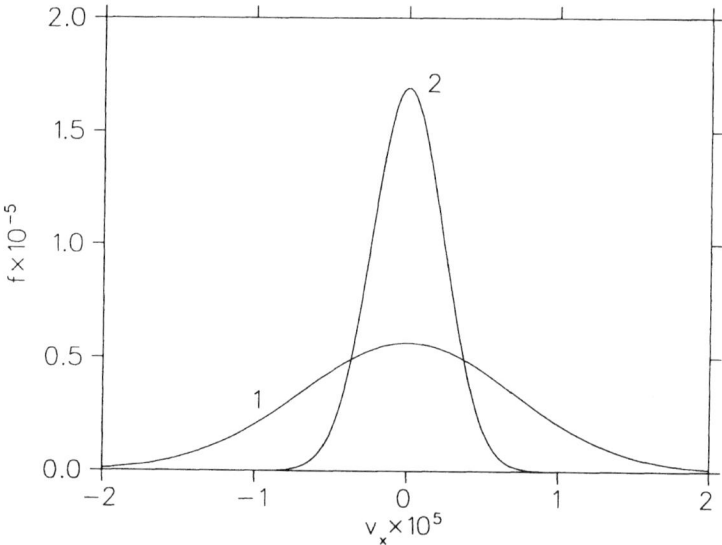

Figure 1.3 The Maxwellian distribution of a component of the molecular velocity at different temperatures: (1) $\alpha = 10^{-10}$, (2) $\alpha = 9 \times 10^{-10}$.

Substituting (1.3.2) into (1.1.9) gives

$$pV = \frac{Nm}{2\alpha}. \qquad (1.3.3)$$

If we deal with one mole of a gas, then N is equal to N_0, the Avogadro number. In this case, the equation $pV = N_0 m/2\alpha$ is equivalent to the equation $pV = RT$, and the required exact correspondence between α and T follows from the identity of the two definitions of one and the same law

$$\frac{N_0 m}{2\alpha} = RT, \qquad \alpha = \frac{m}{2kT}, \qquad (1.3.4)$$

where $k = R/N_0 = 1.38^{-16}$ erg/K. According to Avogadro's law, N_0 is a fundamental physical constant. So k, the ratio of two fundamental constants, is also a fundamental constant—the Boltzmann constant. In view of (1.3.4), we can obtain instead of (1.3.2)

$$\frac{\overline{mv_x^2}}{2} = \frac{\overline{mv_y^2}}{2} = \frac{\overline{mv_z^2}}{2} = \frac{kT}{2}, \qquad \bar{\epsilon} = \frac{m\left(\overline{v_x^2} + \overline{v_y^2} + \overline{v_z^2}\right)}{2} = \frac{3}{2}kT, \qquad (1.3.5)$$

It is seen that the value $3kT/2$ defines the average kinetic energy of translational motion of gas molecules. At $T = 300$ K it is equal to

6.2×10^{-14} erg = 4×10^{-2} eV. For comparison, it should be noted that the bonding energy of an atom in a molecule is greater by two orders of magnitude. The difference between room thermal energy of molecules and the energy of electrons in atoms, or that of accelerated particles is even more impressive. Still, the thermal energy of molecules is sufficient for them to move at rather high velocities. For example, for *oxygen*: $(\overline{v^2})^{1/2} = \sqrt{2\overline{\epsilon}/m} = (12.4 \times 10^{-14}/5.3 \times 10^{-23})^{1/2} = 4.8 \times 10^4$ cm/s while for *hydrogen* the velocity is four times greater.

Finally let us return to the previous discussion and analyze the assumptions made, in particular, the model of an "ideal" gas. Purely qualitative property of a gas—the tendency to disperse in all directions, if no walls prevent the expansion—has been considered as an indication that gas molecules are not linked by intermolecular forces. In such a strong statement this assumption is not quite justified. A gas will disperse provided that the average kinetic energy ϵ exceeds the attraction energy when molecules are at the average distance from one another. In fact, the energy of intermolecular interaction is not negligible, as it is responsible for the condensation of a gas into a liquid or solid. This fact, as well as the finite sizes of molecules which were previously considered as point masses, leads to the conclusion that the equation of state is essentially different from $pV = RT$. For most gases the ideal gas law holds only approximately and only at rather low pressures and high temperatures. However, the limited applicability of the ideal gas model does not remove its advantages. Due to its simplicity, an exact mathematical treatment is possible. The model of noninteracting point masses provides information about the most important features of molecular motion in a gas and sheds light upon the basic distinctions between the classical gas and ideal photon and phonon gases which will be considered later. The theoretical basis of the model is consistent and clear. It may be used as a skeleton upon which to build models suited to reveal the specific features of real objects. Abandoning ideal model simplifications one after another, we shall discover effects or phenomena omitted in the simplest scheme. Transfer phenomena are associated with finite sizes of molecules, chemical reactions—with molecular destruction, electric conduction—with their ionization, and so on. The advantages of the ideal model are elegance and strictness, while the advantage of the real models is their applicability to a variety of phenomena and physical situations.

According to (1.3.3) and (1.3.4), in an ideal gas

$$p = nkT. \qquad (1.3.6)$$

In a real gas nonlinear corrections in n appear which become dominant in condensed phases.

1.4 MAXWELL DISTRIBUTION

Substituting (1.2.9) into (1.2.3) in view of (1.2.10) and (1.2.11), we find

$$dW(\mathbf{v}) = \left(\frac{\alpha}{\pi}\right)^{3/2} e^{-\alpha\left(v_x^2+v_y^2+v_z^2\right)} dv_x dv_y dv_z, \qquad (1.4.1)$$

where $\alpha = m/(2kT)$. The probability of detecting the velocity defined by the value and direction cannot depend on the shape of an elementary volume in velocity space which contains the end of the vector $\mathbf{v}$. Thus (1.4.1) may be recast as

$$dW(\mathbf{v}) = \left(\frac{\alpha}{\pi}\right)^{3/2} e^{-\alpha v^2} d\mathbf{v}, \qquad (1.4.2)$$

where by $d\mathbf{v}$ (or d^3v) we mean the volume of the element of arbitrary shape in the velocity space containing the group of molecules we are interested in. In particular, if we wish to know the probability of finding a molecule with the definite velocity $\mathbf{v}$ whose orientation in the space is specified by spherical coordinates θ and φ, it is necessary to calculate the volume $d\mathbf{v}$ defined by the increments dv, $d\theta$, $d\varphi$. As is seen from Fig. 1.4a, this volume is equal to the product of the height of the spherical layer dv and the area of its base: $(v\sin\theta d\varphi)vd\theta$, so that $d\mathbf{v} = v^2 dv \sin\theta d\theta d\varphi = v^2 dv d\Omega$, where $d\Omega$ is a spherical angle limiting the direction dispersion of velocities. Thus in these coordinates the Maxwell distribution is of the form

$$dW(\mathbf{v}) = \left(\frac{\alpha}{\pi}\right)^{3/2} e^{-\alpha v^2} v^2 dv d\Omega = \left(\frac{\alpha}{\pi}\right)^{3/2} e^{-\alpha v^2} v^2 dv \sin\theta d\theta d\varphi. \qquad (1.4.3)$$

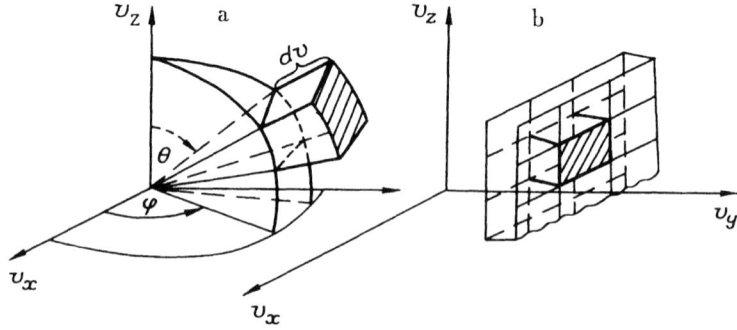

Figure 1.4 Elements of velocity space in (a) spherical and (b) Cartesian coordinates.

Summation of Probabilities

Sometimes, only part of the information obtainable from (1.4.1) or (1.4.3) is necessary to solve a problem. For example, in calculating the pressure, we need only know the velocity of a molecule moving towards the wall along the x axis, independent of its projections on the other axes. What is wanted for the calculation of $\overline{v_x^2}$, and, hence, p, is the distribution $dW(v_x)$. How can this particular distribution be obtained from the general Maxwell distribution (1.4.1)?

The recipe is given by the summation theorem: the probability of observing one of a number of incompatible events, no matter which of them, is equal to the sum of probabilities of these events. To appreciate the physical meaning of the theorem, just remember that the probability of a state is the relative number of particles in this state: $dn(v_x, v_y, v_z)/n = dW(\mathbf{v})$. For the given value v_x other projections are not essential: they may be identical, different, great or little. All molecules with the same v_x which differ only in other projections should be counted. The ratio of their total number to the corpuscular density n gives the desired probability

$$dW(v_x) = \frac{1}{n} \int_{v_y} \int_{v_z} dn(v_x, v_y, v_z) = \int_{v_y} \int_{v_z} dW(v_x, v_y, v_z) = \sqrt{\frac{\alpha}{\pi}} e^{-\alpha v_x^2} dv_x. \tag{1.4.4}$$

Substituting (1.4.1) into (1.4.4), we can readily see that this equality is valid. The summation theorem also has a clear geometrical meaning. The quantity $dW(\mathbf{v})$ is the probability that the end of the velocity vector will find itself in one of a number of equal cubic elements $d\mathbf{v}$ confined between two planes $v_x = \text{const}$ and $v_x + dv_x = \text{const}$, while $dW(v_x)$ is the probability that it will be found in any of these cubes, that is, in any point of the infinite plane layer (Fig. 1.4b).

Reasoning similarly, it is easy to calculate the probability that a molecule will move at the absolute velocity v (which is greater than v and less than $v + dv$), irrespective of the direction. Obviously, all points of a spherical layer confined between two spheres of the radius v and $v + dv$ fit the criterion thus formulated. At whichever point the end of the vector $\mathbf{v}$ finds itself, its modulus will meet the requirement, while the direction may be arbitrary. The probability of moving with the same velocity but in a strictly specified direction—within the limits of the spherical angle $d\Omega$—was defined by the general distribution (1.4.3). To find from it the required probability, it is necessary to integrate over all angles, taking into account the entire volume of the spherical layer:

$$dW(v) = \int_0^{2\pi} d\varphi \int_0^{\pi} d\theta \, dW(\mathbf{v}) = 4\pi \left(\frac{\alpha}{\pi}\right)^{3/2} e^{-\alpha v^2} v^2 dv. \tag{1.4.5}$$

Statistical Weight

Let us consider the origin of the multiplier v^2 which appears in Eq. (1.4.5) but was not involved either in (1.4.1) or in (1.4.4). Through the following analogy we can deduce that it is associated with different definitions of the "states" whose probability is being sought. Compare a many-storied pyramid-shaped building with a box-like skyscraper, assuming that all living units inside are of equal size and quality (Fig. 1.5). If they are inhabited uniformly the probability of finding a tenant in any apartment in the daytime is one and the same regardless of the story or the architecture. However, the probability of finding a person in a particular story is less, the nearer the story to the top of the pyramid house. This is because the greater the number of apartments on the story, the greater the probability of a successful search. As far as a box-like house is concerned, the apartment distribution over stories is uniform, and the probability of finding a person in any of them is the same.

The unit volumes of the velocity space may be associated with "apartments." The number of units in a spherical layer increases with distance from the origin of coordinates as $4\pi v^2 dv$. The probability of finding the end of the vector **v** in the layer of the width dv is proportional to statistical weight $4\pi v^2$. On the other hand, the number of units between two parallel planes remains unchanged when the planes are shifted together either to the right or to the left. Thus the corresponding statistical weight is invariant with respect to the position of a flat layer (see Fig. 1.4).

The previous analogy is not quite precise. The probability of occupancy of equal elements of the volume dv at different spherical "stories" is not the same: it decreases exponentially as $e^{-\epsilon/kT}$. Imagine that most of the rooms in a hotel are vacant and the few visitors occupy apartments as they like. If there is no escalator in the hotel, then the visitors will prefer those on the lower floors, even though the rooms are equal in size. The statistical weight $\rho(v)$ is a measure of the "state" (spherical layer) capacity while the Boltzmann factor $e^{-\epsilon/kT}$ relates the attendance of each unit in the state to its energy.

By the "state" whose probability is given by (1.4.3) we mean the element defined by the increments dv, $d\theta$, $d\varphi$, which is the spherical layer delimited by two close meridional cross-sections. Its volume increases the nearer it is to the equator and is equal to zero at the poles. Correspondingly, the statistical weight is $\rho(v, \theta) = v^2 \sin\theta$. Using the summation theorem, we can find the

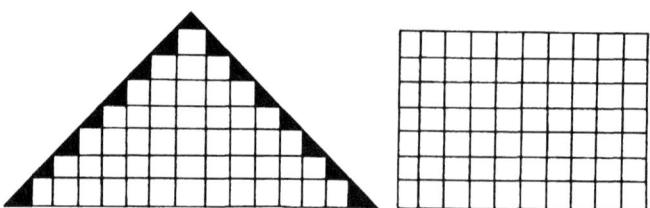

Figure 1.5 Illustration of the concept of statistical weight.

probability of moving in a definite direction irrespective of the absolute value of the velocity:

$$dW(\Omega) = \int_v dW(\mathbf{v}) = \frac{\sin\theta\, d\theta\, d\varphi}{4\pi} = \frac{d\Omega}{4\pi}. \quad (1.4.6)$$

This is just what we expected, due to the space isotropy.

From (1.4.5) we can also derive the probability "to be in a definite energy state," that is, to have kinetic energy greater than ϵ, but less than $\epsilon + d\epsilon$. Substituting $v = \sqrt{2\epsilon/m}$, Eq. (1.4.5) is brought to the form

$$dW(\epsilon) = 2\pi \left(\frac{2\alpha}{\pi m}\right)^{3/2} \exp\left(-\frac{2\alpha\epsilon}{m}\right)\sqrt{\epsilon}\, d\epsilon = \left(\frac{m}{2\pi kT}\right)^{3/2} \exp\left(\frac{\epsilon}{kT}\right)\rho(\epsilon)d\epsilon, \quad (1.4.7)$$

where

$$\rho(\epsilon) = 2\pi \left(\frac{2}{m}\right)^{3/2} \sqrt{\epsilon} \quad (1.4.7a)$$

is the statistical weight.

The general definition of the statistical weight following from the above examples is given by the equality

$$d\mathbf{v} = \rho(q)dq, \quad (1.4.8)$$

where q is any variable or a set of variables expressed in terms of v_x, v_y, v_z, and $\rho(q)$ is the statistical weight.

Mean Values

The probability density that enters the Maxwell distribution $dW(q) = g(q)dq$ is as follows

$$g(q) = \frac{1}{Z} \exp\left[-\alpha\epsilon(q)\right] \rho(q),$$

where Z is the normalization constant (*partition function*). As the exponential factor and statistical weight change inversely with q, $g(q)$ has a sharply defined maximum (Fig. 1.6). The most probable velocity v_e and the most probable energy ϵ_e may be determined from the conventional condition $dg/dq = 0$. Using Eqs. (1.4.5) and (1.4.7) one can get

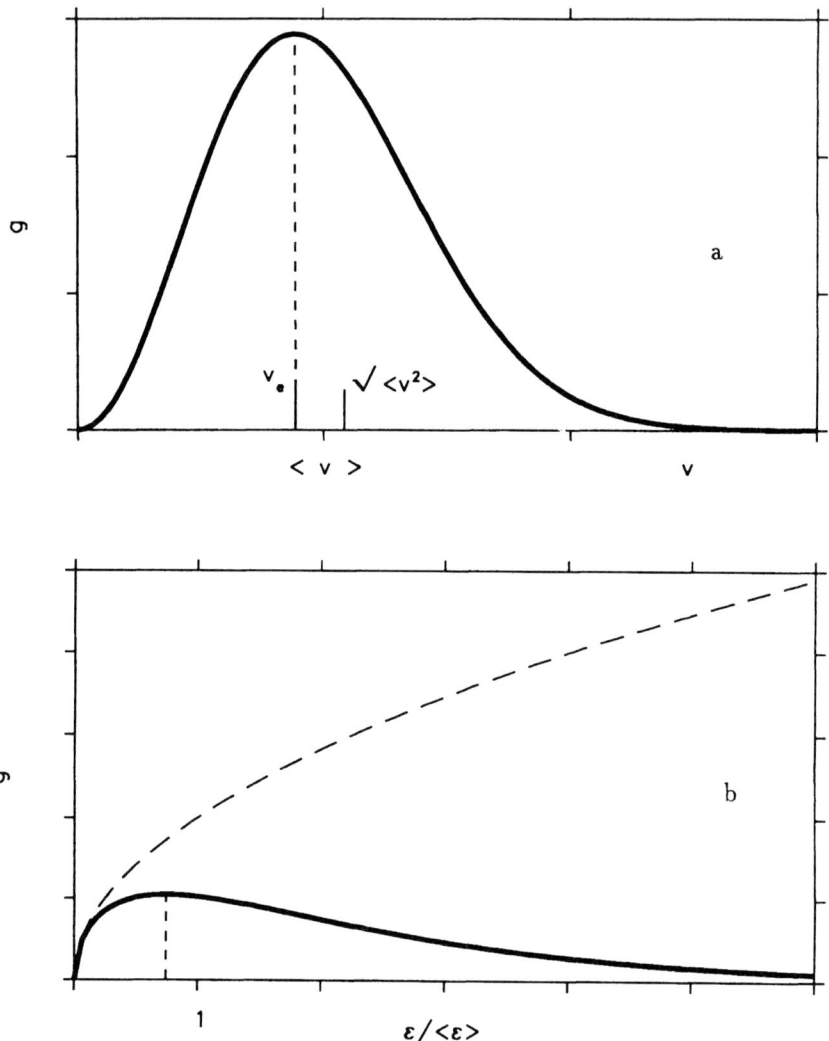

Figure 1.6 Distribution of (a) the molecular speed and (b) kinetic energy.

$$v_e = \sqrt{\frac{2kT}{m}} \quad \text{and} \quad \epsilon_e = \frac{kT}{2}. \tag{1.4.9}$$

Note that $\epsilon_e \neq mv_e^2/2$. This is hardly surprising: an ensemble of molecules with different velocities is poorly described by a simplified model that divides all particles into three groups moving with equal velocities along the coordinate axes. This model is often used for rough kinetic estimates, but it is only within this model that the relations between energy and velocity of a single particle remain valid both for extreme and average values. What actually happens after averaging over the Maxwell distribution is that $\bar{v}$ coincides with rms value $(\overline{v^2})^{1/2}$ as well as with v_e only up to an accuracy of a constant multiplier. Different experiments deal with different averages: $\overline{|v_x|}$, $\bar{v}$, $\overline{v^2}$, and so on. Of course, the order of magnitude of any quantity being measured may be evaluated by replacing v by one assumed value (i.e., the root-mean-square one), but a correct calculation always calls for actual averaging of the required quantity.

For example, let us calculate a number of particles striking a unit area of the wall per unit time. According to Eq. (1.1.4) the elementary flux is

$$dj = dN/S = v_x dn(v_x), \qquad (1.4.10)$$

while the total flux is an integral over positive values of v_x:

$$j = \int_0^\infty v_x dn(v_x) = \frac{1}{2} n \overline{|v_x|}. \qquad (1.4.10a)$$

If all molecules were moving with the same velocity to and fro, then the velocity would be chosen to be $\overline{|v_x|}$, giving exactly the same flux as in the Maxwellian gas. It is obvious that

$$\overline{|v_x|} = 2 \int_0^\infty v_x dW(v_x) = \frac{1}{\sqrt{\pi \alpha}} = \sqrt{\frac{2kT}{\pi m}} \qquad (1.4.11)$$

coincides neither with the root-mean-square nor with the average modulus of the velocity, given by

$$\bar{v} = \int_0^\infty v dW(v) = \frac{2}{\sqrt{\pi \alpha}} = \sqrt{\frac{8kT}{\pi m}}. \qquad (1.4.11a)$$

Using this mean value the expression (1.4.10) can be recast in the form

$$j = \frac{1}{4} n\bar{v}. \qquad (1.4.12)$$

This result coincides with that for a flux of particles moving in all directions but with a single speed $\bar{v}$.

Stern Experiment

The concept of chaotic thermal motion and the shape of Maxwell's distribution were experimentally verified by Otto Stern in 1920. In this experiment, a hollow rotating cylinder, called the vacuum analyzer, is placed in the path of a molecular jet. During the short time interval when the vertical slot of the analyzer is aligned with the jet, molecules can enter the cylinder (Fig. 1.7). These inertially

18   GASES

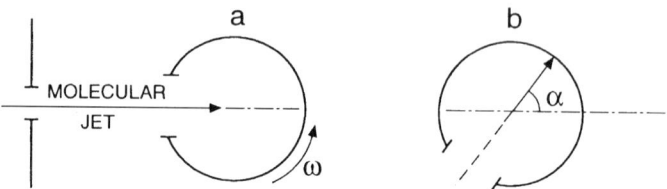

Figure 1.7 Configuration of the analyzer (a) at the time when molecules enter it and (b) at the moment of their adsorption.

moving molecules reach the screen on the opposite side of the analyzer at different times $\Delta t = D/v$, depending on their velocity (D is the cylinder diameter). Due to the fact that during the time Δt the cylinder rotates by a certain angle, the molecules encounter the screen not along the line opposite the slot but at a position which depends on their velocity: the lower their velocity the farther from this line (Fig. 1.7b). At a rather low analyzer temperature almost all the particles are deposited on the screen at the points of collision. Atoms of silver form a stratum on the screen, the thickness of which allows one to determine the number of deposited particles. Knowing the rotation angle of the analyzer $\alpha = \omega \Delta t$ provides a means to judge the velocity of particles which are deflected through this angle: $v = D/\Delta t = D\omega/\alpha$. Therefore one is able to reconstruct the distribution of absolute velocities of atoms which pass through the slot. The experimental determination and study of this distribution were a strong argument in favor of the existence of atoms and verified the validity of the *a priori* assumptions which were of fundamental importance in the derivation of the Maxwell distribution.

Structure of the Flux

However, we have not yet been given sufficient evidence to believe that the distribution thus obtained coincides with the Maxwell one. In the Stern experiment the distribution was reconstructed not for particles in the container but in the jet. The number of rapid molecules leaving the container is greater than the number of slow molecules escaping in the same time. For this reason alone, the distribution obtained by Stern must differ from the Maxwell distribution. The distribution of speeds in a jet is different from that in the volume, which is Maxwellian.

As follows from Eqs. (1.4.10), (1.4.3) and (1.4.5), the flux of particles with a given velocity is

$$dj = v_x n dW(\mathbf{v}) = v \cos \vartheta dW(v) \cdot n \frac{d\Omega}{4\pi}. \qquad (1.4.13)$$

To obtain the normalized velocity distribution of particles in a jet we have only to divide (1.4.13) by its integral value (1.4.12):

$$\frac{dj}{j} = \frac{v}{\bar{v}} dW(v) \frac{\cos \vartheta \, d\Omega}{\pi} = dW_j(v) dW_j(\Omega). \qquad (1.4.14)$$

Here the distribution of the speeds of the particles is

$$dW_j(v) = \frac{v}{\bar{v}} dW(v) = \frac{1}{2} \left(\frac{m}{kT}\right)^2 \exp\left(-\frac{mv^2}{2kT}\right) v^3 dv \qquad (1.4.15)$$

and

$$dW_j(\Omega) = I(\vartheta) \frac{d\Omega}{2\pi}, \qquad I(\vartheta) = 2 \cos \vartheta \qquad (1.4.16)$$

is the distribution of the directions of motion within a semisphere.

In Stern's experiment the velocity distribution of particles moving along the bunch axis ($\vartheta = 0$) in a finite but small spherical angle $\Delta\Omega$ is measured. It was shown to coincide with Eq. (1.4.15). Apart from an unimportant constant this distribution differs from the Maxwell one by the extra factor v. As is clear from Eq. (1.4.13), its origin is associated with the advantage which rapid molecules have over slow molecules in escaping from the container in a given time. That is why both the average velocity and the average energy of particles in the bunch are greater than the Maxwell ones:

$$\bar{v}_j = \int v \, dW_j = \frac{\overline{v^2}}{\bar{v}} = \frac{3\pi}{8} \bar{v}, \qquad \bar{\epsilon}_j = \frac{m}{2} \int v^2 \, dW_j = \frac{m\overline{v^3}}{2\bar{v}} = 2kT = \frac{4}{3} \bar{\epsilon}.$$

1.5 BAROMETRIC FORMULA

In the absence of an external field, it is natural to consider all points of the space filled by gas to be equivalent. Then the probability of entering any element of equal volume is identical:

$$dW(x, y, z) = \frac{dx \, dy \, dz}{V}. \qquad (1.5.1)$$

However, this is not true if the gas in the vessel is under the influence of a gravitational field. In this case, different volume elements differ in the value of the potential energy of molecules contained in them. Generally speaking, one should take

$$dW = f(x, y, z)\,dx dy dz, \qquad (1.5.2)$$

and special considerations are necessary for the determination of $f(x, y, z) \neq$ const. If the gravitational force acts in the z direction, then the variation of $f(z)$ has the result that the gas density $n = n(z)$ is different at different heights. Due to Eq. (1.3.6) this must also affect the pressure.

The variation of pressure with the height is inevitable, at least because the gas confined in the vessel has weight only when the pressure on the bottom exceeds that on the lid. Assuming that the vessel is a weightless box in the form of a parallelepiped with a bottom area S the weight of the gas is

$$\mathcal{P} = S[p(0) - p(H)] = SkT[n(0) - n(H)]. \qquad (1.5.3)$$

On the other hand, the gas weight is the product of the total number of molecules in the vessel and the molecular weight which is mg at reasonably small heights H:

$$\mathcal{P} = mgN = mgS \int_0^H n(z)\,dz. \qquad (1.5.4)$$

Since the identity

$$mg \int_0^H n(z)\,dz = kT[n(0) - n(H)]$$

is valid at any H, $n(z)$ must satisfy the equation

$$\frac{dn}{dz} = -\frac{mg}{kT} n. \qquad (1.5.5)$$

The solution of this equation

$$n(z) = n(0) \exp\left[-\frac{mgz}{kT}\right] \qquad (1.5.6)$$

contains the integration constant $n(0)$ which is the gas density at the bottom of the vessel. Substituting (1.5.6) into the normalization condition $\int_0^H n(z)\,dz = N/S$ determines $n(0)$ to be

$$n(0) = \frac{N}{S}\frac{mg}{kT}\left[1 - \exp\left(-\frac{mgH}{kT}\right)\right]^{-1}. \qquad (1.5.7)$$

With $mgH \ll kT$ (not very tall container or rather high temperatures) the gas density $n(z) \approx n(0)$, that is, it is practically uniform over the whole vessel, just as in our everyday experience. However, when $mgH \gg kT$, the decrease in density with height becomes noticeable.

Conditions of the type

$$\alpha = \frac{mgH}{kT} = \frac{T_c}{T} > 1 \tag{1.5.8}$$

are typical for statistical physics. They represent a competition between deterministic mechanical action (in the given case, the gravitational force) and chaotic thermal motion. The gravitational force tries to make all molecules fall to the bottom of the vessel, while the thermal kinetic energy enables them to overcome the gravitation and move upwards. The dimensionless exponent (1.5.8) defines the relative magnitude of the two effects which is determined by the ratio of the corresponding energies: mgH and kT. Concentration of a gas at the bottom of the vessel may take place only when $T \ll T_c$. At $H = 1$ m the characteristic temperature $T_c = mgH/k \approx 5 \cdot 10^{-23} \times 10^3 \times 10^2 / 1.4 \times 10^{-16} \approx 3 \times 10^{-2}$ K. As a rule, the gaseous phase cannot exist at such low temperatures. Therefore the space inhomogeneity of a gas in a vessel of real size at a usual temperature $T \gg T_c$ is insignificant and should be taken into account just for the correct definition of the gas weight. One may hope to detect the gas inhomogeneity in a taller vessel, but from the same formula it follows that at room temperature a noticeable decrease in density (by a factor of e) is observed at $H = kT/mg = 1.4 \times 10^{-16} \times 300/5 \times 10^{-23} \times 10^3 = 10^6$ cm = 10 km. Obviously, it is impossible to imagine such a vessel but this suggests another possibility: comparison with the Earth's atmosphere.

The decrease in atmospheric density with height is a well-known fact. If it could be described by Eq. (1.5.6), then the corresponding pressure fall would be defined by the "barometric" formula

$$p(z) = p(0) \cdot \exp\left[-\frac{mgz}{kT}\right]. \tag{1.5.9}$$

It is also seen from Eq. (1.5.6) that the separation of gases with height, which actually happens, is due to different molecular weights

$$\frac{n'}{n''} = \frac{n'(0)}{n''(0)} \exp\left[\frac{(m''-m')gz}{kT}\right]. \tag{1.5.10}$$

Good qualitative agreement between these conclusions and well-known facts suggests an attractive idea: to use the above formula for the quantitative description of the composition and density of the atmosphere at different altitudes. However, such an extrapolation is not quite adequate. The approximate law $F = -mg$ used to describe the action of the gravitational force is not valid at high altitudes. It should be replaced by Newton's law of gravitation. According to this law

$$U = f \int_{R_0}^{R} \frac{mM}{r^2} \, dr = fmM \left(\frac{1}{R_0} - \frac{1}{R} \right). \tag{1.5.11}$$

It has not yet been made clear, however, how this can be realized. The distribution of particles far from the Earth's surface is not one-dimensional, but has a spherical symmetry. Thus before employing (1.5.11), it is necessary to generalize the results obtained.

1.6 THE BOLTZMANN DISTRIBUTION

As the gas density at small height z is

$$n(z) = dN(z)S dz = (N/S) \frac{dW}{dz},$$

the one-dimensional distribution of particles may be obtained from Eq. (1.5.6):

$$dW(z) = \frac{n(0)S}{N} \exp\left[-\frac{mgz}{kT}\right] dz = \frac{1}{Z} \exp\left[-\frac{mgz}{kT}\right] dz, \tag{1.6.1}$$

where $Z = N/n(0)S$. This is the probability to find a molecule in a layer of thickness dz. Though this distribution is obtained for the particular potential energy $u(z) = mgz$ the form of the distribution density

$$f(z) = \frac{1}{Z} \exp\left[-\frac{u(z)}{kT}\right] \tag{1.6.2}$$

indicates how to generalize this for other forms of $u(z)$. In particular, it is clear that even in the case considered, the reference system does not necessarily coincide with the most convenient one where the z axis is parallel to the force of gravity. Thus in the general case, the distribution (1.6.1) takes the form

$$dW(x,y,z) = \frac{1}{Z} \exp\left[-\frac{u(x,y,z)}{kT}\right] dx\,dy\,dz = \frac{1}{Z} \exp\left[-\frac{u(\mathbf{r})}{kT}\right] d\mathbf{r}. \tag{1.6.3}$$

It differs from the Maxwell distribution only in that the potential energy plays the role of the kinetic energy and $d\mathbf{r}$ is the element of volume in ordinary coordinate space, not in the space of velocities. This is the Boltzmann distribution.

By analogy with Eq. (1.4.3), for a spherically symmetrical field it may be represented as

$$dW(R,\theta,\varphi) = \frac{1}{Z} \exp\left[-\frac{u(r)}{kT}\right] R^2 dR \sin\theta\, d\theta\, d\varphi, \tag{1.6.4}$$

where $Z = \int \exp[-u(R)/kT]d\mathbf{R}$. If we are interested in the composition of gas in the near-Earth layer of thickness ΔR, it is natural to employ (1.6.4). If the variation of density or pressure is of interest, we pass from the probability distribution to the density distribution and get the analog of the barometric formula

$$n(R) = \frac{NdW(R)}{4\pi R^2 dR} = \frac{N}{Z}\exp\left(-\frac{u}{kT}\right) =$$

$$= n(0)\exp\left[-\frac{fmM}{kT}\left(\frac{1}{R_0} - \frac{1}{R}\right)\right]. \quad (1.6.5)$$

Thus knowing the density near the Earth's surface, we could, in principle, estimate its variation with height. Unfortunately, any conclusions on the Earth's atmosphere density following from (1.6.5) are substantiated only qualitatively. Quantitative agreement is impossible, at least, for the reason that Z diverges. Besides, the atmosphere is not in thermodynamic equilibrium. Its temperature near the Earth's surface differs from that at different altitudes, while in equilibrium it must be the same at all points of space. The difference in temperature is responsible for convectional gas flows (winds), heat exchange among neighboring layers and other phenomena incompatible with equilibrium. So this example is not perfect for a quantitative verification of the theory.

However, nature has indeed provided systems which are, on one hand, much more convenient from the experimental standpoint, and, on the other hand, are adequately modeled by the ideal gas. These are so-called suspensions or emulsions: solutions of noninteracting solid or liquid macroscopic particles, suspended in a transparent solvent. The fact that small particles of this kind perpetually move around in a random manner was first observed by Brown, a botanist, in the nineteenth century. Brownian motion is caused by random collisions of the small particles with molecules of the solvent. They alternately gain or lose momentum due to unbalanced collisions with molecules from all sides. As a result the Brownian particles participate in thermal motion along with the solvent molecules themselves in spite of the difference in size. In studies of such properties as distribution with height this difference is not essential; only the mass of dissolved particles is of importance. The mass may be chosen such that the distribution inhomogeneity can be easily observed in a vessel of moderate size.

Since a particle with mass m and density ρ suspended in a liquid of density ρ_0 experiences a buoyant force its weight in solution is $mg - mg\rho_0/\rho$. Therefore the analog of the barometric formula for Brownian particles is

$$n(z) = n(0)\exp\left[-\frac{mgz}{kT}\left(1 - \frac{\rho_0}{\rho}\right)\right] \quad (1.6.6)$$

Random motion of Brownian particles and their distribution with height has defied any explanation other than that given above. Thus the quantitative study of such systems performed by Perrin, in particular, the experimental verification of barometric formula (1.6.6), were a real triumph of both atomic and statistical theory, and finally led to their acknowledgment as the valid theories. Moreover, owing to Perrin's measurements the Avogadro number $N_0 = R/k$ was first determined to high accuracy using

$$k = \frac{mg}{T}\left(1 - \frac{\rho_0}{\rho}\right) \frac{z_2 - z_1}{\ln\left[n(z_1)\right] - \ln\left[n(z_2)\right]}. \tag{1.6.7}$$

The Boltzmann constant was found from Eq. (1.6.7) by calculating the particle density at different heights, then measuring the particles' mass by several methods to obtain the exact value.

As soon as k is known, the inverted problem may be solved with the same formula (1.6.7). This is the problem of the determination of small masses, which is of practical interest. The smaller the masses, the greater the distances that are required to reveal the difference between them at a given accuracy of density measurements. Since the height of the vessel is restricted, essential progress can only be achieved by enhancing the potential field. This aim is achieved by replacing the gravitational field by that of centrifugal force $F = m\omega^2 r$, which can be extremely strong in modern centrifuges. With this replacement, Perrin's method has become a technique which is extensively employed in chemistry and biology as an effective way of separating substances similar in molecular weight.

With the centrifugal potential

$$u = -\int_0^R m\omega^2 r dr = -\frac{m\omega^2 R^2}{2}, \tag{1.6.8}$$

the Boltzmann distribution is of the form

$$dW(\mathbf{R}) = \frac{1}{Z}\exp\left[\frac{m\omega^2}{2kT}R^2\right]d\mathbf{R}, \tag{1.6.9}$$

or, in cylindrical coordinates,

$$dW(R,\varphi,z) = \frac{1}{Z}\exp\left[BR^2\right]RdRd\varphi dz, \tag{1.6.10}$$

where $B = m\omega^2/2kT$. For example, we centrifuge milk and wish to determine the number of oil particles in a layer of the thickness ΔR. The best way is to integrate the distribution (1.6.10) with respect to φ and z. However, if one is interested in the density of particles it is better to proceed from Eq. (1.6.9) bringing this distribution into a "barometric form"

$$n = \frac{NdW}{d\mathbf{R}} = \frac{N}{Z}\exp[BR^2] = n(0)\exp[BR^2]. \quad (1.6.11)$$

The greater the angular frequency of rotation ω, the more inhomogeneous is the distribution of particles along the radius: the density at the periphery increases and that near the rotation axis decreases.

1.7 CANONICAL DISTRIBUTION

If one wishes to know the velocity of the molecule and at the same time its position, the answer is given by the product of corresponding probabilities—(1.4.1) and (1.6.3):

$$dW(\mathbf{r},\mathbf{v}) = dW(\mathbf{r})dW(\mathbf{v}) = \frac{1}{Z}\exp\left(-\frac{\mathcal{E}}{kT}\right)dxdydzdv_xdv_ydv_z, \quad (1.7.1)$$

where $\mathcal{E} = \epsilon + u = mv^2/2 + u(x,y,z)$ and

$$Z = Z_M Z_B = \int e^{-mv^2/2kT}d\mathbf{v}\int e^{-u/kT}d\mathbf{r} = \left(\frac{m}{2\pi kT}\right)^{3/2}\int e^{-u(x,y,z)/kT}dxdydz. \quad (1.7.1a)$$

This is the Maxwell–Boltzmann distribution defined in so-called μ-space, the six-dimensional *configurational space* parametrized by the six coordinates x, y, z, v_x, v_y, v_z. The Maxwell and Boltzmann distributions are obtained from this distribution as particular cases by the familiar summation procedure

$$dW(\mathbf{v}) = \iiint dW(r,\mathbf{v})d\mathbf{r}, \quad dW(\mathbf{r}) = \iiint dW(\mathbf{r},\mathbf{v})d\mathbf{v}. \quad (1.7.2)$$

The Maxwell–Boltzmann distribution (1.7.1) gives a complete description of the translational degrees of freedom, which is quite sufficient in the case of monoatomic gas. However, for molecules, it is desirable also to know the probability of their rotation and orientation, energy of vibration, and so on. Hence, many other distributions may be required.

Generalizing Eq. (1.7.1) by induction, one can expect that the distribution applicable to various motions of the molecular system is similar in form to (1.7.1):

$$dW = \frac{1}{Z}\exp\left(-\frac{\mathcal{E}}{kT}\right)d\Gamma, \quad (1.7.3)$$

where $\mathcal{E}$ is the total energy. It is necessary only to adjust the space element $d\Gamma$ to the product of the coordinate increments associated with a definite type of motion. However, as we have already seen, the element $d\Gamma$ has different statistical weights depending on the coordinates chosen to describe the motion. In order to determine its value, it is sufficient to calculate the statistical weight in some particular coordinate system. Then, by changing variables, we shall automatically obtain it in other coordinates. For example, one may find a coordinate frame in which the statistical weight is equal to unity, that is, indicate the *phase space* where the probability of entering isoenergy elements of equal size is identical. In other words, one should determine coordinates in which increments constitute equivalent "rooms," but not "stories" of different capacity. Formerly, this problem seemed obvious. Equal elements of Cartesian space $dxdydz$ were naturally considered equivalent, while the similar elements in spherical coordinates $drd\theta d\varphi$, were weighed correspondingly. Nevertheless, this was just a hypothesis; generally speaking it might be quite the reverse. This is our belief that only equal elements of real physical space are indistinguishable and therefore equiprobable.

Unfortunately, obvious things hide misunderstanding and confusion, particularly when they serve as a basis for generalization. What seems unquestionable in the case of ordinary coordinate space is not so evident even in the space of velocities, to say nothing of rotational or vibrational motion. Thus relation (1.7.3) becomes a law only after the element $d\Gamma$ is exactly defined.

For reasons to be clarified in Section 1.10, the probability of entering equal elements is considered to be identical only in the *phase space* of canonical variables q_i and p_i, which are generalized coordinates and momenta. It is known from mechanics that when the generalized coordinates q_i and the corresponding velocities $\dot{q}_i = dq_i/dt$ are chosen, the generalized momenta may be found by the formula

$$p_i = \frac{\partial L(q_i, \dot{q}_i)}{\partial \dot{q}_i} = \frac{\partial [\epsilon(q_i, \dot{q}_i) - u(q_i, \dot{q}_i)]}{\partial \dot{q}_i}. \tag{1.7.4}$$

Here ϵ is the kinetic energy, u is the potential energy, and $L = \epsilon - u$ is the Lagrange function.

The value $d\Gamma$ in the canonical Gibbs distribution (1.7.3) is

$$d\Gamma = d\mathbf{p}d\mathbf{q} = \prod_i dq_i dp_i, \tag{1.7.5}$$

so the statistical weight is equal to unity only in coordinate-momenta space. Hence, for a system of noninteracting point masses the Gibbs distribution is of the form

$$dW = \frac{1}{Z}\exp\left[-\frac{\epsilon + u}{kT}\right] d\mathbf{q}d\mathbf{p}, \tag{1.7.6}$$

where $\int dW = 1$, so

$$Z = \int qp e^{-\epsilon+u/kT}\, d\mathbf{q}d\mathbf{p} . \tag{1.7.7}$$

Substitution of $p_i = m\dot{q}_i = mv_i$ reveals the identity of this distribution and the Maxwell–Boltzmann one. The constant statistical weight m^3 is not essential, because it may be excluded by renormalization of the partition function Z.

The formula

$$d\Gamma = d\mathbf{p}d\mathbf{q} = |I|d\mathbf{a}d\mathbf{b} . \tag{1.7.8}$$

enables one to transform the canonical variables used in the Gibbs distribution to any other system. The Jacobian I of this transformation defines the statistical weight in the new variables. Should we need a more specific form of the distribution free of unnecessary variables, the particular expression for the statistical weight is obtained by integration with respect to "unwanted" variables.

1.8 DIELECTRIC PROPERTIES OF GASES

Generalized Coordinates and Momenta

To illustrate the application of the canonical distribution to some system, we consider a linear rotator—a rigid diatomic molecule whose atoms carry alternative charges $\pm e$. Such a molecular dumb-bell (Fig. 1.8) is characterized (apart from the mass $M = m_1 + m_2$) by two more constants—the moment of inertia in the center-of-mass system $I = m_1 r_1^2 + m_2 r_2^2$ and the permanent dipole moment $\mathbf{q} = e(\mathbf{r}_2 - \mathbf{r}_1) = e(r_2 + r_1)\mathbf{a}$ where $\mathbf{a} = \mathbf{r}_2/r_2 = -\mathbf{r}_1/r_1$. Accordingly, its kinetic energy along with the energy of translational motion involves the kinetic energy of rotation

$$\epsilon_{rot} = \frac{m_1 \dot{\mathbf{r}}_1^2}{2} + \frac{m_2 \dot{\mathbf{r}}_2^2}{2} = \frac{I}{2}\dot{\mathbf{a}}^2 \tag{1.8.1}$$

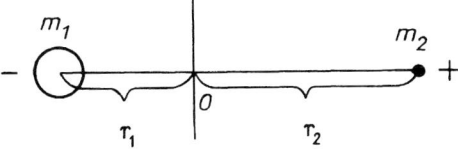

Figure 1.8 Configuration of masses and charges in a diatomic molecule relative to the center of mass.

and potential energy depending on the orientation of the dipole in an external electric field **E**,

$$u = -\mathbf{q}\mathbf{E} = -q_z E = -qE\cos\theta . \tag{1.8.2}$$

Since the potential energy is easily expressed via the angle θ between the dipole and the field, it is natural to choose as generalized coordinates the two spherical coordinates θ and φ with respect to the molecular axis. In this case the kinetic energy should be expressed as a function of the generalized velocities $\dot\theta$ and $\dot\varphi$. The coordinate space of the linear rotator and the space of its generalized velocities are two-dimensional.

Let us bring the origin of coordinates into coincidence with the molecular center-of-mass. In this frame the translational motion vanishes The radius-vector

$$\mathbf{a} = (a\sin\theta\cos\varphi,\quad a\sin\theta\sin\varphi,\quad a\cos\theta) \tag{1.8.3}$$

completely represents a molecule in this reference system. It circumscribes a sphere about the center-of-mass, each point of which is a possible state of the system. To be more exact, this is a set of states coincident in position but differing in the rotational velocity

$$\dot{\mathbf{a}} = a\left(\dot\theta\cos\theta\cos\varphi - \dot\varphi\sin\theta\sin\varphi,\ \dot\theta\cos\theta\sin\varphi + \dot\varphi\sin\theta\cos\varphi,\ -\dot\theta\sin\theta\right)$$

Squaring this expression and substituting it into (1.8.1), we have

$$\epsilon_{rot} = \frac{I}{2}\left[\dot\theta^2 + \sin^2\theta\,\dot\varphi^2\right] . \tag{1.8.4}$$

Using this result in (1.7.4) and bearing in mind that the potential energy does not depend on $\dot\theta$ and $\dot\varphi$, we derive

$$p_\varphi = I\sin^2\theta\,\dot\varphi,\qquad p_\theta = I\dot\theta . \tag{1.8.5}$$

The generalized momentum p_φ corresponds to rotation about the z axis, and p_θ to rotation about the axis perpendicular to the plane passing through the z axis and the instantaneous position of the molecular axis. As is seen from Eq. (1.8.5) and Fig. 1.9, at a given velocity $\dot\varphi$ the angular momentum p_φ is the function of θ, and at $\theta = 0$ it is equal to zero. Thus, unlike p_θ and $\dot\theta$, the ratio between generalized momentum p_φ and generalized velocity $\dot\varphi$ is not a constant.

1.8 DIELECTRIC PROPERTIES OF GASES

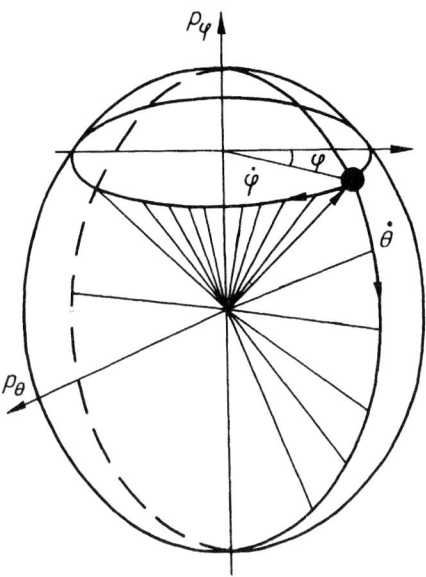

Figure 1.9 Components of angular velocity and momentum in spherical coordinates.

Probability Distribution

According to the Gibbs canonical distribution it is essential that

$$d\Gamma = d\theta d\varphi dp_\theta dp_\varphi , \qquad (1.8.6)$$

but not $d\theta d\varphi d\dot\theta d\dot\varphi$. In view of (1.8.2), (1.8.4) and (1.8.5), the total energy in canonical variables is as follows

$$\mathcal{E}(\theta, \varphi, p_\theta, p_\varphi) = \frac{p_\theta^2}{2I} + \frac{p_\varphi^2}{2I \sin^2 \theta} - qE \cos \theta . \qquad (1.8.7)$$

Now the distribution (1.7.3) can easily be made more specific. Using (1.8.6) and (1.8.7), we put it into the form

$$dW(\theta,\varphi,p_\varphi,p_\theta) = \frac{1}{Z} \exp\left[-\frac{p_\theta^2}{2IkT} - \frac{p_\varphi^2}{2IkT \sin^2 \theta} + \frac{qE}{kT} \cos \theta\right] dp_\theta dp_\varphi d\theta d\varphi . \qquad (1.8.8)$$

The partition function is

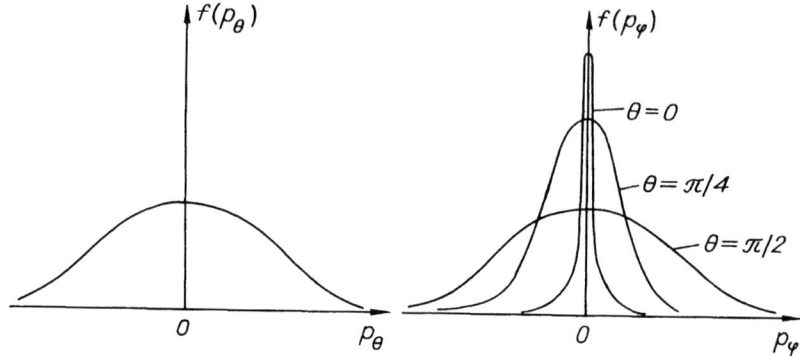

Figure 1.10 Distribution of angular momentum components at different orientations of the molecular axis given by Θ.

$$Z = \frac{8\pi^2 I(kT)^2}{qE} \sinh\left(\frac{qE}{kT}\right).$$

According to (1.8.8), the probability of finding a certain value p_φ depends on θ. This is quite natural: the smaller the angle θ, the less probable are large values of p_φ. At $\theta = 0$, all p_φ are improbable, except zero (Fig. 1.10).

Now we express the distribution of molecules in terms of generalized coordinates and velocities. In view of (1.8.5)

$$dW(\theta, \varphi, \dot\theta, \dot\varphi) = \frac{1}{Z} \exp\left[-\frac{I\dot\theta^2}{2kT} - \frac{I\sin^2\theta\dot\varphi^2}{2kT} + \frac{qE}{kT}\cos\theta\right] I^2 \sin^2\theta\, d\theta d\varphi d\dot\theta d\dot\varphi, \qquad (1.8.9)$$

where $\rho(\theta, \varphi, \dot\theta, \dot\varphi) = I^2 \sin^2\theta$ is the statistical weight, which is not equal to unity in these variables (in contrast to canonical ones) and could not be inferred beforehand.

If we are interested in the dipole orientation rather than in the rotation rate, then, following the routine procedure, we have

$$dW(\theta, \varphi) = \int_{p_\theta} \int_{p_\varphi} dW(\theta, \varphi, p_\theta, p_\varphi) = \frac{\alpha}{4\pi \sh \alpha} e^{\alpha \cos\theta} \sin\theta d\theta d\varphi, \qquad (1.8.10)$$

where $\alpha = qE/kT$. For $\alpha \gg 1$, significant deflections of the molecular axis from the z axis are highly improbable, while at $\alpha \ll 1$, all directions are almost equiprobable (Fig. 1.11).

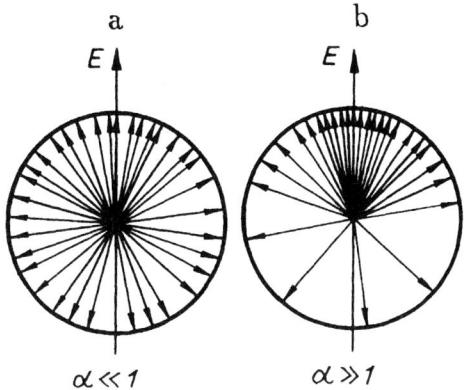

Figure 1.11 Distribution of dipole moments related to the external field E at (a) high and (b) low temperatures.

Electric Polarization

Let us calculate the equilibrium polarization of a gas in an electric field, which is the average dipole moment per unit volume

$$\mathbf{Q} = \int \mathbf{q}\, dn(\mathbf{q}) . \qquad (1.8.11)$$

The average number of molecules with a given orientation of the dipole moment **q** contained in the unit volume is

$$dn(\mathbf{q}) = n\, dW(\theta, \varphi) . \qquad (1.8.12)$$

Due to the axial symmetry of the dipole moment distribution (1.8.10), the mean values of the transversal components of the dipole moment are equal to zero, and the problem reduces to the averaging of the longitudinal component

$$Q = nq \int_\theta \int_\varphi \cos\theta\, dW(\theta,\varphi) = nqL(\alpha) , \qquad (1.8.13)$$

where $L(\alpha)$ is the so-called Langevin function

$$L(\alpha) = \frac{ch\,\alpha}{sh\,\alpha} - \frac{1}{\alpha} = \frac{e^\alpha + e^{-\alpha}}{e^\alpha - e^{-\alpha}} - \frac{1}{\alpha} \simeq \begin{cases} 1 - \dfrac{1}{\alpha} & \alpha \gg 1 \\ \dfrac{\alpha}{3} & \alpha \ll 1 \end{cases} . \qquad (1.8.14)$$

Thus the equilibrium polarization of the gas is parallel to the applied field, and its absolute value is a function of the field and temperature. So the two limiting cases may be distinguished

$$Q = qn\frac{\alpha}{3} = \frac{1}{3}\frac{q^2 n}{kT} E = \chi \cdot E \quad \text{at} \quad \alpha \ll 1; \quad (1.8.15\text{a})$$

$$Q = qn \quad \text{at} \quad \alpha \gg 1. \quad (1.8.15\text{b})$$

At rather high temperatures, or, correspondingly, low fields, the polarization is linear in the field and the quantity

$$\chi = \frac{q^2}{3kT} n \quad (1.8.16)$$

is called the electric susceptibility per unit volume. At low fields the excess of dipoles lined up with the field is fairly insignificant, but increases in direct proportion to E. In this case, the linear field theory, or linear optics, is said to be valid, if we deal with alternating fields. Retaining the components quadratic in E in a power expansion of the Langevin function gives rise to the effects of nonlinear optics. Finally, at very strong fields ($\alpha \gg 1$), almost all molecules become orientated with the field, and further increase of E has practically no effect on the polarization ("saturation effect"). In ordinary fields and at room temperatures this case is realized only rarely. As a rule, $\theta = \chi E$, where χ, according to (1.8.16), increases in inverse proportion to the absolute temperature (the Curie law).

1.9 REAL GAS

With the advent of the Gibbs distribution, the problem with the ideal statement may be considered to be solved. All properties of a gas consisting of noninteracting particles are rigorously described by it, however complicated the structure of an isolated molecule.

Intermolecular Interaction

However, as soon as intermolecular interaction is taken into account, the situation becomes different. The ideal gas equation of state $pV = RT$ is not observed to be a universal law. Even real gases do not strictly obey it. In fact, there are as many equations of state as there are gases, since the interaction is specific for each particular kind of molecule. Any general regularity can be inferred only from a rough, idealized model of actual intermolecular interaction. The simplest and most general model of an angle-independent interaction is that involving only two parameters: one for mutual attraction, and the other, for repulsion of molecules.

The mere existence of condensed phases of a substance—liquids and solids—suggests that at low temperature molecules are bound together by attraction forces. However, mutual attraction prevailing at remote distances gives way to repulsion at very close proximity. The origin of repulsion is even easier to understand than that of attraction. When molecules approach one another up to their "eigenvolume," further approach encounters stiff resistance, since it can result in the deformation of the molecules' structure or atomic shells. A quantitative measure of this resistance is the potential energy of interaction $u(r)$. It abruptly increases after passing through the minimum at $r = a_0$ where attraction and repulsion forces balance, then changes sign at $r = d$ and tends to infinity (Fig. 1.12). The approach of one molecule towards the other becomes energy-consuming as soon as their centers come closer than d. Thus the interaction law can be simplified assuming that at the distance d a molecule encounters a potential wall. This is equivalent to assuming that the molecules can be modeled as rigid impenetrable sphere of radius $d/2$. The only advantage of such a simplification is that we can easily allow for the "eigenvolume" (or excluded volume) of molecules $v_d = \frac{4}{3}(d/2)^3$, neglecting

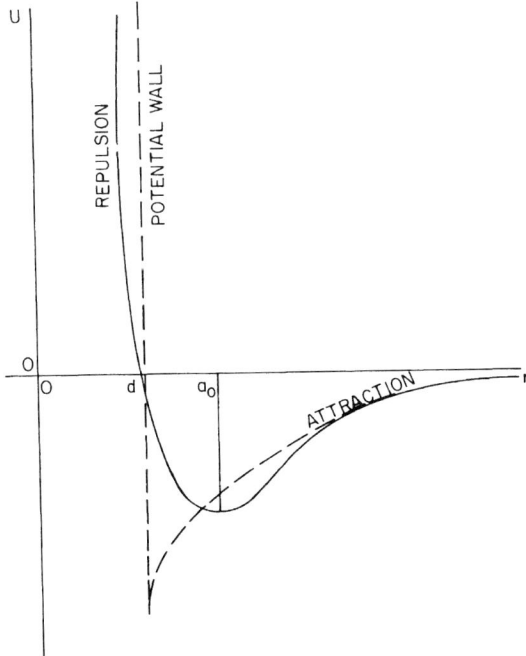

Figure 1.12 Actual intermolecular potential (solid line) and its approximation within the model of attracting hard spheres (dashed line).

the details. This is just the acknowledgment of the fact that the molecules are particles of finite size rather than point masses. The derivation of the van der Waals equation is based exactly on this idealization of molecules as hard spheres attracting one another.

Repulsion

First let us see what we obtain by making allowance for the molecules' eigenvolume. Consider one molecule as a point mass, and all the other molecules as hard spheres of radius d (Fig. 1.13). In our previous discussion, a point mass could be found at any point in space, while now a part of the space is excluded, equivalent to a total volume of spheres of doubled radius. Since the volume of such a sphere is $8v_d$, then in any cubic centimeter very remote from the vessel walls the fraction of excluded volume is $8v_d n$, while that of accessible volume is $1 - 8v_d n$. Thus in the vessel of a volume V, the total free volume is $V_f = V(1 - 8v_d n)$. In the elementary volume dV the free part is $dV_f = dV(1 - 8v_d n)$.

Since penetration into the occupied part of the space is impossible, only equal elements of the free part are occupied equiprobably. In other words, the probability of finding a molecule in any space element free of other molecules is dV_f/V_f. So far, this makes no difference, since the total volume and its unoccupied part are related in the same way: $dW = dV_f/V_f = dV/V$. However, for elements in the immediate vicinity of the walls the relationship between dV_f and dV is somewhat different. Since the molecules are actually hard spheres with the radius $d/2$, none of them including the chosen one can approach the wall closer than $d/2$. Therefore only semispheres presented inside the vessel constitute the excluded volume inaccessible near the wall while the opposite semispheres are of no importance. Each molecule in contact with the wall reduces the free part of the space by half as much as the molecule inside the volume (Fig. 1.14), namely, $4v_d$. If the corpuscular density in the neighborhood of the wall is n_s, in each near-surface unit volume just $4n_s v_d$ is filled by the matter, while the remaining part is a free volume $dV_s = dV(1 - 4v_d n)$.

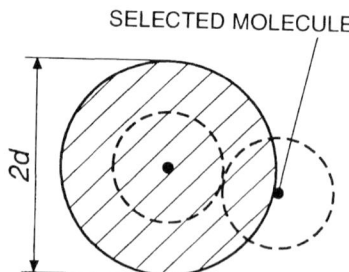

Figure 1.13 Excluded volume in a collision of hard spheres (shaded).

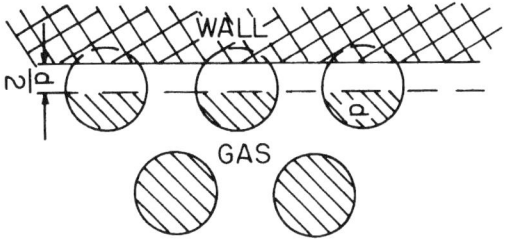

Figure 1.14 Excluded volume near the wall and in the interior of a gas container. The layer of width $d/2$ is unavailable to the center of any molecule.

Accordingly, the probability of finding a molecule near the surface is somewhat greater than that away from it.

$$dW_s = \frac{dV_s}{V_f} = \frac{1 - 4v_d n_s}{1 - 8v_d n}\frac{dV}{V}.$$

For this reason we have to distinguish between the near-surface density n_s and the average density $n = N/V$. Obviously,

$$n_s = \frac{NdW_s}{dV} = \frac{1 - 4v_d n_s}{1 - 8v_d n} n,$$

so

$$n_s = \frac{n}{1 - 4v_d n}. \tag{1.9.1}$$

The near-wall increase in the particles density also affects the pressure applied to the wall

$$p = n_s kT = \frac{nkT}{1 - 4v_d n}. \tag{1.9.2}$$

This is essentially the simplest estimate of *thermal* pressure of hard-sphere gas $p_t = \Gamma nkT$. Since $\Gamma = 1/(1 - 4v_d n) > 1$ at $v_d \neq 0$ the real gas pressure is greater than that in the ideal gas (point mass) model.

Attraction

If one takes into account now the mutual attraction of molecules, the opposite results are obtained. Consider a molecule far from the wall that is being attracted by each molecule nearby. Such a molecule is thrown from side to side, since the resultant of the attractive forces varies randomly in magnitude

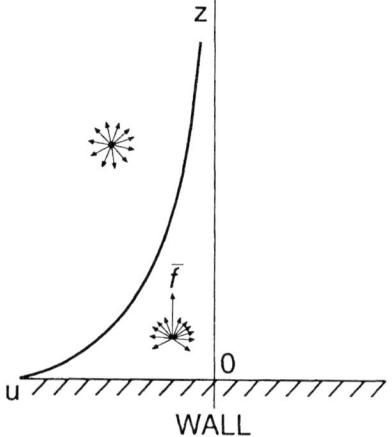

Figure 1.15 The resultant attractive force (near the wall and inside the gas container) and its potential $u(z)$.

and direction. However, the time-averaged force is equal to zero, because the molecule is pulled in different directions and none of them is preferred (Fig. 1.15).

However, a molecule near the surface experiences attraction only from the inside of the vessel, because the wall is still considered as ideal. Thus, despite fluctuations, the time-average resultant of attractive forces differs from zero. Due to symmetry, this average force must be perpendicular to the wall and directed inward. As a molecule moves away from the wall, the direction of the force remains the same, while its magnitude decreases and, eventually, goes to zero. At any point the average force is the increasing function of the density of molecules whose collective efforts draw a periphery molecule inside the vessel:

$$\bar{f} = \gamma(z)n . \qquad (1.9.3)$$

The coefficient $\gamma(z)$ must be equal to zero deep inside the vessel (with $z \to \infty$), but increases monotonically when the wall is approached, reaching its maximum at $z = 0$. As with any other field, this field of average force may be associated with a certain potential going to zero away from the walls

$$u(z) = n \int_z^\infty \gamma(z)dz . \qquad (1.9.4)$$

The distribution of molecules in such a field is described by the barometric formula:

$$n(z) = n \exp\left(-\frac{n}{kT} \int_z^\infty \gamma(z) dz\right). \tag{1.9.5}$$

Naturally, the gas density decreases approaching the wall, though the precise dependence is unknown due to the lack of knowledge of $\gamma(z)$. Fortunately, this is of minor importance. To find the pressure, we need only know the density in the vicinity of the wall

$$n(0) = n \exp\left(-\frac{n\bar{\gamma}}{kT}\right), \tag{1.9.6}$$

which is expressed in terms of the single unknown parameter $\bar{\gamma} = \int_0^\infty \gamma(z) dz$. The quantity $\bar{\gamma}n$ has the meaning of the "work expended on escaping from the gas," that is, the energy spent by a "surface molecule" in overcoming the coupling forces that bind it to other molecules inside the gas. The parameter $\bar{\gamma}$ describes the attraction of molecules, just as v_d determines the repulsion effect. It is seen from Eq. (1.9.6) that the pressure on the wall is

$$p = n(0)kT = nkT \exp\left(-\frac{n\bar{\gamma}}{kT}\right). \tag{1.9.7}$$

It is not surprising that the pressure is less than that of the ideal gas: attraction forces acting from inside partly prevent the boundary molecules from striking the walls.

Equation of State

Equation (1.9.2) is valid for noninteracting hard spheres and Eq. (1.9.7) for point particles attracting each other. These models can be synthesized so that both repulsion and attraction will be taken into account. For this purpose it is sufficient to generalize Eq. (1.9.6) by considering the near-border gas compression, as was done before for Eq. (1.9.1). Finally, we arrive at

$$n_s = \frac{n(0)}{1 - 4v_d n} = \frac{n}{1 - 4v_d n} \exp\left(-\frac{n\bar{\gamma}}{kT}\right) \tag{1.9.8}$$

and, consequently:

$$p = \frac{nkT}{1 - 4v_d n} \exp\left(-\frac{n\bar{\gamma}}{kT}\right). \tag{1.9.9}$$

This is the so-called Dieterici equation of state. When the exponent is much less than unity, one can easily derive, by power expansion of Eq. (1.9.9), the following equation for one mole of gas:

$$p = \frac{RT}{V-B} - \frac{A}{V^2}, \qquad (1.9.10)$$

where $B = 4N_0 v_d$ and $A = N_0^2 \bar{\gamma}$. This is the famous van der Waals equation. It may be represented as

$$p = p_t - p_i,$$

where thermal pressure p_t is the same as in Eq. (1.9.2), while the *internal* pressure $p_i = A/V^2$ accounts for attraction of particles. We will see that even two parameters, used to take roughly into account repulsion and attraction, are sufficient to reveal important qualitative features of real gases, including their capability to condense into a liquid state.

This phase transition occurs in the region where the van der Waals isotherm makes a "loop" (Fig. 1.16). Above and below the loop, at any given pressure there is a single equilibrium state on the isotherm, while within the loop there are three of them. The side-states correspond to the equilibrium states of liquid (point) and vapor (circle). The states marked with crosses are in the interior of the loop, between the minimum and maximum, which is characterized by a positive sign of the derivative $(\partial p/\partial V)_T$. These states are unstable. If any volume fluctuation occurs there, it would tend to develop, thus returning the system to one of the stable branches of the isotherm (either left or right).

The instability of intermediate states prevents a gradual liquid–vapor transition, which would correspond to a successive movement along the

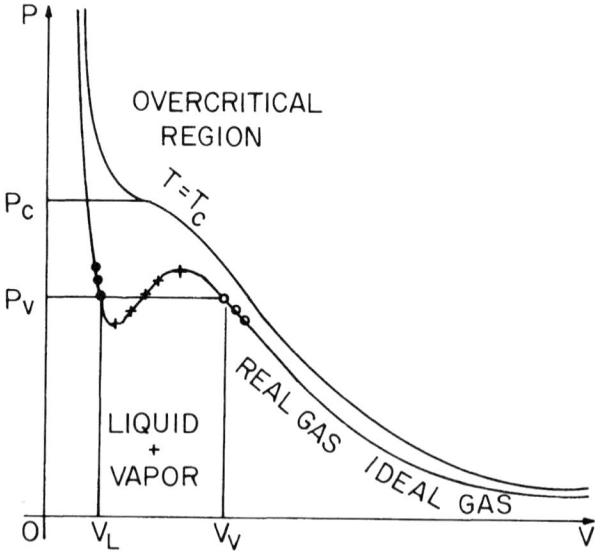

Figure 1.16 Liquid (•), unstable (+), and gaseous (○) states on the van der Waals isotherm.

loop, including its inner region. Instead, a homogeneous phase decomposes into liquid and vapor and, thus, transforms into a binary system. When the pressure exerted on a dense homogeneous fluid tends to decrease at constant temperature, both phases and an interface appear simultaneously at $p = p_V$. The pressure of the saturated vapor p_V remains constant until all liquid is evaporated. Only after the liquid–vapor transition is complete does the system again becomes homogeneous and the gas pressure starts to fall upon expansion.

The homogeneous systems may not be distinguished if they are disordered. One can distinguish between the gas and liquid if only they are the integral parts of the binary system. The criterion is the density, which makes a "jump" at the interface. In Fig. 1.16 this jump is characterized by the difference between molar volumes, V_L and V_V. This difference decreases with rising temperature and goes to zero once the critical temperature T_c is achieved. For all $T > T_c$ (overcritical region) the existence of a binary system is impossible and, consequently, liquid and gas are indistinguishable.

For any temperature below critical the liquid is in equilibrium with its vapor at a distinct pressure p_V, kept constant on a horizontal "Maxwell shelf," that joins the corresponding states on Fig. 1.16. The vapor pressure may be estimated on the basis of the so-called Maxwell rule (see Eq. (5.6.24) in Chapter 5):

$$p_V(V_V - V_L) = \int_{V_L}^{V_V} p(V, T) dV. \qquad (1.9.11)$$

Thus, p_V is determined from the equality of the areas below the Maxwell shelf and the van der Waals loop joining the same states.

The extent to which the van der Waals theory represents the qualitative features of the behavior of real gases can be judged from Fig. 1.17, which depicts a family of nitrogen isotherms. They are plotted according to an empirical equation of state, which is far more complex than the van der Waals equation but approximates experimental data within the error. Parameters of the equation are adjusted using data related to stable isotherm branches (liquid and gas). In the unstable region, one can only extrapolate these data to it, relying on analytical continuation of well established isotherms $p(V)$. If the extrapolation is reliable, both branches are linked within the unstable region, thus demonstrating a behavior typical for the van der Waals loop. This was shown to be the case at a few different temperatures below critical one. Moreover, when estimating the vapor pressure using the Maxwell rule applied to the restored loops it was verified that p_v so obtained coincides with the experimental one. This is evidence that the van der Waals loop is something real.

Points connected by the Maxwell shelves lie on curve a (Fig. 1.17), called a binodal. An ascending branch of the binodal is called a liquid line and a descending branch a gas line. By carefully expanding a very pure liquid, it is

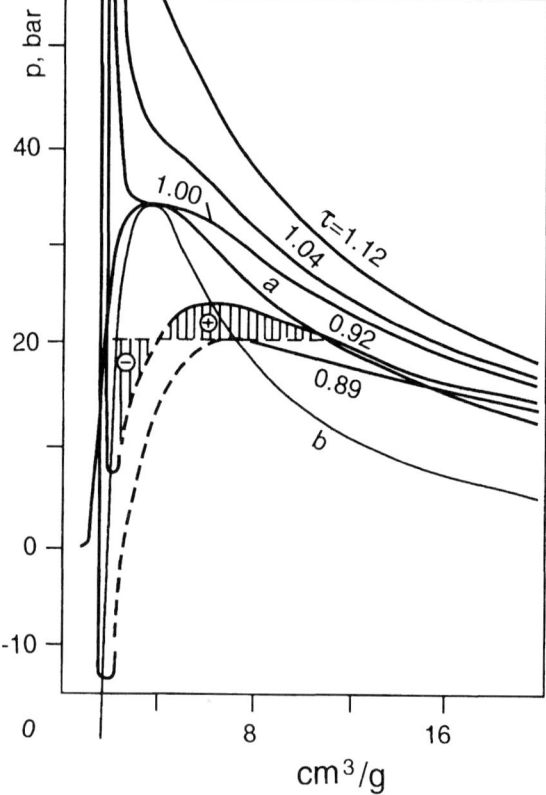

Figure 1.17 Nitrogen isotherms at corresponding temperatures $\tau = T/T_c$: (a) binodal (coexistence line), (b) spinodal. The loop for $\tau = 0.92$ is hatched. (From A. A. Vasserman *Russ. J. Phys. Chem.* **38**, 1289 (1964).)

possible to cross the binodal and move down along the isotherm to approach its minimum, unless it lands in the negative pressure region. Similarly, it is possible to advance upwards along the gas branch of the isotherm to its maximum. In these cases the homogeneous metastable states of the *overexpanded* liquid or *supersaturated* vapor are attained. Although unstable with respect to division into two phases they are still realizable. Totally unstable are the inner parts of the isotherms between the extrema situated on curve b, which is called spinodal (see Fig. 1.17). While its ascending branch corresponds to a maximally expanded condensed state, its descending branch corresponds to a maximally condensed gas state. Metastable states located between the binodal and spinodal may be also related either to an *overheated* liquid or to an *overcooled* vapor, provided they are compared with stable phases at the same pressure. The binodal and spinodal have a common apex at the critical point. The critical isotherm is bent at this point, and at higher temperatures $(\partial p/\partial V)_T < 0$ all along the isotherms.

The whole picture is described semiquantitatively by the van der Waals equation. Positions of isotherm extrema that constitute the van der Waals spinodal are located by the conventional condition:

$$\left(\frac{\partial p}{\partial V}\right)_T = -\frac{RT}{(V-B)^2} + \frac{2A}{V^3} = 0. \tag{1.9.12a}$$

As the temperature increases, the extrema come closer to each other and eventually merge at $T = T_c$ creating an inflection point, which is determined by an additional condition:

$$\left(\frac{\partial^2 p}{\partial V^2}\right)_T = \frac{2RT}{(V-B)^3} - \frac{6A}{V^4} = 0. \tag{1.9.12b}$$

Using both equations (1.9.12) one can find, within the van der Waals theory, the coordinates of the critical state:

$$V_c = 3B, \quad T_c = \frac{8}{27}\frac{A}{RB}, \quad p_c = \frac{A}{27B^2}. \tag{1.9.13}$$

The last coordinate was determined by substituting V_c and T_c into Eq. (1.9.10). All the parameters taken together determine the van der Waals gas compressibility factor at the critical point: $p_c V_c / RT_c = F_c = 3/8 = 0.375$. It turns out to be much higher than the experimental value. For example, for a series of saturated hydrocarbons $F_c = 0.267$. This value is in much better agreement with the value found by the Dieterici equation ($F_c = 0.2706$). However, any equations that use only two parameters can hardly claim to provide good quantitative agreement with experiment over wide ranges of p and V. Still, from the heuristic viewpoint these equations are very valuable owing to their simplicity and transparency as regards the physical sense.

The difference in size of molecules and in the cohesive force between them has quite a strong effect on the van der Waals parameters and, eventually, on critical characteristics of real gases (1.9.13). To facilitate and unify comparison of different molecular media, van der Waals put forward the idea of their "corresponding states," which are determined by the following reduced variables:

$$\pi = \frac{p}{p_c}, \quad \omega = \frac{V_c}{V}, \quad \tau = \frac{T}{T_c}. \tag{1.9.14}$$

With these variables, the van der Waals equation (1.9.10) acquires universality, independent of the nature of the gas:

$$\pi + 3\omega^2 = \frac{8\omega\tau}{3-\omega}. \qquad (1.9.15)$$

This universality is a useful guide for estimating the behavior of dense gases and liquids. It was formulated as a "principle of corresponding states": all substances taken at the corresponding temperatures and pressures also have the same corresponding density.

As found, the principle of corresponding states is in good agreement with experimental data despite the fact that the van der Waals equation itself is not very accurate. In other words, the principle has far wider applications than the equation it is based on. Isotherms of various gases plotted as functions of the reduced variables (1.9.14) are, in fact, very close to each other and are nearly the same for gases having intermolecular interactions of the same type.

Compressibility Factor

The Hougen and Watson chart shown in Fig. 1.18 was prepared by averaging data for seven gases (H_2, N_2, CO, NH_3, CH_4, C_3H_8 and C_5H_{12}). It depicts the isothermal pressure dependence of the gas compressibility factor, $F = pV/RT$. A modern analog of this chart for N_2 is shown in Fig. 1.19. The latter gives an idea of isothermal behavior of $F(p)$ over much wider ranges and, in addition, for both gas and liquid states.

On these charts the ideal gas is presented by a single point: $p = 0$, $F = 1$. All isotherms begin from this point corresponding to an infinitely rarefied state, but behave differently as the pressure increases. For high-temperature isotherms the compressibility factor increases from the beginning. For low-temperature isotherms it first decreases and undergoes a sharp jump downwards upon gas liquefaction and only then tends upwards, returns to its starting value $F = 1$ and eventually exceeds it. This is because attraction makes compressibility smaller, while repulsion enhances it. With increasing temperature or pressure the role of repulsion increases and finally becomes dominating even if initially it was relatively low.

To make these arguments more concrete let us present the van der Waals compressibility factor as density expansion in the vicinity of the ideal gas state:

$$F = \frac{1}{1 - B/V} - \frac{A}{RTV} = 1 + \frac{B - A/RT}{V} + \left(\frac{B}{V}\right)^2 + \left(\frac{B}{V}\right)^3 + \ldots \qquad (1.9.16)$$

This form of the van der Waals equation can be regarded as a particular case of the virial expansion:

$$F = 1 + \frac{B_0}{V} + \frac{C_0}{V^2} + \frac{D_0}{V^3} + \ldots, \qquad (1.9.17)$$

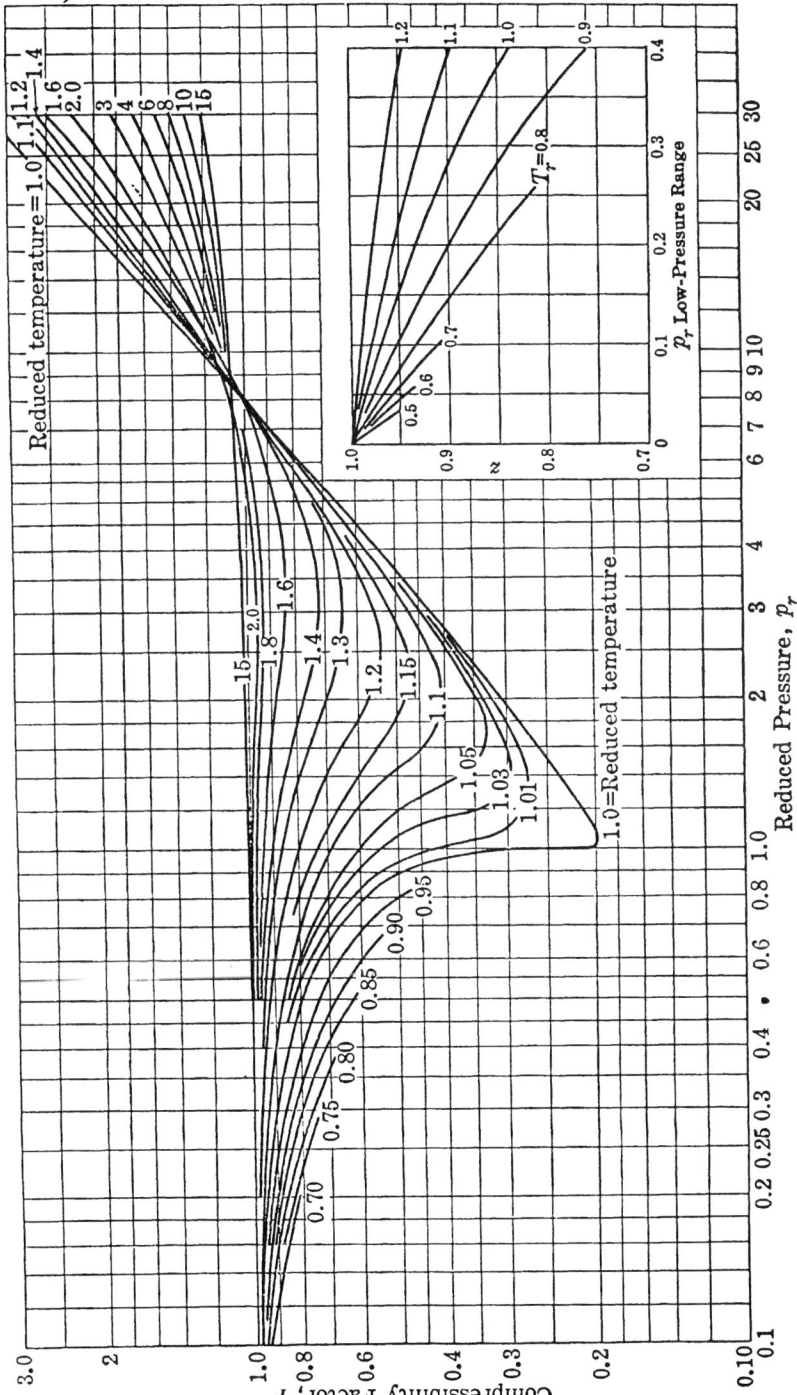

Figure 1.18 The compressibility factor as a function of the reduced pressure and the reduced temperature. (From O. A. Hougen and K. M. Watson, *Chemical Process Principles*, John Wiley (1947), Part II; see also Fig. 4.1–2 in *Molecular Theory of Gases and Liquids* by J. O. Hirschfelder, C. F. Curtiss, and B. R. Bird.)

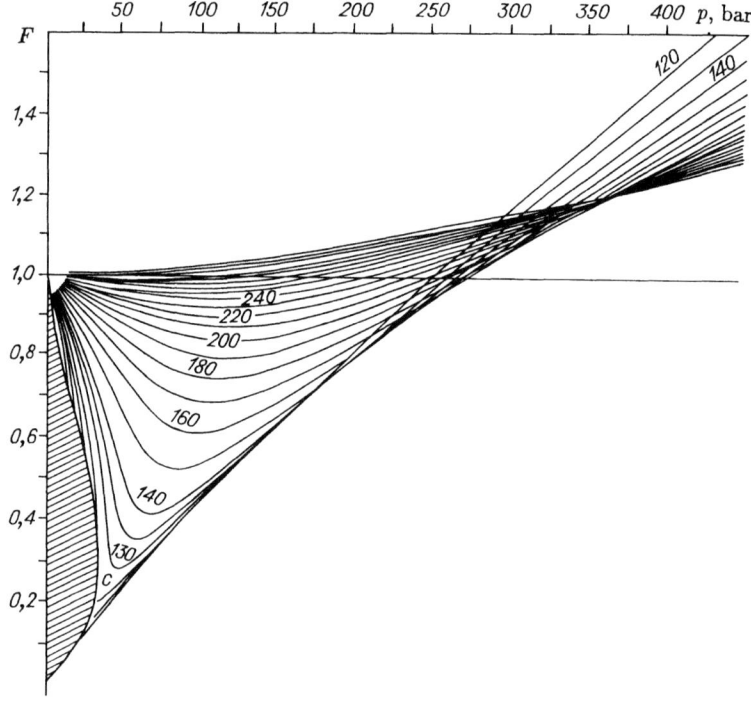

Figure 1.19 The contemporary Hougen–Watson chart for nitrogen.

whose coefficients are expressed in terms of intermolecular forces (Latin *vires*). Generally speaking, each of the coefficients depend on the forces of attraction and repulsion. However, within the approximate van der Waals model it is the second virial coefficient alone that takes care of both forces. This coefficient is defined as:

$$B_0 = B - \frac{A}{RT}. \tag{1.9.18}$$

The competition between attraction and repulsion affects not only the value, but also the sign of the coefficient, which determines the isotherm slope at $p = 0$:

$$F \approx 1 + \frac{B_0}{V} \approx 1 + \frac{B_0}{RT} p \quad \text{at} \quad |1 - F| \ll 1. \tag{1.9.19}$$

The second virial coefficient is equal to zero at the so-called Boyle's temperature. Within the van der Waals theory this is:

$$T_B = \frac{A}{BR}. \tag{1.9.20}$$

Below Boyle's temperature, upon isothermal compression of a real gas the compressibility factor first falls. Above Boyle's temperature the compressibility factor increases from the very beginning to the very end.

On inspecting Figs. 1.18 and 1.19 one can easily see that below Boyle's temperature the compressibility factor equals unity at two points of any isotherm. One of the points, which is common for all isotherms, corresponds to the true ideal gas state at $p = RT/V \to 0$, when both van der Waals corrections are equal to zero. The second point corresponds to a "pseudoideal" state. In this state both corrections are significant, but are equal and opposite in sign and so cancel one another. As a result, one obtains again $pV = RT$. Pseudoideal states correspond to the points where isotherms cut the line $F = 1$. Their abscissae depend on the temperatures (both on pressure and density scales).

The locus where gas becomes pseudoideal is the "ideal gas curve," or the "orthometric curve." The latter notion was originally introduced by Batschinski. Assuming $F = 1$ and taking into account the definition of Boyle's temperature one can readily derive from the van der Waals equation the corresponding equation of the orthometric curve:

$$1 - B/V = T_B/T.$$

From this it follows that density on the gas orthometric line falls linearly with increasing temperature:

$$n = \frac{N_0}{B}\left(1 - \frac{T}{T_B}\right) = \frac{1}{4v_d}\left(1 - \frac{T}{T_B}\right). \tag{1.9.21}$$

This surprising and impressive linearity (see Fig. 4.16) discovered by Batschinski in 1906 was a subject of intensive studies made in the last century's sixties and seventies by Vasserman, Nedostup, Burshtein and Shokhirev. Quite recently this phenomenon was subjected to a new investigation by Ferschbach et al. who gave it a new name: "Zeno line".

Along this line the pressure changes parabolically:

$$p = nkT = \frac{RT}{B}\left(1 - \frac{T}{T_B}\right). \tag{1.9.22}$$

It is seen that the orthometric curve $p(T)$ is a parabola with an apex at the point $T_B/2$. The maximum orthometric pressure is

$$p_0 = \frac{RT_B}{4B} = \frac{A}{4B^2}. \tag{1.9.23}$$

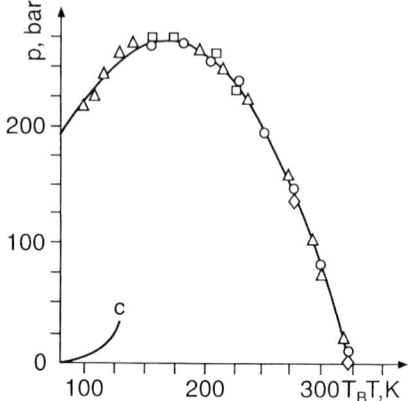

Figure 1.20 Nitrogen orthometric ("ideal gas") curve. The coexistence line (binodal) shown in the lower part of the graph comes to an end at the critical point C.

These conclusions of the van der Waals theory are in excellent agreement with experimental studies of orthometric curves of noble and molecular gases, which condense in "simple liquids" (see Chapter 4). Fig. 1.20 shows a nitrogen orthometric curve, plotted using the most reliable experimental data of different authors. The curve is, indeed, parabolic along the whole length. Everywhere under this curve attraction forces dominate over repulsion, while above the situation is opposite. The liquid–vapor coexistence curve that ends with a critical point is deep inside the region where attraction plays a governing role.

The fact that the pressure, where pseudoidealization occurs, first increases with temperature and then falls is seen on close inspection of Fig. 1.19. Taking into account that near the orthometric maximum the pressure changes only slightly, one can expect that a few isotherms having temperatures close to $T_B/2$ will pass through pseudoideal states under nearly the same pressure. Due to the selection of exactly such isotherms for the Hougen–Watson chart (Fig. 1.18), it appears that they cross in at a common point. Apparently, the abscissa of the intersection point is nothing else but p_0—the maximum orthometric pressure.

With relation (1.9.13) it is possible to obtain the following estimates of typical orthometric curve parameters:

$$T_B = \frac{27}{8} T_c, \quad p_0 = \frac{27}{4} p_c. \qquad (1.9.24)$$

However, empirically, the actual T_B/T_c ratio is close to 2.64 rather than to 3.37 as follows from (1.9.24). It is exactly for this reason that the temperature of the orthometric maximum is very close to the critical temperature: $T_B/2 = 1.3\, T_c$. Similarly, the maximum pressure, which according to the Hougen–Watson chart is equal to $8p_c$, is notably higher than the value obtained from (1.9.24): $p_0 = 6.8\, p_c$. However, one should not expect better from the van der Waals

1.10 BASIC IDEAS OF STATISTICAL THERMODYNAMICS

equation, which uses only two parameters and is quantitatively valid for only low pressures.

1.10 BASIC IDEAS OF STATISTICAL THERMODYNAMICS

The above derivation of the van der Waals equation is possible due to the essential simplifications of the intermolecular potential. In any attempt to do it more rigorously and accurately we have to deal with actual interactions of particles, in which it is difficult or even impossible to separate attraction and repulsion. Fortunately, there is an alternative method which leads to a standard recipe for the calculation of thermodynamic properties of macroscopic systems with arbitrary interparticle interactions. This method was advanced by Gibbs.

The Gibbs Ensemble

The main idea was to consider N interacting particles as a unit rather than separately. To define the "gas state" as a whole, we need now to specify the positions and momenta of all N molecules. Formerly, the "state" was defined by the six variables (coordinates and velocities or coordinates and momenta) of a single molecule; now we need the $6N$ variables of all molecules in the given volume. To specify $6N$ numbers means to indicate the point in the $6N$-dimensional configurational space ("Γ- space") that defines the state of the entire gas, while the point in the ordinary 6-dimensional "μ-space" specifies that of one particle only.

In fact, in moving to the Γ-space terminology we can define by one point in Γ-space the distribution of N points in μ-space. This distribution is not, however, the Maxwell Boltzmann distribution. For an arbitrarily chosen point in Γ-space, the distribution of particles corresponds to an instantaneously "photographed" gas state. Some time later, the state is changed and is therefore associated with another point in Γ-space or with another distribution of particles in μ-space. The gas as a mechanical system develops in time: the particles move, change their position, and, upon collisions, their velocity. The system passes through a series of states lying on a trajectory in Γ-space (Fig. 1.21a). In μ-space distributions permanently change from one to another. To illustrate this, consider, for example, the energy distribution in μ-space (Fig. 1.21b). A given point in Γ-space corresponds to one distribution, another point to another distribution, and thus Maxwell's $\bar{f}$ may be found only by exhausting all trajectory points, that is, by averaging over all possible distributions. It is easily seen that this procedure is similar to the determination of the average pressure (see Section 1.1). Each state of the system is associated with a certain pressure $F(t)/S$ that oscillates randomly in time about its mean value (Fig. 1.21c). Similarly, the instantaneous distribution of molecules in a gas differs

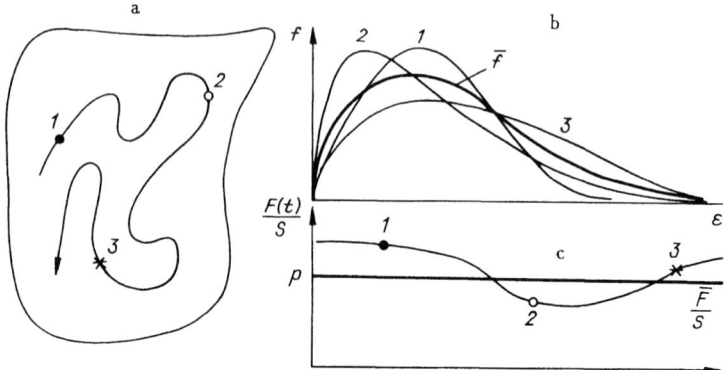

Figure 1.21 (a) The points of the trajectory in Γ-space, (b) the corresponding energy distributions, and (c) pressure acting on the wall at successive moments of time 1, 2, and 3.

from the time-averaged distribution which is the only appropriate one for a description of the system's equilibrium properties.

How can we find the time-averaged distribution? Before averaging the distributions, it is necessary to know the frequency of their realization, that is, the probability that the system will find itself at one or another point in Γ-space. This is the statement of the main problem of statistical mechanics. To clarify, let us imagine, as suggested by Gibbs, a collection of identical systems—an "ensemble" of identical boxes containing the same number of identical particles. Obviously, while the macroscopic gas state is the same in each of them, the microscopic states of the ensemble's systems do not necessarily coincide: each is associated with a different point in Γ-space. How can one find the equilibrium distribution of these points in Γ-space?

Microcanonical Distribution

Some information on the form of this distribution may be obtained from the fact that all points representing these systems in Γ-space move along the corresponding phase trajectories. We want the result of any averaging over the ensemble to be time-independent, as this is just what happens under equilibrium conditions. Therefore, we should require that the density of points in any Γ-space element be constant, despite motion. In other words, the probability of any microstate realization must be time-invariant. If all systems of the ensemble are isolated from heat and mass exchange with the environment, their internal motion conserves energy and they are described by ordinary mechanical laws. By these laws, when in motion, the points in the Γ-space of *canonical variables* (coordinates and momenta) behave as marked points of incompressible liquid: their density in a flow does not vary with time ("Liouville's theorem"). For everything to be time-invariant, the points must be distributed uniformly along the trajectory. Then the number of points entering any Γ-space volume element will coincide with the number of points leaving it in the same period of time. Thus it is the Liouville theorem that distinguishes the *phase space* of canonical variables as unique (between other configurational spaces). In this space and this space only the distribution over equal isoenergetic elements is assumed to be equiprobable (see Section 1.7). However, the Liouville theorem does not define the distribution of points in the phase space in a unique fashion. To meet the condition of equilibrium, we need only distribute the points uniformly along each trajectory. However, many trajectories have the same energy, and it is unknown how the point density will vary from one trajectory to another.

However, all trajectories are indistinguishable from any point of view. If isolated, the system has the same energy for any accessible point of Γ-space, that is, all points of Γ-space lie on a hypersurface of equal energy. Thus the only possible distinguishing feature disappears and the only way out is to assume that all trajectories are equivalent. Therefore, the equal volume elements of different trajectories must also be equivalent and consequently equally populated. In other words, the distribution of systems in Γ-space may be reasonably assumed to be completely uniform. Of course, such an extension of the conclusion following from the Liouville theorem should be considered as a principle. Its final statement is known as Gibbs' microcanonical distribution: an isolated system may be found equiprobably at any accessible state of the prescribed energy $\mathcal{E}$ (between E and $E + \Delta E$), while the probability of finding it outside this energy layer is equal to zero. As usual, the probability is proportional to the volume of the element $d\Gamma$ where the system is sought:

$$dW = \begin{cases} \dfrac{d\Gamma}{\Delta\Gamma} = \Phi d\Gamma & E \leq \mathcal{E} \leq E + \Delta E, \\ 0 & \mathcal{E} < E \quad \mathcal{E} > E + \Delta E. \end{cases} \quad (1.10.1)$$

Here $\Delta\Gamma$ is the complete volume of the Γ-space accessible to the system at the given energy.

Gibbs' Canonical Distribution

The Gibbs canonical ensemble may be thought of as an ensemble of systems which are "Gibbs' boxes" disposed in a heat reservoir. Such systems are no longer isolated, but have a heat contact with the environment. In each system the temperature is held constant (equal to that of the reservoir), while the energy may vary. Due to the energy exchange with the reservoir, the system is able either to gain or to lose energy, thus moving from one isoenergy surface to another. As for a microcanonical ensemble, each system always remains on a single defined surface (within the layer of the thickness ΔE).

Therefore the microcanonical and canonical distributions differ in that the former has no exponential multiplier, while the latter includes this factor:

$$dW = \frac{1}{Z} \exp\left(-\frac{\mathcal{E}}{kT}\right) d\Gamma = \Phi(\mathcal{E}) d\Gamma, \quad 0 \leq \mathcal{E} < \infty \quad (1.10.2)$$

This has nothing to do with the system's size. The canonical distribution is applicable both to an isolated molecule and to the gas on the whole.

In fact the availability of different energies leads to the introduction of a probability density $\Phi(\mathcal{E})$ in (1.10.2), which is the exponentially decreasing function of energy. Even considering it unknown, we can write

$$dW = \Phi(\mathcal{E}) d\Gamma.$$

However, if the system consisted of two weakly interacting parts, so that $\mathcal{E} = \mathcal{E}_1 + \mathcal{E}_2 + \mathcal{E}_{int}$, $N = N_1 + N_2$, then the same relation would obviously be valid for each of them. Assuming them to be independent by virtue of weak interaction: $\mathcal{E}_{int} \ll \mathcal{E}_1, \mathcal{E}_2$, it is possible to determine the probability of finding the entire system in a certain state by the multiplication theorem:

$$dW(\mathcal{E}) = dW(\mathcal{E}_1) dW(\mathcal{E}_2).$$

Therefore

$$\Phi(\mathcal{E}) d\Gamma = \Phi(\mathcal{E}_1) \Phi(\mathcal{E}_2) d\Gamma_1 d\Gamma_2.$$

As $d\Gamma = d\Gamma_1 d\Gamma_2$ and $\mathcal{E}_{int}$ is negligible, $\Phi(\mathcal{E}_1 + \mathcal{E}_2) = \Phi(\mathcal{E}_1)\Phi(\mathcal{E}_2)$. By analogy with (1.2.6), this equation immediately gives

$$\Phi(\mathcal{E}) = \frac{1}{Z} e^{-\alpha \mathcal{E}},$$

where $\alpha = 1/kT$ and

$$Z = \int \exp(-\mathcal{E}/kT) d\Gamma, \quad (1.10.2a)$$

where $d\Gamma = dq_1 dp_1 dq_2 dp_2 ... dq_N dp_N$. A similar distribution for an isolated molecule may be obtained only for an ideal classical gas when $\mathcal{E} = \sum_{i=1}^{N} \epsilon_i$. Only in this case $dW = \prod_{i=1}^{N} dW_i$, so simply summing probabilities we get from (1.10.2)

$$dW(q_1, p_1) = \int_2 \cdots \int_N dW(1, 2, \ldots, N) = \frac{e^{-\epsilon_1/kT} dq_1 dp_1}{Z^{1/N}}, \quad (1.10.3)$$

that is, the conventional Maxwell–Boltzmann distribution. This result seems quite natural, since any classical molecule is a subsystem that interacts with the gas as with a thermostat, and the Gibbs distribution is well applicable to it.

In the case of an ideal (but not classical!) gas, passing from Γ-space to μ-space requires much greater effort and leads to distributions differing from the Maxwell–Boltzmann one. The reason is that even noninteracting quantum particles cannot be considered as independent subsystems.

For a classical real gas, a reduction of the type (1.10.3) is impossible for a different reason. If in total energy

$$\mathcal{E} = \Sigma \epsilon_i + U(q_1, q_2, \ldots, q_N) \tag{1.10.4}$$

U is not ignored, then $dW \neq \prod_i dW_i$, and it is impossible to describe a real gas by any distribution in μ-space.

Entropy

So we return to the problem posed at the very beginning. Now the distributions in Γ-space are known, but we still do not know how to use them. Even in the derivation of the ideal gas equation of state, the problem was divided into two parts: firstly to find the distribution $dW(\mathbf{v})$, and then to establish the relationship between this distribution and the observed quantity, for example, pressure: $p = nm \int v_x^2 dW(\mathbf{v})$. This second part until now has remained uncertain. For an isolated system, we know the volume and internal energy (V and $\mathcal{E}$) of the gas, along with (1.10.1). In the case of an open system, its volume V and temperature T as well as distribution (1.10.2) are known. To define the state of the gas completely, the two known variables must be complemented by a third one. This need not be the pressure as determined earlier from T and V. But, whatever the third parameter is, it should be calculated from distributions related to the gas as a whole: either (1.10.1) or (1.10.2) according to the situation.

Let us vary the gas volume V with either $\mathcal{E} = $ const (isolated system), or $T = $ const (open system). This will cause a change in the gas pressure and other characteristic quantities. However, in this process the distributions (1.10.1) and (1.10.2) will remain unchanged, except for the normalization constants $\Delta \Gamma$ and Z. This suggests that these quantities contain the required information.

For example, for the ideal gas $\Delta \Gamma = \int d\Gamma \propto V^N$. Consequently, a twofold increase in the vessel's volume will lead to an increase in accessible phase volume $\Delta \Gamma$ by a factor of 2^N and, correspondingly, to a decrease in the probability density Φ of the same factor. The greater the number of accessible states, the smaller the normalized probability of attaining any one. For an isolated system the quantity

$$\Delta\Gamma = \int d\Gamma = \left(\underbrace{\int\cdots\int}_{q}\underbrace{\int\cdots\int}_{p} dq_1\ldots dp_N\right)_{V,\mathcal{E}=\text{const}}$$

is a measure of the macroscopic state degeneration, that is, the number of ways N molecules with any velocity may be arranged within the vessel, provided that the total energy of the system is kept within the specified range ΔE.

If an ideal gas is constituted of molecules of two types, both may be considered as subsystems with particle number N_1 and N_2 ($N = N_1 + N_2$). Each subsystem may be in any one of the states of its phase volume $\Delta\Gamma_1$ or $\Delta\Gamma_2$, respectively. The total number of all possible states of the whole system is determined by the product of these volumes $\Delta\Gamma = \int d\Gamma_1 \int d\Gamma_2 = \Delta\Gamma_1\Delta\Gamma_2$. Hence, the relation between $\Delta\Gamma$ and any additive macroscopic characteristic of the system must be only of the form

$$S = k \ln \Delta\Gamma. \tag{1.10.5}$$

Only a logarithmic relationship ensures the additivity of

$$S = k \ln \Delta\Gamma_1 + k \ln \Delta\Gamma_2 = S_1 + S_2$$

when $\Delta\Gamma = \Delta\Gamma_1 \Delta\Gamma_2$ is multiplicative. However, the realization that S in the relationship (1.10.5), is the *entropy* was reached first by Boltzmann. He recognized that entropy is the measure of the system's disorder. The greater the number of distinguishable microstates which are compatible with the particular value of V and $\mathcal{E}$, the greater the entropy.

Fluctuations

It should be noted that not all states of an isolated system can be identified with the macroscopic equilibrium. For example, if all ideal gas molecules are concentrated in a small portion of the vessel, with the equilibrium velocity distribution remaining unchanged, this state should be considered as a fluctuation although it is accessible (belongs to $\Delta\Gamma$). The state when all molecules are uniformly distributed over the vessel and move with the same velocity $v = (2\mathcal{E}/Nm)^{1/2}$ is also a fluctuation.

Fortunately, such states which differ drastically from the average one are very few in number. For the majority of states, the corresponding distributions of molecules in the space of coordinates and velocities differ little from the equilibrium, time-averaged distribution. Due to the huge number of particles in an isolated system, the number of states of the system which are actually distinguishable from the equilibrium is a minute fraction of all states contained in $\Delta\Gamma$. Thus it makes no difference how the entropy of an isolated system is defined: as a logarithm of the number of equilibrium states, or as a logarithm of all states compatible with the specified V and $\mathcal{E}$, as in Eq. (1.10.5).

Free Energy

In the case of an open system (canonical ensemble) the situation is quite different. Such a system can find itself in a state with arbitrary energy, that is, the whole volume of Γ-space is accessible to it. Naturally, the entropy cannot be expressed in terms of the total volume $\Delta\Gamma$ which is infinite, but solely in terms of the finite part $\Delta\Gamma_e$ that actually corresponds to equilibrium. Despite the possibility of an energy exchange with the environment, the open system is much more frequently found in this part of space than in the rest of it.

The reason for this is that the statistical weight $\rho(\mathcal{E})$ increases sharply with energy. In the case of one particle it increases as $\mathcal{E}^{1/2}$, while in a system of N noninteracting particles the degeneration of states with the same total energy increases as $\mathcal{E}^{N/2}$. Such an abrupt rise of $\rho(\mathcal{E})$ competes with the equally dramatic decrease of the exponential factor $\exp(-\mathcal{E}/kT)$. Consequently, there is a sharp maximum near $\mathcal{E} = \bar{\mathcal{E}}$, and any significant deviations of energy from $\bar{\mathcal{E}}$ are improbable (Fig. 1.22). So, by order of magnitude

$$\int dW = \int \Phi(\mathcal{E})\, d\Gamma \approx \Phi(\bar{\mathcal{E}})\, \Delta\Gamma_e = 1,$$

where $\Delta\Gamma_e = \rho(\bar{\mathcal{E}})\Delta E = 1/\Phi(\bar{\mathcal{E}})$ is the volume of that part of phase space where $dW/d\mathcal{E} = \Phi(\mathcal{E})\rho(\mathcal{E})$ is essentially different from zero. The definition of

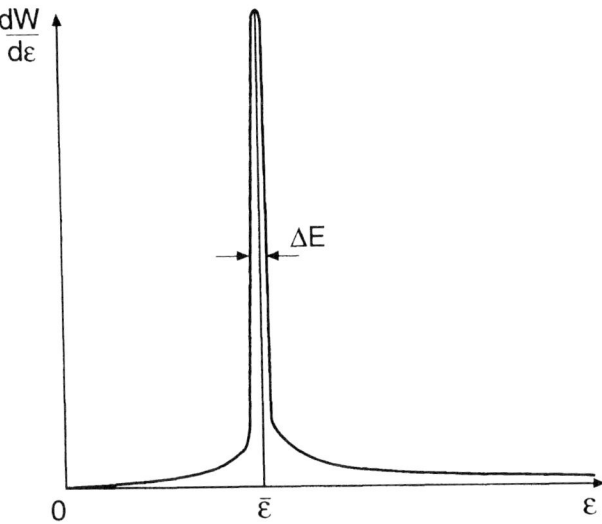

Figure 1.22 The canonical energy distribution for large system.

of entropy which remains the measure of equilibrium state degeneracy must be generalized as follows:

$$S = k \ln \Delta\Gamma_e = -k \ln \Phi(\bar{\mathcal{E}}), \qquad (1.10.5a)$$

For isolated systems the definition of entropy thus generalized is identical to the previous one, since $\Phi(\bar{\mathcal{E}}) \equiv \Phi(\mathcal{E}) = 1/\Delta\Gamma$. As for the canonical ensemble (thermostated systems), substituting $\Phi(\bar{\mathcal{E}}) = \exp(-\bar{\mathcal{E}}/kT)/Z$ yields

$$\mathcal{F} = \bar{\mathcal{E}} - TS = -kT \ln Z. \qquad (1.10.6)$$

Here $\mathcal{F}$ is the Helmholtz free energy, and the average energy is exactly the internal energy of the system $\mathcal{E}$ (the line above it is omitted here and below).

Computational Scheme

From Eqs. (1.10.5) and (1.10.6) we are able to calculate either the entropy or the free energy of the system, provided that the corresponding normalization constants $\Delta\Gamma$ or Z are known. With the above formulae we can calculate the macroscopic characteristics of the substance without determining the distributions in μ-space. That is why this procedure is applicable to both ideal and real gases as well as to condensed media. In outline it is as follows.

In the case of a microcanonical ensemble, the main problem is the calculation of the accessible phase volume $\Delta\Gamma(V, \mathcal{E})$ for a multiparticle mechanical system, provided the number and mass of molecules as well as their total energy $\mathcal{E}$ and the vessel's volume V are known. If this problem is solved, formula (1.10.5) immediately gives the entropy as a function of the given variables: $S(V, \mathcal{E})$. Using equilibrium thermodynamics any other macroscopic parameters may be obtained from this relation which carries the full information about the system including the equation of state. In particular, from Eq. (5.2.36)

$$TdS = d\mathcal{E} + pdV \qquad (1.10.7)$$

it follows that

$$p = -\left(\frac{\partial \mathcal{E}}{\partial V}\right)_S, \quad T = \left(\frac{\partial \mathcal{E}}{\partial S}\right)_V; \qquad (1.10.7a)$$

thus both the pressure and the temperature are determined. As is customary in thermodynamics, the state parameters to be considered constant in differentiation are noted by indices to the right of the brackets.

If the ensemble is described by the canonical distribution, the main difficulties lie in the calculation of the statistical sum Z as a function of the volume V and the temperature T. This problem is usually a little easier than the previous one. Once it is solved, Eq. (1.10.6) immediately yields the free energy $\mathcal{F}(V,T)$ which satisfies Eq. (5.5.8)

$$d\mathcal{F} = -pdV - SdT, \qquad (1.10.8)$$

This then gives information about the rest of the parameters:

$$p = -\left(\frac{\partial \mathcal{F}}{\partial V}\right)_T \quad S = -\left(\frac{\partial \mathcal{F}}{\partial T}\right)_V. \qquad (1.10.8\text{a})$$

The results obtained from the microcanonical or canonical distributions are identical, because the equilibrium properties do not depend on whether the system is thermally isolated or not. However, mathematically, estimation of the statistical sum Z is often more convenient than that of $\Delta\Gamma$.

Ideal Gas

The validity of these general recipes may be easily verified using the simplest model of an ideal gas which is well known. For example, let us make use of the canonical distribution. As $\mathcal{E} = \sum_{i=1}^{N} p_i^2/2m$, all variables in (1.10.2a) separate, and upon integration we have

$$Z = V^N (2\pi mkT)^{3N/2}.$$

In view of (1.10.6), we find

$$\mathcal{F} = -NkT \ln\left[(2\pi mkT)^{3/2} V\right], \qquad (1.10.9)$$

and (1.10.8a) gives

$$p = \frac{NkT}{V}, \quad S = kN \ln\left[V(2\pi mkT)^{3/2}\right] + \frac{3}{2}kN. \qquad (1.10.10)$$

In particular, we have attained the aim of the derivation of the equation of state. The general method gives the required result: $p = RT/V$. However, neither free energy nor entropy are proportional to N as must be the case as when the system's volume increases with constant density. This is quite a surprise, because both $\mathcal{F}$ and S are additive quantities. One can always run into difficulties in developing a new theory, but they are particularly undesirable at the end of a road which has appeared to be right! Naturally, it is tempting to overcome this impediment by making some minimal changes.

Identity of Microparticles

To get an idea of the required modifications, compare S from Eq. (1.10.10) with the Sackur–Tetrode formula (identical to Eq. (5.8.2) of Chapter 5) which is free of this problem:

$$S = kN\ln\left[\frac{V}{N}(2\pi mkT)^{3/2}\right] + \frac{5}{2}kN, \qquad (1.10.11)$$

This differs from Eq. (1.10.10) only by the term $-kN \ln N/e$, that would appear in (1.10.10) if Z were divided by $N!$:

$$Z = \frac{1}{N!}(2\pi mkT)^{3N/2} V^N. \qquad (1.10.11a)$$

Using the well-known Stirling formula $\ln N! \approx N \ln N/e$, the substitution of (1.10.11a) into (1.10.6) leads to the additive quantity

$$\mathcal{F} = -NkT \ln\left[(2\pi mkT)^{3/2} \frac{eV}{N}\right], \qquad (1.10.12)$$

which in turn yields the correct result (1.10.11). The redefined $\mathcal{F}$ as well as S depend on specific volume $v = V/N$, but not on the volume itself, so the paradox is resolved.

However, the multiplier $N!$ may enter Z solely from the canonical distribution which, therefore, must differ from (1.10.2):

$$dW = \frac{1}{Z}\exp\left(-\frac{\mathcal{E}}{kT}\right)\frac{d\Gamma}{N!}. \qquad (1.10.13)$$

Similarly the microcanonical distribution should be modified

$$dW = \frac{d\Gamma}{\Delta\Gamma \cdot N!} \qquad E \leq \mathcal{E} \leq E + \Delta E. \qquad (1.10.14)$$

This innovation is well justified. The distributions themselves do not suffer any significant change; only the absolute weight of each state is redefined. Therefore if the problems are successfully remedied by this redefinition, the price is not too high. Moreover, innovations that are useful often prove to be necessary as well. Here the multiplier $N!$ in fact is a result of the fundamentally important property of microparticles, discussed in Chapter 5 (Section 5.8): their "identity" or indistinguishability. Due to this property, there is no meaning in discriminating between states of the system which differ only by permutation of the particles. These are all a single state, because identical particles cannot be numbered or marked.

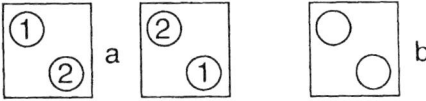

Figure 1.23 (a) Different states of distinguishable particles obtained by permutation and (b) a single state of indistinguishable particles.

This peculiarity manifests itself even in a system of two particles. If the particles were distinguishable (like billiard balls), then each configuration would be counted twice (owing to possible permutation), while if not, then only once (Fig. 1.23). Another example: after an exchange of apartments the tenants can always be identified, at least by fingerprints, if not by marked individuality. However, it is impossible to differentiate among electrons, protons, or identical atoms. We can say: there are two atoms occupying these two positions, but not: the first atom is here, and the second there. In calculating the number of states in Γ-space, this specific property of microparticles has so far been ignored: all states differing by the permutation of N particles (in all, $N!$) were considered in $d\Gamma$ as different. To correct this mistake, we now have to divide the result by $N!$ This is just what was done in (1.10.13) and (1.10.14).

Real Gas

Although in general the procedure is clear, the calculation of particular partition functions presents considerable problems once we pass to the real gas. The energy of any real system is composed of kinetic and potential energies. In the simplest case, the latter is additive with respect to the pair intermolecular interaction $u_{ik} = u(\mathbf{q}_i - \mathbf{q}_k)$:

$$U = \sum_{k>i} u_{ik}. \tag{1.10.15}$$

The partition function of distribution (1.10.13) is

$$Z = \frac{1}{N!}\int\cdots\int \exp\left\{-\sum_{i=1}^{N} p_i^2/2mkT - U/kT\right\} d\mathbf{q}_1 \dots d\mathbf{p}_N = Z_{id} \cdot Q^*. \tag{1.10.16}$$

This differs from its ideal analogue Z_{id} defined in (1.10.11a) in the presence of the extra multiplier, the dimensionless configurational integral

$$Q^* = \int_1 \cdots \int_N e^{-U/kT} \frac{d\mathbf{q}_1}{V} \cdot \frac{d\mathbf{q}_2}{V} \cdots \frac{d\mathbf{q}_N}{V}, \tag{1.10.17}$$

that is difficult to calculate. This may be considered as the result of averaging the integrand over a uniform distribution of N particles in the volume V.

Denoting the averaging by a bar and taking account of (1.10.15), we represent the configuration integral as

$$Q^* = \int_1 \cdots \int_N \prod_{i=1}^{N-1} \Theta_i \frac{d\mathbf{q}_1}{V} \cdot \frac{d\mathbf{q}_2}{V} \cdots \frac{d\mathbf{q}_N}{V} = \overline{\prod_{i=1}^{N-1} \Theta_i}, \qquad (1.10.18)$$

where $\Theta_i = \exp\left[-\sum_{k=i+1}^{N} u(\mathbf{q}_k - \mathbf{q}_i)/kT\right]$ describes the interaction of the ith particle only with the $N-i$ particles with higher ordinal number. All Θ_i are of the same nature: they define the effect produced on the ith particle by its surroundings. As the arrangement of particles in the space changes, all Θ_i fluctuate over the entire range of possible values. At low gas densities the action of the surroundings on any particle is naturally assumed to be independent, that is, the mean of the product is equal to the product of means:

$$Q^* = \overline{\prod_{i=1}^{N-1} \Theta_i} = \prod_{i=1}^{N-1} \overline{\Theta_i}. \qquad (1.10.19)$$

This assumption simplifies further calculations of the configurational integral, however, it restricts the region of applicability to relatively rarefied gases.

Changing variables to $\mathbf{r}_k = \mathbf{q}_k - \mathbf{q}_i$ we have

$$\overline{\Theta_i} = \int_i \cdots \int_N \exp\left[-\sum_{k=i+1}^{N} u(\mathbf{q}_k - \mathbf{q}_i)/kT\right] \frac{d\mathbf{q}_i}{V} \cdots \frac{d\mathbf{q}_N}{V}$$

$$= \int_{i+1} \cdots \int_N \exp\left[-\sum_{k=i+1}^{N} u(\mathbf{r}_k)/kT\right] \frac{d\mathbf{r}_{i+1}}{V} \cdots \frac{d\mathbf{r}_N}{V}. \qquad (1.10.20)$$

This expression may be presented as the product of equal multipliers:

$$\overline{\Theta_i} = \prod_{k=i+1}^{N} \int e^{-u(\mathbf{r}_k)/kT} \frac{d\vec{\mathbf{r}}_k}{V} = \left[\int e^{-u(\mathbf{r}_k)/kT} \frac{d\mathbf{r}_k}{V}\right]^{N-i}.$$

For scalar interactions depending solely on the distance between particles, further simplifications are possible:

$$\overline{\Theta_i} = \left[\int_0^R e^{-u(r)/kT} \frac{4\pi r^2 dr}{V}\right]^{N-i} = \left[1 + \int_0^R \left(e^{-u(r)/kT} - 1\right) \frac{4\pi r^2 dr}{V}\right]^{V(n-i)/V}.$$

$$(1.10.21)$$

Here R is the radius of the volume filled with gas. For simplicity, it is considered spherical. The last expression is given in a form convenient for passing to the limit $V \to \infty$ ($R \to \infty$) at constant gas density. This is quite justified, since the number of particles in the system is macroscopically large. The final result must depend solely on the density as a universal parameter, and this is just what we have upon passing to the limit

$$\overline{\Theta_i} = \exp\left[\frac{N-i}{V}\int_0^\infty \left[\exp\left(-\frac{u(r)}{kT}\right) - 1\right] 4\pi r^2 dr\right]. \tag{1.10.22}$$

Substituting this result into (1.10.19) yields

$$Q^* = \exp\left[\frac{N(N-1)}{2}\mu\right], \qquad \mu = \int_0^\infty f(r)\frac{4\pi r^2 dr}{V}.$$

As N is large

$$Q^* \approx \exp\left(\frac{N^2}{2}\mu\right) = \exp\left[\frac{N^2}{2V}\int_0^\infty f(r) 4\pi r^2 dr\right], \tag{1.10.23}$$

where

$$f(r) = \exp\left(-\frac{u(r)}{kT}\right) - 1. \tag{1.10.24}$$

As the distance between molecules increases, the function $f(r)$ tends rapidly to zero (Fig. 1.24), thus providing the convergence of the integral in (1.10.23).

Substituting (1.10.23) in (1.10.16) and calculating the free energy by formula (1.10.6), we find

$$\mathcal{F} = \mathcal{F}_{id} - \frac{N_0^2 kT}{2V}\int_0^\infty f(r) 4\pi r^2 dr, \tag{1.10.25}$$

where $\mathcal{F}_{id}$ is the free energy of an ideal gas (1.10.12). Therefore, the pressure

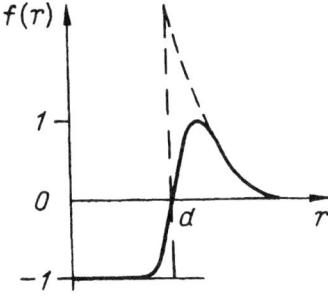

Figure 1.24 A qualitative view of the function $f(r)$ and its approximation in a model of attracting hard spheres (dashed line).

$$p = p_{id} - \frac{N_0^2 kT}{2V^2} \int_0^\infty f(r)\, 4\pi r^2\, dr \qquad (1.10.26)$$

proves to be different from $p_{id} = RT/V$. In virial form this equation of state appears as

$$F = 1 - n\frac{\int_0^\infty f(r)\, 4\pi r^2 dr}{2} = 1 + \frac{B_0}{V}. \qquad (1.10.27)$$

Here we can easily recognize the reduced form of Eq. (1.9.19) with the second virial coefficient

$$B_0 = \frac{N_0}{2}\int_0^\infty \left[1 - e^{-u(r)/kT}\right] 4\pi r^2 dr, \qquad (1.10.28)$$

expressed in terms of the intermolecular interaction potential.

In the van der Waals model of "hard spheres attracting one another," the repulsive branch of the interaction is replaced by a potential wall at the distance of a molecular diameter: $u(d - 0) = \infty$ (dashed line in Fig. 1.24). Taking this model and assuming that the average kinetic energy of particles exceeds the potential well depth, we get from (1.10.28) in the first approximation with respect to u/kT

$$B_0 \approx \frac{N_0}{2}\left[\frac{4}{3}\pi d^3 + \int_d^\infty \frac{u(r)}{kT} 4\pi r^2 dr\right] = B - \frac{A}{RT}. \qquad (1.10.28a)$$

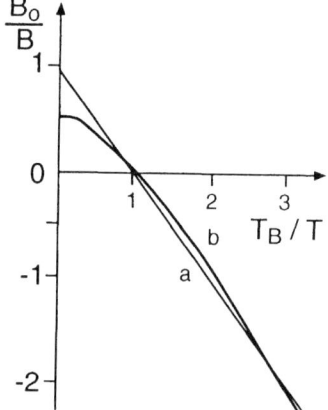

Figure 1.25 The second virial coefficient (a) in the van der Waals theory and (b) in the Lennard-Jones approximation of the interparticle potential.

From a formal standpoint, this result coincides with expression (1.9.18) obtained in the van der Waals theory, moreover, here the parameter $4N_0 v_d = B$ is defined identically, and A is expressed in terms of the intermolecular interaction potential:

$$A = 2\pi N_0^2 \int_0^\infty |u(r)| r^2 dr. \qquad (1.10.29)$$

Agreement with the van der Waals theory is achieved at the cost of considering the interaction potential in a rough approximation. Obviously, this introduces an error in the theoretical estimation of the second virial coefficient. It can be avoided by using the actual interaction potential in a rather rigorous and general formula (1.10.28). For noble and simple gases, the so-called "6–12 Lennard-Jones" potential

$$u(r) = 4\epsilon \left[\left(\frac{\sigma}{r}\right)^{12} - \left(\frac{\sigma}{r}\right)^6 \right], \qquad (1.10.30)$$

is most often used. Here σ and ϵ are molecular constants

$$\sigma = 3 \div 6 \times 10^{-8} \text{ cm}, \quad \epsilon = 1 \div 30 \times 10^{-15} \text{ erg}.$$

The first term in (1.10.30) corresponds to repulsion, and the second to attraction of two particles separated by a distance r. The temperature-dependence of the second virial coefficient determined from (1.10.28) with this potential is given in Fig. 1.25. It is seen that in the high-temperature region it deviates essentially from the linear, van der Waals dependence. The reason is that for particles with high kinetic energy of particles, even a repulsive potential decreasing as abruptly as r^{-12} departs noticeably from a vertical wall.

On the other hand, the original van der Waals equation (1.9.16) describes the state of condensed media much better than Eqs. (1.10.27) and (1.9.19), however rigorous the definition of the second virial coefficient may be. Although using a rough potential model the van der Waals equation does allow for all orders of virial expansion, while here we have only succeeded in the correct determination of the first, linear in density term of the virial expansion (1.10.27). Although more sophisticated statistical methods allow one to express all virial coefficients in terms of $u(r)$ this does not improve the situation very much. It is impossible to sum up the series or reduce it to any expression similar to the van der Waals equation. As the density increases, the description of the substance requires not only linear but also quadratic and cubic corrections to be taken into account. For condensed matter the entire series is necessary, but even this is deficient, because the virial expansion diverges as the critical region is approached. Thus the virial expansion cannot in principle be applied to media of higher than critical densities.

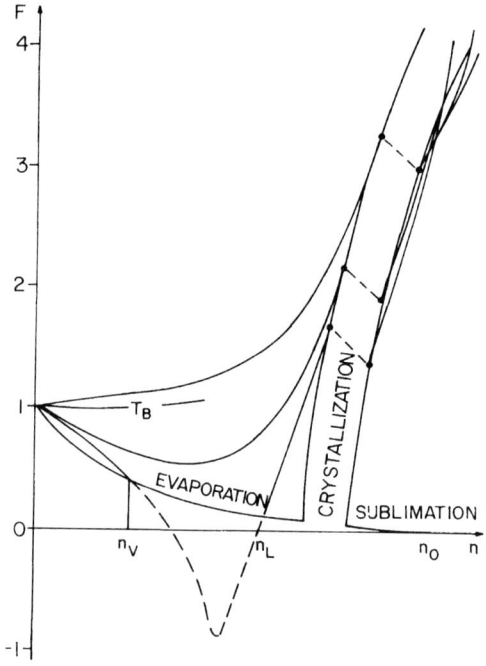

Figure 1.26 The dependence of the compressibility factor $F = pV/RT$ on the corpuscular density n. The dashed sections of the isotherms are their extrapolation to the density region intermediate between the vapor and the liquid or between the liquid and the crystal. n_V and n_L are the densities of the saturated vapor and liquid at boiling point while n_0 is the density of the tensionless state of the crystal.

To clear up this important point, consider the variation of the compressibility factor with density as schematically given in Fig. 1.26. In fact, the virial equation of state is equivalent to the Taylor series expansion about zero density. Its linear variant (1.10.27) is obviously inapplicable even to describe gas isotherms in the range $T_c < T < T_B$ to the right of the indicated minimum. For liquids, the entire series is not sufficient, because the curve has discontinuities associated with condensation, and the behavior of $F(n)$ is open to speculation (dashed line in Fig. 1.26). This is also valid for the solid phase. For this reason, it is useful to have an alternative to virial expansion, that is, an approximation intended for highly condensed media. This is the free volume theory of Lennard-Jones and Devonshire to be outlined below.

Free Volume Theory

The starting point in the calculation of configurational integral (1.10.17) of a condensed system is the model of dense packing of particles (Fig. 1.27a). Surrounded by neighbors on all sides, such particles move in a potential well

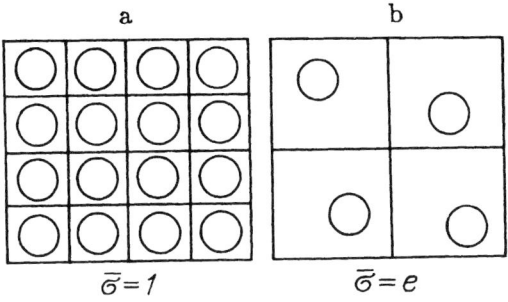

Figure 1.27 (a) A perfect simple lattice and (b) the expanded substance leaving room for a multiple occupation of cages.

with the effective potential $\Phi(\mathbf{q}_i)$. As a rule they do not leave the well, and the center-of-mass of each particle is always within the bounded *free volume*. Thus the molecules can be considered separately, with integration over spatial coordinates being restricted to a small cell, a region of order Δ. If the center of cell is placed at the origin of coordinates, its depth is $\Phi(0)$, and the total potential energy can be represented as

$$U = \frac{N}{2} \Phi(0) + \sum_{i=1}^{N} \Delta\Phi(\mathbf{q}_i) . \qquad (1.10.31)$$

Here $\Delta\Phi(\mathbf{q}) = \Phi(\mathbf{q}) - \Phi(0)$ is the cage potential delimiting the free volume, while the first term is the total potential energy of the substance when all particles are in the center of their cells. It is half as large as $N\Phi(0)$, because the interaction of a molecule with any other is counted twice in the sum $\sum_{i=1}^{N} \Phi_i(0) = N\Phi(0)$. This is taken into account by dividing the sum by 2.

Using the model (1.10.31), we can easily calculate the configuration integral (1.10.17) which is the product of the integrals over different cells summed over all possible permutations of particles:

$$V^N Q_N^* = \exp\left(-\frac{N\Phi(0)}{2kT}\right) \int_V \cdots \int_V e^{-\sum_i \Delta\Phi(\mathbf{q}_i)/kT} d\mathbf{q}_1 \ldots d\mathbf{q}_N =$$

$$= N! v_f^N \exp\left(-\frac{N\Phi(0)}{2kT}\right) . \qquad (1.10.32)$$

Here

$$v_f = \int_\Delta \exp\left[-\frac{\Delta\Phi(\mathbf{q})}{kT}\right] d\mathbf{q} \qquad (1.10.33)$$

is the free volume of the cell formed by summing the space elements accessible to a particle weighted by the probability of finding the particle in them.

Using Stirling's formula, substituting (1.10.32) into (1.10.16) and the result into (1.10.6) yields

$$\mathcal{F} = \mathcal{F}_{id} + NkT - NkT \ln \frac{v_f}{v} + N \frac{\Phi(0)}{2}, \qquad (1.10.34)$$

where $v = V/N$ is the specific volume, and $\mathcal{F}_{id}$ is defined in (1.10.12). This result is quite reliable for crystals, where the migration of particles between cells is very unlikely. However, in a liquid the neighborhood of the particle is not impenetrable, and under further rarefaction the notion of a cell becomes a mere conventionality. This limits the application of formula (1.10.34).

Its drawback is revealed in passing to the lowest densities. Although potential barriers are removed ($\Delta\Phi \to 0$ at $v \to \infty$) and almost the entire cage volume (see Fig. 1.27b) becomes accessible to the particle ($v_f \to v$), the sample free energy $\mathcal{F} \to \mathcal{F}_{id} + NkT$ still differs from the ideal gas value by NkT. The ideal gas pressure calculated by this formula proves to be correct, but the entropy differs from the ideal value: $S = S_{id} - Nk$. Obviously, this failing of (1.10.34) is due to the excessive order of the imposed cell structure, which underestimates the disorder arising in a real system under rarefaction. This drawback may be eliminated by introducing an additional parameter, the so-called "collective entropy" $\bar{\sigma}$ which varies from 1 to e with increasing v, thus correcting the asymptotic behavior of the free energy

$$\mathcal{F} = \mathcal{F}_{id} + NkT \ln \frac{e}{\bar{\sigma}} - NkT \ln \frac{v_f}{v} + N \frac{\Phi(0)}{2}. \qquad (1.10.35)$$

Unfortunately, the explicit form of the $\bar{\sigma}(v)$ dependence remains unknown. This leads to some uncertainty in the equation of state

$$p = \frac{1}{N}\left(\frac{\partial \mathcal{F}}{\partial v}\right)_T = p_t - p_i, \qquad (1.10.36)$$

where

$$p_i = \frac{1}{2}\frac{d\Phi(0)}{dv} \qquad (1.10.37a)$$

is the so-called *internal* pressure, while

$$p_t = nkT\left(\frac{\partial \ln(\bar{\sigma} v_f)}{\partial \ln v}\right)_T = \frac{RT}{V}\left(\frac{\partial \ln(\bar{\sigma} v_f)}{\partial \ln v}\right)_T \qquad (1.10.37b)$$

has the meaning of *motional* or thermal pressure. The former is analogous to the van der Waals term A/V^2 in Eq. (1.9.10) and depends on the volume only. However, the latter is created by particle motion, which is thermal in origin at real gas temperatures. From a phenomenological point of view the equation

$$p + p_i = p_t \qquad (1.10.38)$$

is valid for all phases of the substance, although the particular form of the components is different.

The free-volume theory was first advanced by Lennard-Jones and Devonshire in 1937. It became immediately evident that despite all simplifications of the model, it was still too complicated for particular calculations. This is partly due to the fact that the potential of a cell resulting from an actual crystallographic arrangement of the neighboring molecules proved to be too cumbersome for analytical calculation of the free volume. Thus the authors of the theory and their adherents preferred to use the angle averaged potential

$$\Delta\Phi(q) = c\left[\frac{1}{4\pi}\int\int u\left(\sqrt{a^2 + q^2 - 2aq\cos\theta}\right)\sin\theta d\theta d\varphi - u(0)\right]. \qquad (1.10.39)$$

This implies that particles of the first coordination sphere equally distant from the cell center are uniformly distributed over the sphere of radius a.

Even for hard spheres of diameter σ it is customary to employ a "smeared" or "sphericalized" free volume. The structures of the first shell in simple liquids and their crystals are similar. Each molecule in a cell of a face-centered cubic lattice has 12 nearest neighbors, a distance of a away. The volume per molecule is $v = a^3/\sqrt{2}$ and the cell corresponding to this volume is dodecahedron. If $v* = v/\sigma^3 \gg 1 (a \gg \sigma)$ the full volume of a cell is available for a center of molecule ("wanderer") inside it, but the cell borders are just conditional, because they are transparent. At higher densities each face is partly protected from penetration of a wanderer by a spherical body of corresponding neighbor. The emigration is possible only through the chinks between them at the tops of dodecahedron (Fig. 1.28). At even higher densities ($v < 2\sigma^3$) the wanderer can not escape from the "cage" formed by its nearest neighbors. So the region of applicability of the free volume theory to hard sphere model is actually

$$2\sigma^3 > v > \sigma^3/\sqrt{2},$$

where the lowest limit corresponds to the tightest possible packing. At such densities the shape of a cage becomes rather whimsical and its volume, which is free for wanderer center, is hardly available for calculation. On the contrary, the smeared free volume is the largest sphere which can fit inside the exact cages shown in Fig. 1.28. This approximation is appropriate for untransparent

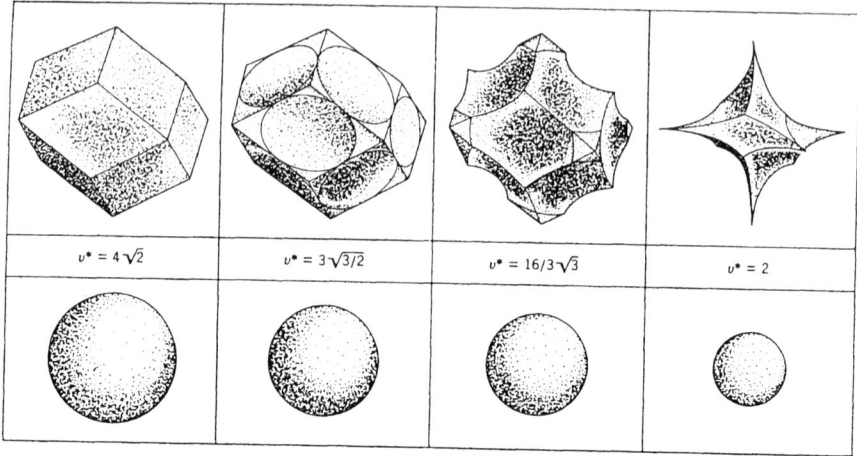

Figure 1.28 The figures in the top row show the shape of the exact free volumes at different densities for face-centered lattice ($v^* = v/\sigma^3$). The corresponding "smeared" free volumes are shown in the bottom row. (From R. J. Buehler, R. H. Wentorf, J. O. Hirschfelder, and C. F. Curtiss, *J. Chem. Phys.*, **19**, 61 (1951).)

cages but becomes questionable at lower densities when the wanderer is no longer confined to the cage formed by the nearest neighbors.

In (1.10.39) only the contribution from the nearest neighbors is taken into account but the contribution of remote particles has to be also averaged in a similar fashion. The best version of the theory took account of the three nearest spheres which turned out to be quite sufficient for the short-range Lennard-Jones potential. The results of the calculations are illustrated in the original diagram presented in Fig. 1.29. Qualitatively they reproduce the behavior of the isotherms given in Fig. 1.26, but the breaks due to evaporation and crystallization of the substance are absent. The reason is that the collective entropy was taken to be equal to unity because the recipe for its calculation had not yet been found. Within the framework of this theory, thermodynamic properties of solids may be described semiquantitatively, while those of liquids only qualitatively. In fact, the theory is applicable only in the neighborhood of the knot point shown in Figs. 1.26 and 1.29. Conversely the virial expansion is valid in the region of quasilinear variation of $F(n)$ near the ideal gas state which is also the knot point of the compressibility factor isotherms.

1.11 HEAT CAPACITY OF GASES

Sometimes it happens in science that the most radical innovations appear under most prosaic circumstances. One of the problems facing statistical physics was the calculation of the molecular heat capacity of ideal gases. No

1.11 HEAT CAPACITY OF GASES

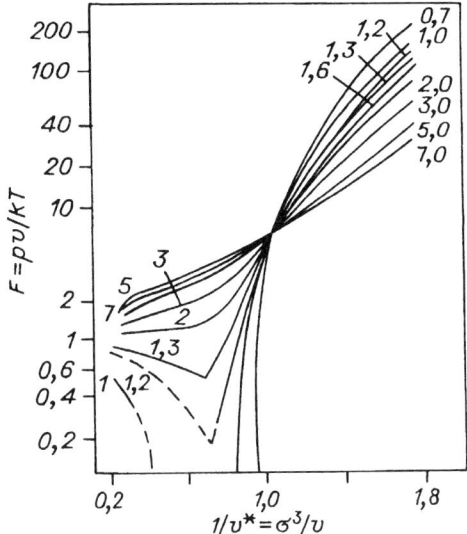

Figure 1.29 The compressibility factor F as a function of the reduced density $1/v^*$. The curves are isotherms labeled by the values of the reduced temperature $T^* = kT/\epsilon$. (From R. H. Wentorf, R. G. Buehler, J. O. Hirschfelder and C. F. Curtiss, *J. Chem. Phys.* **18**, 1484 (1950); see also Fig. 4.7-3 in *Molecular Theory of Gases and Liquids* by J. O. Hirschfelder, C. F. Curtiss, B. R. Bird.)

fundamental difficulties were expected. However, this problem proved to be one of the "hard nuts to crack" which served to verify the newly born physical theory.

According to the formal definition, the molar specific heat at constant volume is given by

$$C_V = \left(\frac{\partial \bar{\mathcal{E}}}{\partial T}\right)_V = N_0 \frac{d\bar{E}}{dT}. \quad (1.11.1)$$

For an ideal gas it is expressed via the total heat energy per molecule $\bar{E}$ including all possible types of intramolecular motion. In the simplest case of an atomic gas, the particles are capable of nothing but translational motion, that is, $\bar{E} \equiv \bar{\epsilon} = \frac{3}{2}kT$, and $C_V = \frac{3}{2}R \approx 3$ cal/mole · deg. The validity of this conclusion is verified experimentally: the heat capacity of single-atom molecules is temperature-independent and is equal to three. However, this is the only point of excellent agreement between the theory and experiment. Extension of the calculation to diatomic molecules capable not only of translation, but also of rotation and vibration leads to the conclusions which do not show even qualitative agreement with experiment.

To clarify the essence of the problem, let us note that in the most general case $\bar{E}$ is uniquely expressed in terms of Z:

$$\overline{E} = \frac{1}{Z}\int E\exp\left(-\frac{E}{kT}\right)dpdq = -\frac{\partial(\ln Z)}{\partial\left(\frac{1}{kT}\right)}. \quad (1.11.2)$$

Thus the calculation of heat capacity is reduced to that of the partition function:

$$Z = \int_p\int_q \exp\left(-\frac{E}{kT}\right)dpdq. \quad (1.11.3)$$

For translational, rotational and vibrational motion (in the harmonic approximation) calculations are similar and give identical results. Let us see it for ourselves.

Translational Motion

In the absence of external fields $E = \sum_{i=x,y,z}(p_i^2/2m)$, and

$$Z = V(2\pi mkT)^{3/2}. \quad (1.11.4)$$

There is one translational degree of freedom per axis, each associated with the multiplier $(2\pi mkT)^{1/2}$ in Z. The average energy $kT/2$ accounts for any of them according to Eq. (1.11.2). Since there are three translational degrees of freedom, the total heat energy $\overline{E} = \frac{3}{2}kT$. It is remarkable that it does not depend on the molecular mass but is completely determined by the temperature. As has already been shown, the corresponding specific heat is $\frac{3}{2}R$.

Rotational Motion

In the absence of orienting fields the calculation is similar to the previous one. The Gibbs distribution obtained from (1.8.8) with $E = 0$ takes the form

$$dW(p_\theta, p_\varphi, \theta, \varphi) = \frac{1}{Z}\exp\left[-\frac{p_\theta^2}{2IkT} - \frac{p_\varphi^2}{2IkT\sin^2\theta}\right]dp_\theta\, dp_\varphi\, d\theta\, d\varphi.$$

The substitution $p_1 = p_\theta$, $p_2 = p_\varphi/\sin\theta$ brings the distribution to a more conventional form:

$$dW(p_1, p_2, \theta, \varphi) = \frac{1}{Z}\exp\left[-\frac{p_1^2}{2IkT} - \frac{p_2^2}{2IkT}\right]dp_1\, dp_2\, \sin\theta\, d\theta\, d\varphi. \quad (1.11.5)$$

In these variables, the energy is a coordinate-independent quadratic function of momenta. Thus

$$Z = \int \exp\left(-\frac{p_1^2}{2IkT}\right) dp_1 \int \exp\left(-\frac{p_2^2}{2IkT}\right) dp_2 \int d\Omega = 8\pi^2 IkT, \quad (1.11.6)$$

which involves two equal multipliers $\sqrt{2\pi IkT}$ which yield $kT/2$ each, if substituted into (1.11.2). Therefore, $\overline{E} = kT/2 + kT/2$ and $C_V = R \approx 2$ cal/mole · deg. As in the previous case, the average energy does not depend on the parameters of the molecule itself such as the moment of inertia.

The presence of only two components $kT/2$ in the expression for average rotational energy is due to the specific geometry of linear molecules. Such molecules have only two rotational degrees of freedom. The molecule's rotation about its own axis is neglected for reasons which will be clarified later. Only nonlinear molecules have three rotational degrees of freedom. As before, each contributes one $kT/2$, whatever the magnitude of the corresponding moment of inertia. In this case, the total energy $\overline{E} = 3(kT/2)$, and $C_V = \frac{3}{2}R$.

Vibrational Motion

It is reasonable to choose the quantity $q = x_2 - x_1 - a$ as one of canonical variables describing the vibrational motion of diatomic molecules (a is the equilibrium distance between atoms located at the points x_1 and x_2). The greater the bond stretching q, the greater the effort required to return the atoms to their equilibrium position. In the harmonic approximation they are proportional and opposite in sign

$$F = -\beta q. \quad (1.11.7)$$

If linear dependence holds within the whole range of q, the oscillator is called harmonic. The corresponding equation of motion

$$m\ddot{q} = F = -\beta q \quad (1.11.8)$$

shows that the vibrational frequency $\omega_0 = \sqrt{\beta/m}$ is the basic characteristic of the oscillator. The total energy of harmonic vibrations

$$E = \frac{p^2}{2m} + \int \beta q \, dq = \frac{p^2}{2m} + \frac{m\omega_0^2 q^2}{2}, \quad (1.11.9)$$

where $p = m\dot{q}$. Thus

$$Z = \int \exp\left(-\frac{p^2}{2mkT}\right) dp \int \exp\left(-\frac{q^2 m\omega_0^2}{2kT}\right) dq = \frac{2\pi kT}{\omega_0}. \quad (1.11.10)$$

Although in this case the multipliers appearing on integration over different variables (coordinates and momenta) are not equal to each other, the difference is insignificant. On substitution into (1.11.2) and differentiation, all T-independent parameters disappear. Thus we arrive at the same result

$$\overline{E} = \frac{kT}{2} + \frac{kT}{2} \quad \text{and} \quad C_V = R.$$

Equipartition Law

It is seen that identity of all estimates of the mean heat energy and its invariance to the type of motion are due to similar dependence of mechanical energy on the corresponding coordinates. In all examples given above the energy proved to be a quadratic form of coordinates or momenta. Thus the following generalization is justified: if the energy takes the form

$$E = \sum_{\ell=1}^{i} \gamma_\ell \xi_\ell^2, \tag{1.11.11}$$

then its average equilibrium value at any γ_ℓ is

$$\overline{E} = \sum_{\ell=1}^{i} \gamma_\ell \overline{\xi_\ell^2} = \sum_{\ell=1}^{i} \frac{kT}{2} = i \frac{kT}{2}, \tag{1.11.12}$$

where i is the number of quadratic terms in Eq. (1.11.11).

Sometimes this statement is formulated as a law or *the principle of equipartition of energy* among the various degrees of freedom. It is implied that each quadratic term in (1.11.11) is associated with one or another degree of freedom of mechanical motion. Therefore the corresponding mean energy is equal to $kT/2$, whatever the value of γ_ℓ. Thus for translational motion there are always three degrees of freedom, for rotational motion, either two or three, depending on the molecule's shape. As for vibrational motion, note that, according to (1.11.9), the expression for energy involves two (not one) quadratic terms corresponding to vibrational motion: one term is related to the kinetic part of the energy, the other, to the potential part. For each of the terms there is one $kT/2$. Hence, the above formulation of the law will remain valid, provided that not one but two degrees of freedom are assigned to each vibrational mode of a molecule.

Unfortunately, an artificial "doubling" of vibrational degrees is not the only correction to the "equipartition" law that should be taken into account. In the presence of external fields translational and rotational degrees of freedom also have not only kinetic but potential energy as well. However, potential energy terms like mgz or $qE\cos\theta$ are not quadratic in the coordinates, so they do not

Temperature Anomalies

If the total number of the gas molecules' degrees of freedom is i, then, according to (1.11.12) and (1.11.1), $C_V = iR/2 = i\,\text{cal/mole} \cdot \text{deg}$, that is, the absolute value of specific heat coincides numerically with i.

This consequence of the equipartition law, and, in fact, the Gibbs distribution itself, did not agree with experimental findings concerning polyatomic molecules. In particular, the experience shows that for diatomic molecules at ordinary temperatures $C_V = \frac{5}{2}R$, not $\frac{7}{2}R$, as follows from the equipartition law taking that there are three translational, two rotational and two vibrational degrees of freedom. Even if we assume that the molecule does not oscillate, the agreement is violated as soon as the temperature falls, because the heat capacity falls as well approaching the atomic value $\frac{3}{2}$ (Fig. 1.30). For molecular hydrogen, for example, this limit is reached at temperatures below 80°C. On the other hand, the specific heat of diatomic and polyatomic gases increases with rising temperature, and, eventually, becomes equal to $iR/2$ (unless dissociation takes place) in excellent agreement with the equipartition law, but in contrast with their room temperature values which are $5R/2$ and $3R$, respectively, as if vibrations were absent.

We run into a paradoxical situation: though the deductions of the theory do not completely agree with experiment, they are by no means senseless and

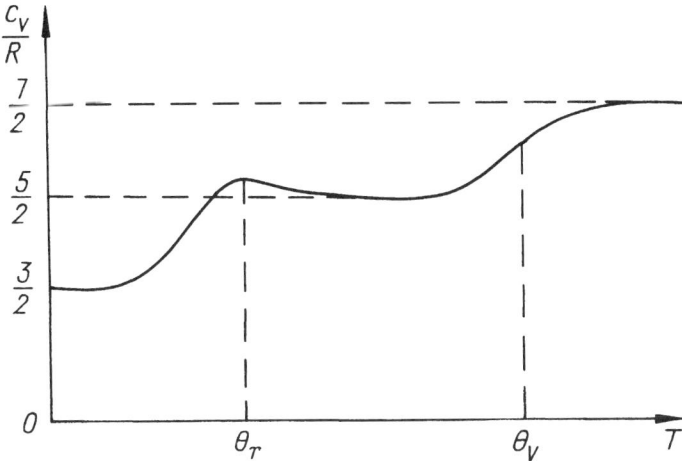

Figure 1.30 A schematic plot of the diatomic specific heat as a function of temperature. Characteristic temperatures for rotation and vibration are Θ_r and Θ_v correspondingly.

adequately represent physical reality in a restricted temperature range. Obviously, the theory should be not abandoned, but modified somehow so that we can predict which degrees of freedom and what temperature range should be taken into account, and which degree must be considered to be not involved in the heat motion. The problem is: can it happen that some type of motion contributing to the heat capacity under certain conditions will prove to be inessential in another circumstances? To see that this situation is quite possible, let us consider the potential part of translational and rotational motion energy in external field and its contribution to heat capacity.

(a) Assume that a vessel with a gas is in a uniform gravitational field. Then integrating over coordinates, we can see that the multiplier V in Eq. (1.11.4) is replaced by the following partition function

$$Z = \int\int dx dy \int_0^H \exp\left(-\frac{mgz}{kT}\right) dz = \frac{SkT}{mg}\left[1 - \exp\left(-\frac{mgH}{kT}\right)\right]. \tag{1.11.13}$$

The corresponding part of the heat energy, naturally separated from other (kinetic) components, is

$$\overline{U} = \mathbf{mg}\bar{z} = kT - \frac{mgH}{\exp\left(\frac{mgH}{kT}\right) - 1}. \tag{1.11.14}$$

With $mgH \ll kT$, the total center of gravity is approximately in the center of the vessel and $\overline{U} = \frac{1}{2}mgH - \frac{1}{12}[(mgH)^2/kT]$. Naturally, the heat capacity is close to zero, since the distribution of molecules in height, and, therefore, their total potential energy are essentially unaffected by the temperature variation: $C_V = N_0(d\overline{U}/dT) = (R/12)[(mgH)/kT]^2 \ll R$. As the temperature rises, $C_V \propto 1/T^2 \to 0$, because the distribution becomes more uniform.

On the contrary, at $mgH \gg kT$, $\overline{U} \approx kT$. Similarly to the equipartition law, the corresponding specific heat is constant although $C_V = R$, instead of $R/2$. The heat capacity is height-independent because the gas is at the bottom of the vessel and the position of the lid is of no importance. As the temperature increases, energy is expended in shifting the center of gravity of the gas upwards, R for each degree.

In general, the dependence of C_V on T may be inferred from Fig. 1.31: the boundary between the region of constant C_V and hyperbolic disappearance of the specific heat is governed by the characteristic temperature $T_c = mgH/k$.

(b) Now take the case that molecules with a dipole moment are placed in a uniform electric field. The potential energy related to a rotation by the angle θ may be determined by the procedure described above. However, this is quite unnecessary, as $\overline{U} = -\bar{q}_z E$, and $\bar{q}_z = qL(\alpha)$ has already been calculated in (1.8.13). So we immediately obtain

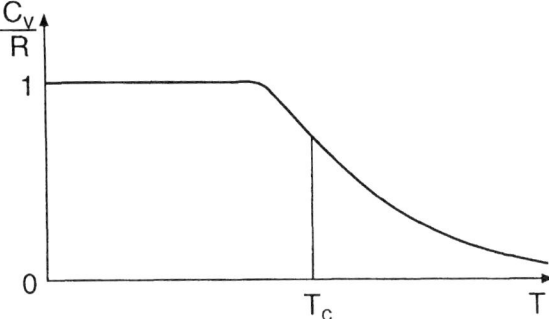

Figure 1.31 The ideal gas heat capacity in a homogeneous gravitational field.

$$C_V = N_0 \frac{d\overline{U}}{dT} = R\left(1 - \frac{\alpha^2}{\text{sh}^2\alpha}\right),$$

where $\alpha = qE/kT$ as before. An analysis of this expression shows that under saturation conditions $(qE \gg kT)$, the heat capacity is equal to R, while at high temperatures $(qE \ll kT)$ it tends to zero by a hyperbolic law: $C_V = (R/3)(qE/kT)^2$, just as in the previous example (see Fig. 1.31).

The nature of the extra heat capacity of a dielectric in an external field is rather obvious. The increase in temperature turns some dipoles opposite to the field increasing their potential energy at the expense of the heat taken out of the thermostat. As soon as a uniform distribution of dipoles in all directions is attained, further rise of temperature is no longer of any importance, and the heat supply ceases.

In both examples the heat capacity goes to zero above a characteristic temperature. Of course, this cannot be attributed to the fact that the energy is nonlinear in the coordinate. The decrease in the heat capacity is due to a limited phase space. The first example demonstrates this most clearly: if the vessel were infinitely tall $(H = \infty)$, that is, z varied from 0 to ∞, then the specific heat would remain constant (equal to R) over the whole temperature range. In the second case, we have the same situation: the projection of the moment on the z axis affecting the energy $U = -q_z \cdot E$ is bounded from above ($|q_z| < q$). If one had $q = \infty$, the characteristic temperature $T_c = qE/k$ could also never be reached.

In the case of translational, rotational and vibrational motions, the corresponding canonical variables vary in an infinite range. Besides, a comparison of Figs. 1.30 and 1.31 shows that the anomalous behavior of $C_V(T)$ corresponding to these types of motion is opposite in character. At high temperatures excellent agreement with the equipartition law is observed, and only at fairly low temperatures does the heat capacity decrease, tending to zero. An explanation could be found if the phase space of these variables proved to be

74 GASES

bounded from below rather than from above. This important point will enable us to clear up the cause of the heat capacity paradox.

1.12 HARMONIC OSCILLATOR QUANTIZATION

Freezing out of Vibrations

The problems of the classical theory of heat capacity cannot be eliminated by perfecting the gas model. They are fundamental. The equipartition law is a direct consequence of the Gibbs distribution. Its failure means that either the distribution itself is invalid, or the use of classical mechanics is not justified; and at first glance one will tend to blame the former.

However, the reason proved to be different. It lies in the unjustified extension of Newton's mechanics to elementary particles. The mechanical properties of atoms and molecules could not be predicted *a priori*. This unsuccessful attempt to describe them classically brought about innovations in principle which helped to resolve the paradox of heat capacities and formed the basis for the first quantum postulates. The most revolutionary principle of quantum mechanics had been formulated long before direct experimental study of elementary particles became feasible.

This principle was established by Planck in order to eliminate the so-called "ultraviolet catastrophe"—another consequence of the equipartition law (see Chapter 2, Section 2.3). In outline, it is as follows: a radiation field may be treated as an ensemble of harmonic oscillators which are described in the same fashion as harmonic vibrations of molecules. If the field is in equilibrium with the substance, then, according to the equipartition law, the thermal energy per oscillator is equal to kT. However, as the number of field oscillators is infinitely large (with frequencies between 0 and ∞), their total energy is to be ∞! Of course, this is not the case. By analyzing the experimental energy spectrum of equilibrium radiation Planck discovered that the mean thermal energy per oscillator is

$$\overline{E} = \frac{h\nu}{\exp(h\nu/kT) - 1}, \qquad (1.12.1)$$

where $h = 6.6\,10^{-27}\,\text{erg}\cdot\text{s}$ is the Planck constant. At low frequencies ($\nu \ll kT/h$) this expression reduces to the classical result $\overline{E} = kT$, while at high frequencies it goes to zero. The absence (freezing out) of high frequency vibrations resolves the ultraviolet paradox, but is completely inconsistent with the equipartition law, according to which the mean energy of vibrations does not depend on their frequency.

The experimentally obtained formula (1.12.1) also eliminates the difficulties in the theory of heat capacities related to the vibrational energy of molecules. It is seen that for each oscillator with vibration frequency ν there is a characteristic temperature $\Theta = h\nu/k$. At higher temperatures $\overline{E} = kT$, $C_V = R$, and at lower ones $\overline{E} = h\nu\exp(-h\nu/kT)$ and $C_V \to 0$ with $T \to 0$ (vibrations are frozen out). However, we still do not know what drawback in the previous calculations must be rectified to derive this formula theoretically.

Energy Quantization

With a contradiction of this kind, it is useful to examine it from different viewpoints. Sometimes this helps to reveal the cause of the difficulty. In the case under discussion, it is profitable to reconsider the problem in energy space. To this end, it is necessary to carry out the change of variables from p and q to $E = p^2/2m + (m\omega^2 q^2/2)$ and $\varphi = arctg(p/q)$ (energy and phase of vibration) with the subsequent integration over phases (from 0 to 2π). As usual, the first step is to calculate the Jacobian

$$I = \frac{\partial(E, \varphi)}{\partial(p, q)} = \frac{m^2\omega^2 + tg^2\varphi}{m(1 + tg^2\varphi)} .$$

The element of the phase space in the new variables is as follows

$$d\Gamma = dpdq = \frac{dEd\varphi}{I} = \frac{m}{m^2\omega^2 + tg^2\varphi} \frac{dEd\varphi}{\cos^2\varphi} . \quad (1.12.2)$$

The isoenergy states constitute an ellipsoid-shaped strip (Fig. 1.32), the area of which is found by direct integration over φ:

$$d\Gamma = \oint_\varphi d\Gamma = dE \oint \frac{md\varphi}{(m^2\omega^2 + tg^2\varphi)\cos^2\varphi} = \frac{2\pi}{\omega} dE = \frac{dE}{\nu} . \quad (1.12.3)$$

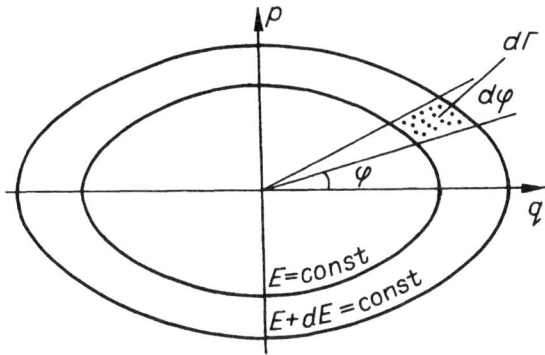

Figure 1.32 An isoenergy strip of the phase space of the harmonic oscillator.

76 GASES

With this in mind, we can write two equivalent definitions for Z

$$Z = \iint e^{-E/kT} \, dp \, dq = \int_0^\infty e^{-E/kT} \frac{dE}{\nu} . \tag{1.12.4}$$

Either of them yields the same result: (1.11.10). However, the second expression has an advantage: its structure gives a clue to a possible way to modify the theory.

Let us state the question in the following way: what value of Z, if substituted into (1.11.2), will lead to the correct result (1.12.1)? Accordingly, we equate (1.11.2) to (1.12.1). Then the desired value of Z is defined by a simple differential equation

$$-\frac{d \ln Z}{d\alpha} = \frac{1}{\exp \alpha - 1} ,$$

where $\alpha = h\nu/kT$. Its solution is

$$Z = \frac{C}{1 - \exp(-\alpha)} = \frac{C}{1 - \exp(-h\nu/kT)} . \tag{1.12.5}$$

Now the problem is laid bare and brought to the point where making the next step requires outstanding shrewdness. The question arises: is there any analogy between the expression (1.12.5) obtained from experimental data and its theoretical analog (1.12.4)? It is not so easy to reveal the analogy, but to use it for a radical modification of the theory is even more difficult. Making such generalizations calls for extraordinary courage. However, this is just what enabled Planck to derive his famous quantum principle.

Planck has noted that expression (1.12.5) may be rewritten using the familiar geometric progression formula

$$\frac{Z}{C} = \sum_{N=0}^{\infty} e^{-\alpha N} = \sum_{N=0}^{\infty} e^{-E_N/kT} . \tag{1.12.6}$$

This result resembles the last version of expression (1.12.4), if we assume (without proof!) that integration should be replaced by summation, taking

$$E_N = Nh\nu, \qquad N = 0, 1, 2, \ldots, \infty . \tag{1.12.7}$$

However, in this case we have to hold that the system's energy cannot take any values, but only those rigorously specified, that the energy changes discretely rather than continuously, and that one allowed value is separated from the next by an energy interval, or *quantum*, of magnitude $h\nu$. Thus the oscillator cannot be found in any state as before, but only in some definite states of motion with energies as specified by Eq. (1.12.7).

1.12 HARMONIC OSCILLATOR QUANTIZATION

So the introduction of formula (1.12.7) is not simply an ordinary working hypothesis, or an improved model of the phenomenon: the quantization of energy reveals the qualitative inadequacy of classical mechanics in its failure to describe the mechanical behavior of atoms and molecules.

High Temperature Limit

The new physical concept does not exclude the previous one, but just reveals its narrowness. This is a brief formulation of the "correspondence principle," according to which a general physical theory must reduce to a particular one in a region where the latter is applicable. To verify the results of the classical theory, we need to show that at high temperatures the sum over states (1.12.6) reduces to the partition function (1.12.4). The procedure is rather simple. For

$$\frac{h\nu}{kT} \ll 1 \quad (1.12.8)$$

the summation in (1.12.6) may be replaced by integration

$$Z = C \sum_{N=0}^{\infty} e^{-Nh\nu/kT} \rightarrow C \int_0^{\infty} e^{-Nh\nu/kT} dN = C \int_0^{\infty} e^{-E/kT} \frac{dE}{h\nu}. \quad (1.12.9)$$

It can be easily seen that this asymptotic formula is similar or even identical to the estimate of Z given by expression (1.12.4). The only difference, the constant multiplier, is eliminated by an appropriate choice of C.

Quantum Cells

However, this should be done without haste. The simplest choice is to put $C = h$, but the requirements of the correspondence principle can also be met in another way. In fact, previously any distribution was determined to be accurate up to a constant multiplier G

$$dW = \frac{1}{Z} \exp\left(-\frac{E}{kT}\right) G d\Gamma, \qquad Z = G \int \exp\left(-\frac{E}{kT}\right) d\Gamma.$$

Nothing depended on this constant, except the setting of a zero for the thermodynamic functions (1.10.5) and (1.10.6). So we have always considered $G = 1$. However, now it is time to make another choice: we put $G = 1/h$ and $C = 1$, thus making (1.12.9) correspond to (1.12.4). With this correction, the classical canonical distribution appears as

$$dW = \frac{1}{Z} \exp\left(-\frac{E}{kT}\right) \frac{d\Gamma}{h}, \qquad Z = \frac{1}{h} \int \exp\left(-\frac{E}{kT}\right) d\Gamma. \quad (1.12.10)$$

On the other hand, the quantum distribution takes the conventional form

$$W_N = \frac{1}{Z}\exp\left(-\frac{E_N}{kT}\right), \qquad Z = \sum_{N=0}^{\infty}\exp\left(-\frac{E_N}{kT}\right), \qquad (1.12.11)$$

where W_N is the dimensionless probability of a state N.

The question of the definition of the constants C and G, that is, whether the Planck constant should appear in the classical distribution or in the quantum one, may seem pointless: at first sight, nothing depends on the answer. However, there are reasons to prefer the solution which leads to (1.12.10) and (1.12.11). If the phase space volume is measured in units of h, as in (1.12.10), then

$$\frac{d\Gamma(E)}{h} = \frac{dE}{h\nu} = dN. \qquad (1.12.12)$$

This equality following from (1.12.3) and (1.12.7) establishes a common terminology, that is, a correspondence between the classical "phase volume" and the quantum "number of states." Now the statement that the probability of finding an oscillator with energy E is proportional to the volume $d\Gamma(E)$ is not quite correct. It is better to say that this probability is measured by the number dN of possible quantum states falling within this interval (it is implied that $dE \gg h\nu$).

In fact, when one has a discrete set of states, the question of how they can be counted does not arise. Nonetheless, if calculations are done in the space of continuously varying variables p and q, the same result may be obtained by dividing it into cells of volume h, with each cell containing one real state. In the case of the harmonic oscillator, the cells are concentric ellipsoidal strips separated by permitted phase trajectories.

Bohr's Postulate

So far the above reasoning has referred to the harmonic oscillator only. Now let us generalize. Rewriting (1.12.12) in integral form, and bearing in mind that the area bounded by the phase trajectory of the system is to be found, we get

$$\Gamma(E) = \oint p\,dq = Nh. \qquad (1.12.13)$$

This relation carries double information: the classical oscillator property (1.12.3) which can be represented as $\Gamma(E) = \int_0^E d\Gamma = E/\nu$ and the quantization condition (1.12.7) selecting solely allowed values $E = Nh\nu$. Remarkable that the frequency of vibration ν, which is the special parameter of the oscillator, is absent in the final expression (1.12.13) although it appears in both premises. When applied to the oscillator, Eqs. (1.12.7) and (1.12.13) serve equally well for energy quantization. However, from the viewpoint of generalization, Eq.

(1.12.13) is preferable, because this expression does not depend on specific features of the system. The idea of applying it to the quantization of translational and rotational motion, as well as the motion of an electron inside an atom, is very attractive. This idea belonged to Bohr who applied Eq. (1.12.13) to find permitted orbits, and, therefore, energies of electrons in hydrogen-like atoms. The great success of the Bohr theory in the explanation of atomic spectra has verified the general character of this quantization rule. Without going into detail (otherwise, we will find ourselves deep inside the field of atomic physics), we shall simply note that before the emergence of rigorous quantum mechanics, Bohr's quasiclassical quantization rule (1.12.13) was the only reliable basis for interpreting and calculating line spectra of elements—the stumbling block of classical prequantum physics.

1.13 FREEZING OUT OF HEAT MOTIONS

Though in principle any type of motion is subjected to quantization, it is not always necessary. At temperatures $T \gg \Theta = h\nu/k$, even an oscillator may be described classically. Each type of motion has its own characteristic temperature $\Theta = \Delta E_1/k$ where ΔE_1 is the energy difference between the lowest (ground state) and the first excited state of the discrete energy spectrum. The effect of quantization is significant only in the range $0 \leq T \leq \Theta$, but Θ takes different values for different types of motion. For molecular vibrations with a typical frequency $\nu \approx 10^{13}$ Hz, the characteristic temperature

$$\Theta = \frac{h\nu}{k} = \frac{6.6 \cdot 10^{-27} \cdot 10^{13}}{1.4 \cdot 10^{-16}} = 471 \text{ K} \approx 200°C$$

is so high that at room temperatures vibrations are frozen out (most molecules do not vibrate). On the other hand, the characteristic temperature for translation is so close to absolute zero that the classical approximation is applicable at practically all temperatures.

An example will be illuminating. Placing a particle into a cubic box with the edge $\mathcal{L}$, we see that the phase trajectory of translational motion between the opposite walls is a rectangle (Fig. 1.33) of the area

$$\oint p\,dq = 2\int_{-\mathcal{L}/2}^{\mathcal{L}/2} p\,dq = 2p\mathcal{L}.$$

According to the quantization rule (1.12.13), the momentum along any axis is strictly specified:

$$p = N\frac{h}{2\mathcal{L}}, \quad N = 0, 1, 2, \ldots \quad (1.13.1)$$

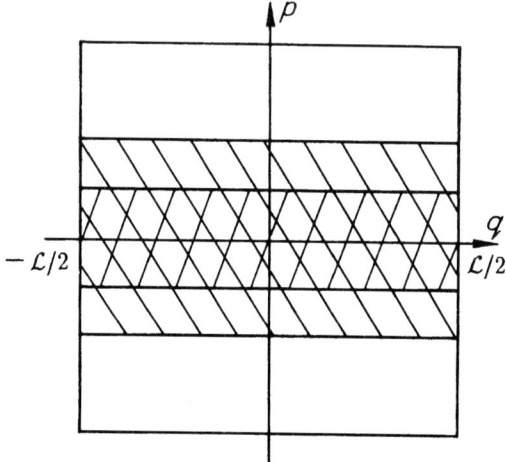

Figure 1.33 Quantum cages of a particle freely moving between the walls of potential well. Hatched regions are onefold and twofold areas of a cage.

Here q is any of x, y, z, and p is any of p_x, p_y, p_z, so

$$p_x = N_x \frac{h}{2\mathcal{L}}, \quad p_y = N_y \frac{h}{2\mathcal{L}}, \quad p_z = N_z \frac{h}{2\mathcal{L}},$$

and

$$p^2 = \frac{h^2}{4\mathcal{L}^2} N^2, \quad N^2 = N_x^2 + N_y^2 + N_z^2 = 0, 1, \ldots \quad (1.13.2)$$

Thus $\Delta E_1 = h^2/8m\mathcal{L}^2$. In a vessel of the size 1 cm

$$\Theta = \frac{h^2}{8m\mathcal{L}^2 k} = \frac{43 \cdot 10^{-54}}{8 \cdot 4 \cdot 10^{-23} \cdot 1.4 \cdot 10^{-16}} = 10^{-15} \text{ K}. \quad (1.13.2a)$$

This is an extremely low, unattainable temperature. Nearly all gases condense long before they are cooled to such temperatures. Therefore, in the gaseous phase (at $T \gg \Theta$) translation is of classical character. This explains the successful interpretation of gas properties on the basis of the Maxwell and Boltzmann distributions.

However, the determination of Θ by comparing kT with the first quantum from below may seem questionable. Unlike a harmonic oscillator, the quantized spectrum of translational energy $\epsilon = p^2/2m$ is not equidistant. The energy difference between the neighboring levels $\Delta \epsilon_N = \epsilon_N - \epsilon_{N-1} = h^2(2N-1)/8\mathcal{L}^2 m$ increases with increasing N. Even if kT is large as compared with $\Delta\epsilon_1$, we may be quite sure that it is less than some rather large $\Delta\epsilon_N$. However, this is not important. In the classical limit $\bar{\epsilon} = \frac{3}{2}kT = \overline{p^2}/2m$, so the quantum number of the corresponding energy level is $\bar{N} \approx \sqrt{3kTm}\,(2\mathcal{L}/h) = \sqrt{3T/2\Theta}$. As the temperature rises, this number increases along with the distance between neighboring levels $\Delta\epsilon_{\bar{N}} \approx h^2\bar{N}/4\mathcal{L}^2 m$. However, the ratio $\Delta\epsilon_{\bar{N}}/\epsilon_{\bar{N}} = 2\bar{N}/\bar{N}^2 = 2/\bar{N} \ll 1$ decreases monotonically with temperature. It can exceed unity at $\bar{N} \ll 1$ only, that is, at $T \ll \Theta$. This is just the case where the first quantum may be compared with kT, since all higher levels are already empty.

Hence, vibrational motion at room temperatures is practically frozen out, while translational motion is classical. Rotational motion occupies an intermediate position. Applying the same quantization rule (1.12.13) to angular momentum projections, we get for each

$$\oint p_\varphi d\varphi = 2\pi p_\varphi = Nh; \qquad p_\varphi = N\hbar. \qquad (1.13.3)$$

Thus angular momentum takes values of integer multiplies of $\hbar = h/2\pi$. Similarly, for a molecule rotating about all three axes, we obtain

$$\epsilon = \frac{N_1^2 \hbar^2}{2I_1} + \frac{N_2^2 \hbar^2}{2I_2} + \frac{N_3^2 \hbar^2}{2I_3}. \qquad (1.13.4)$$

In general, the moments of inertia I_i are different for different axes. At $I_i = 10^{-40}$ g cm^2, the characteristic temperature is

$$\Theta_i = \frac{\hbar^2}{2I_i k} \sim \frac{10^{-54}}{2 \cdot 10^{-40} \cdot 1.4 \cdot 10^{-16}} \approx 30\,\text{K}. \qquad (1.13.5)$$

Rotational motion of molecular hydrogen is frozen out at a temperature near 80 K, because it has the lowest moment of inertia, and correspondingly the largest Θ. The larger a molecule, the lower the freezing temperature of rotational motion.

If we have a linear molecule, the moment of inertia for rotation around the molecular axis is obviously equal to zero, and the characteristic temperature corresponding to this degree of freedom is infinitely high. That is why the mean energy of such a rotation is equal to zero, and it makes no contribution to the heat capacity. Thus the fact that linear molecules have two instead of three rotational degrees of freedom is exclusively of quantum origin. Consider a planar triatomic molecule of H$_2$O type. This is an isosceles triangle rotating in all three directions (Fig. 1.34). The minimum moment of inertia is associated with the axis parallel to the base $H - H$ of the triangle, since the vertex angle is obtuse. Now imagine that this angle increases: the inertia moment under discussion will decrease to become zero at the angle 180°, that is, when the

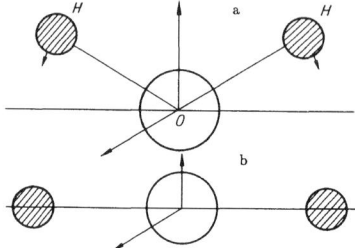

Figure 1.34 The geometrical structure of (a) a water molecule and (b) a linear molecule like CO_2.

molecule is linear. The corresponding Θ will increase gradually and become infinite in a linear molecule. In this limit the related degree of freedom may be excluded from consideration when calculating the heat capacity.

Summing up, we can say that at extremely low temperatures gas molecules execute only translational heat motion, at room temperatures, translational and rotational motions, and only if heated considerably will there be all three types including vibrational motion. This explains the general pattern of the variation of heat capacity with temperature as shown in Fig. 1.30. However, the mechanisms of the unfreezing of rotational and vibrational motions are quite different. When vibrational degrees of freedom appear, the heat capacity increases monotonically with rising temperature, gradually approaching the classical limit. In the case of rotational motion the specific heat of linear molecules passes through the maximum before the classical limit is reached (see Fig. 1.30). These subtle details of the behavior of $C_V(T)$ behavior near characteristic temperatures cannot be understood in the context of Bohr's quasiclassical theory. Although effective in estimating the magnitude of quanta, it does not allow one to correctly determine the positions of levels in the energy spectrum.

To deal with this difficulty, let us take proper account of the results of a consistent quantum theory. According to (1.13.4), the rotational energy of a diatomic molecule is $\epsilon = (\hbar^2/2I) N^2$ where $N = (N_x^2 + N_y^2)^{1/2}$. However, actually

$$\epsilon = \frac{\hbar^2}{2I} N(N+1). \qquad (1.13.6)$$

Each energy level, except the lowest one, is degenerate, that is, consists of $2N+1$ sublevels differing only in the spatial orientation of the angular momentum (Fig. 1.35a). As for the harmonic oscillator, its spectrum, if rigorously calculated, proves to be similar to the quasiclassical spectrum (1.12.7), but differs from it in that all levels are shifted upward along the energy axis by a half quantum:

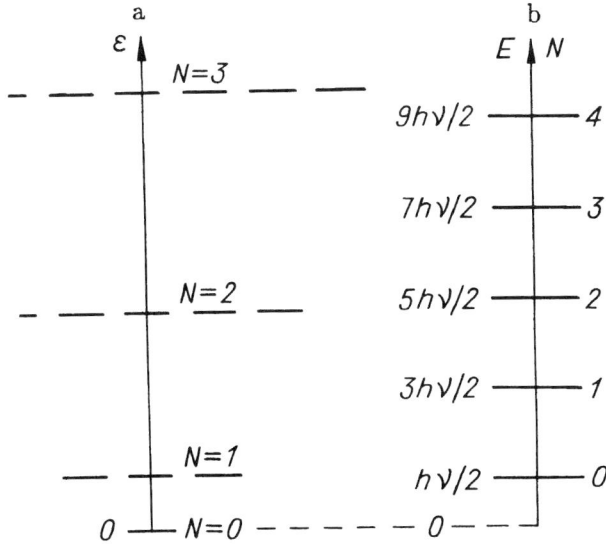

Figure 1.35 The energy spectra of (a) a linear rotator and (b) an oscillator.

$$E = (N + \tfrac{1}{2}) h\nu. \tag{1.13.7}$$

This difference implies that even at $T = 0$, when thermal motion is completely frozen out, the energy of vibrations is not zero. In the lowest state, the oscillator is not at rest, but executes purely quantum motion (*zero vibrations*) with energy equal to $h\nu/2$ (Fig. 1.35b).

Taking this into consideration let us calculate the real state sums of the quantum rotator and oscillator to disclose the origin of the abovementioned difference in the temperature dependence of their heat capacities. For rotator with spectrum (1.13.6), the state sum

$$Z = \sum_{N=0}^{\infty} (2N+1)\, e^{-(\Theta/T)N(N+1)} \tag{1.13.8}$$

may be transformed using the Euler-Macloren formula

$$\sum_{N=0}^{\infty} f(N) = \int_0^{\infty} f(x)dx + \frac{f(0)}{2} - \frac{1}{12}f'(0) + \frac{1}{720}f'''(0)$$

into the following series in Θ/T:

$$Z = \frac{T}{\Theta}\left[1 + \frac{1}{3}\frac{\Theta}{T} + \frac{1}{15}\left(\frac{\Theta}{T}\right)^2 + \ldots\right]. \qquad (1.13.9)$$

Thus the mean rotational energy of the molecule in the high temperature limit may be estimated as follows:

$$\frac{\bar{\epsilon}}{k\Theta} = -\frac{\partial \ln Z}{\partial(\Theta/T)} = \frac{T}{\Theta} - \frac{1}{3} - \frac{1}{45}\frac{\Theta}{T}, \qquad \Theta \ll T \qquad (1.13.10)$$

It differs from its classical limit $\bar{\epsilon} = kT$. As the temperature rises, this deviation decreases to a constant value equal to $k\Theta/3$, but does not disappear completely (Fig. 1.36a). Besides, there is a point of inflection on the curve $\bar{\epsilon}(T)$ at which the specific heat $C_V = N_0 d\bar{\epsilon}/dT$ is at its maximum shown in Fig. 1.30. This follows from the fact that the specific heat calculated from Eq. (1.13.10) approaches its limit from above

$$C_V = R\left[1 + \frac{1}{45}\left(\frac{\Theta}{T}\right)^2\right], \qquad \Theta \ll T.$$

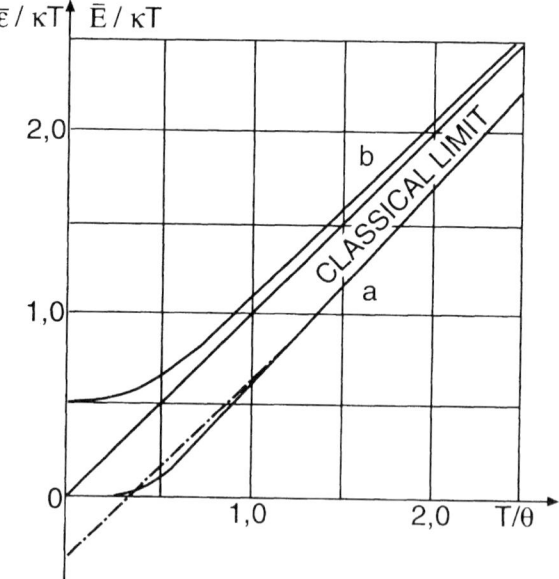

Figure 1.36 Average thermal energies of (a) the linear rotator and (b) the oscillator as a function of temperature.

A different situation arises with the harmonic oscillator. Its mean energy differs from the quasiclassical value found earlier only by a constant $h\nu/2$:

$$\overline{E} = \left(\overline{N} + \frac{1}{2}\right) h\nu = k\Theta \left[\frac{1}{\exp\left(\dfrac{\Theta}{T}\right) - 1} + \frac{1}{2}\right]. \qquad (1.13.11)$$

In the high temperature limit the energy approaches its asymptotic value from above (see Fig. 1.36b):

$$\overline{E} \approx kT \left[1 + \frac{1}{12} \left(\frac{\Theta}{T}\right)^2\right], \qquad \Theta \ll T. \qquad (1.13.12)$$

Thus, the heat capacity increases monotonically as shown in Fig. 1.30.

In a real theory of heat capacities it is necessary to allow for the fact that actually molecular vibrations are never purely harmonic: the vibrational levels are equidistant only deep in the potential well where it has a near-parabolic shape. In reality, as the energy increases, the difference between levels decreases. However, more importantly, at high energies the molecules become unstable and dissociate. This leads to an increase in the total number of particles and a qualitative change in their degrees of freedom.

At significantly high temperatures one must also take account of the excitation of electrons in molecules and atoms to higher energy levels. These excitations are associated with quanta of the order of electronvolts, so the electrons motion remains frozen out up to $10^3 \div 10^4$ K. However, with enough heating, the electrons become involved in thermal motion; first they are excited to higher levels, and then leave the particles ("heat ionization").

Finally, at temperatures of millions, in fact dozens of millions of degrees, when all molecules have already dissociated and all atoms have been ionized, it is time for atomic nuclei excitation and fission will occur. This is just what takes place deep inside the Sun and other stars (thermonuclear reactions).

1.14 GAS PARAMAGNETISM

The angular momentum of a molecule or an atom is due to the rotation of their constituents, ions and/or electrons. As they are charged particles, simultaneously, there arises a magnetic moment. Due to their common origin, the two moments are directly proportional with coefficient γ called the gyromagnetic ratio. The potential energy of a magnetic moment in a constant magnetic field H,

$$U = -\vec{\mu} \cdot \vec{H} = -\mu_z H, \qquad (1.14.1)$$

depends on the angle between the moment and the field, just as the energy of a dipole in an electric field. One might then infer that the picture of magnetic polarization must be similar to the electric one described in detail in Section 1.8.

Actually, this is not true. According to the quantum interpretation of charge rotation, the magnetic moment is quantized in the same way as the corresponding angular momentum (1.13.3)

$$\mu_z = \gamma p_\varphi = \gamma \hbar N, \qquad (1.14.2)$$

where N is the quantum number, and $\gamma \hbar \approx 10^{-20}$ erg/G. The quantization of magnetic moment brings about the quantization of the potential energy

$$U_N = -\mu_z H = -\gamma \hbar H N. \qquad (1.14.3)$$

As the z-projection of magnetic dipole cannot exceed its magnitude ($|\mu_z| \leq \mu = \gamma H J$), N varies within the finite limits

$$N = -J, -J+1, ..., J-1, J. \qquad (1.14.4)$$

All in all, there are $2J + 1$ values of the moment projection (Fig. 1.37). Thus paramagnetic polarization is a purely quantum phenomenon disappearing in

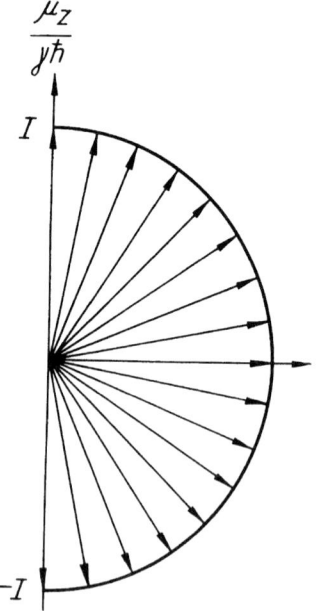

Figure 1.37 The quantized states of a magnetic moment in an external field.

the classical limit attained at $\hbar \to 0$. The quantum number J defining the absolute value of the moment can be either integer or semiinteger. If the magnetic moment is created solely by the orbital movement of electrons in atoms, J is an integer. At the same time, the electrons themselves have a moment—"spin"—different from zero and equal to $\frac{1}{2}$. As a result, the total moment of the atom, composed of the orbital and the spin moments, is integer if the number of electrons is even, and semiinteger if it is odd.

This brief excursus into quantum mechanics is quite sufficient for the statistical calculation of paramagnetic polarization. From (1.14.3) we see that for fields of attainable strength $\sim 10^4 \cdot G$,

$$\Delta U = U_N - U_{N-1} = \gamma \hbar H \sim 10^{-16} \text{ erg}; \tag{1.14.5}$$

that is at room temperatures, ΔU is considerably less than kT. Although magnetic quanta take an intermediate position between rotational and translational quanta, they are not as small as the latter and may not be ignored. Like dielectric polarization, paramagnetic polarization saturates at rather low temperatures: $\Theta = \Delta E/k \sim 1K$.

Spin $\frac{1}{2}$

To be specific, first consider pure spin magnetism with $J = \frac{1}{2}$. In this case, the calculation of magnetization is even more simple than in the classical situation. According to (1.14.14), now we have $N = \pm \frac{1}{2}$, that is, only two states instead of a continuous set. The spins are aligned either with the field or against it. Correspondingly, they have the energy of either the lower or the upper level (Fig. 1.38):

$$U_1 = -\gamma \frac{\hbar}{2} H \quad \text{or} \quad U_2 = \gamma \frac{\hbar}{2} H. \tag{1.14.6}$$

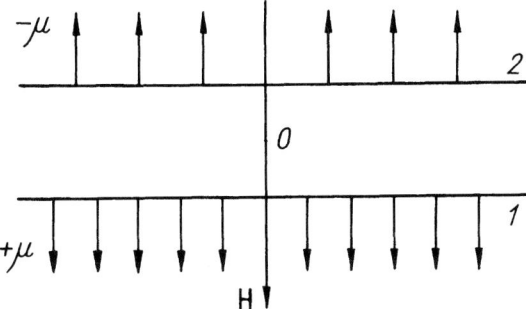

Figure 1.38 Distribution in a two-level system of projections of spin 1/2 in a magnetic field H.

As there are only two states the whole distribution (1.12.11) reduces to two probabilities

$$W_1 = \frac{\exp(\alpha/2)}{\exp(\alpha/2) + \exp(-\alpha/2)}, \qquad W_2 = \frac{\exp(-\alpha/2)}{\exp(\alpha/2) + \exp(-\alpha/2)}, \qquad (1.14.7)$$

where $\alpha = \gamma \hbar H/kT$. These are also the probabilities of the corresponding orientation of magnetic moment. Under equilibrium conditions, the spins are distributed over projections in accordance with these probabilities

$$N_1 = nW_1, \qquad N_2 = nW_2, \qquad (1.14.8)$$

where n is the total number of paramagnetic particles in unit volume. Naturally, their average magnetic moment

$$\overline{M} = \mu N_1 - \mu N_2 = \mu n(W_1 - W_2), \qquad (1.14.9)$$

where $\mu = \gamma \hbar / 2$. It is determined merely by the excess of spins parallel to the field over those aligned against it. According to (1.14.9) and (1.14.7), the mean magnetic moment of a unit volume is

$$\overline{M} = \mu n \frac{1 - e^{-\alpha}}{1 + e^{-\alpha}}. \qquad (1.14.10)$$

In the high temperature limit, with $\alpha = 2\mu H/kT \ll 1$, we can obtain the Curie law simply by expanding the exponents in Eq. (1.14.10)

$$\overline{M} = \frac{\mu^2 n}{kT} H = \frac{\gamma^2 \hbar^2 n}{4kT} H = \chi H. \qquad (1.14.11a)$$

When the magnetization is proportional to the applied field, the paramagnetic susceptibility decreases in inverse proportion to temperature. The close similarity between magnetic and dielectric polarization in the linear region is emphasized by the formal resemblance of (1.14.11a) to (1.8.15a). In the low temperature range where $\alpha \gg 1$ the situation is quite different. In this case, as in (1.8.15b), the saturation effect is observed:

$$\overline{M} = \mu \cdot n. \qquad (1.14.11b)$$

Under intense cooling, almost all spins are orientated parallel to the field, and at absolute zero those aligned against it disappear as completely as dissidents in a totalitarian system. However, the ordering in paramagnetics and dielectrics proceeds variously.

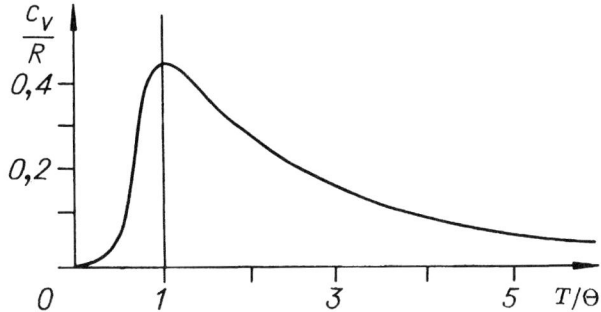

Figure 1.39 Temperature-dependence of two-level system heat capacity.

This difference is particularly pronounced in the temperature dependence of the heat capacity related to the polarization effect. Obviously, the polarization energy of a unit volume is $n\overline{U} = -\overline{M}H$, while the molar specific heat is

$$C_V = -N_0 \frac{d\overline{U}}{dT} = -\frac{d\overline{M}}{dT} N_0 H/n. \qquad (1.14.12)$$

As is seen from this formula and Eq. (1.14.10), the heat capacity associated with magnetic polarization is nowhere constant (Fig. 1.39). In the high temperature (classical) limit it is not equal to "R" for the same reason as the specific heat of polarized dielectric is not: the phase space is bounded above. Comparison of Fig. 1.31 and Fig. 1.39 shows that in this region the heat capacities vary according to the hyperbolic law $C_V = R(\mu H/kT)^2$. However, in the opposite case, where saturation takes place, the dielectric specific heat was found to be constant in qualitative agreement with the classical equipartition law. On the contrary, the specific heat of paramagnetics departs from classical result $C_V = R$ and goes to zero.

This is due to the fact that the spin phase space is bounded both from above *and* from below. All intermediate energies between U_2 and U_1 are forbidden by quantization; thus $C_V \to 0$ as soon as kT becomes less than $\Delta U = U_2 - U_1$. So, if there is a continuous set of orientations (as in the dielectric), the system's ordering by cooling requires that R calories be removed for each degree bringing the gas closer to absolute zero. In the case of two orientations, the closer to zero, the less energy is needed.

General Case

The difference between magnetic and dielectric polarization is particularly pronounced in the case of the spin equal to $\frac{1}{2}$ because only two possible orientations are compared with the continuous set. In fact, the electric dipole orientations are also quantized, but this fact was neglected because of the great number and small value of the corresponding energy quanta. The difference between discretely and continuously oriented dipoles must disappear with increasing absolute value of the moment J and the number of its orientations, provided that the energy interval which is being divided into the increasing number of quanta remains constant.

According to (1.14.2), (1.14.3), (1.14.4), and (1.12.11), the magnetic polarization of unit volume containing molecules with moment J is the following mean value:

$$\overline{M} = n \sum_{N=-J}^{J} \gamma \hbar N W_N = n \gamma \hbar \frac{d \ln Z}{d \alpha} . \qquad (1.14.13)$$

Direct calculation of the state sum, which is a finite geometric progression, yields

$$Z = \sum_{N=-J}^{J} e^{\alpha N} = \frac{e^{\alpha J} - e^{-\alpha(J+1)}}{1 - e^{-\alpha}} . \qquad (1.14.14)$$

On substituting (1.14.14) into (1.14.13) and differentiating, we derive

$$\overline{M} = \gamma \hbar n \left[\frac{J e^{\alpha J} + (J+1) e^{-\alpha(J+1)}}{e^{\alpha J} - e^{-\alpha(J+1)}} - \frac{e^{-\alpha}}{1 - e^{-\alpha}} \right] . \qquad (1.14.15)$$

This result is the most general estimate of magnetization suitable both for $J = 1/2$ when it reduces to (1.14.10) and for any other J, either integer or semiinteger. In this formula the purely quantum nature of paramagnetism manifests itself in the fact that at $\hbar \to 0$ $\overline{M}$ becomes zero as well. However, it is also possible to pass to another limit:

$$J \to \infty, \quad \hbar \to 0 \quad \text{at} \quad \mu = \gamma \hbar J = \text{const} , \qquad (1.14.16)$$

This results in an infinite increase in the number of quanta within the fixed energy range $(-\mu H, \mu H)$. In this case, Eq. (1.14.15) gives

$$\overline{M} = \mu n \left[\frac{e^{\mu H/kT} - e^{-\mu H/kT}}{e^{\mu H/kT} - e^{-\mu H/kT}} - \frac{kT}{\mu H} \right] = \mu n L \left(\frac{\mu H}{kT} \right) . \qquad (1.14.17)$$

This formula differs only in notations from Eqs. (1.8.13) and (1.8.14). Passing to the limit (1.14.16) makes the energy space so finely divided that it is practically continuous, so that the result of the classical theory of orientational polarization is again valid.

1.15 COLLISIONS

Though the equilibrium state of matter was studied and described without any connection with its origin, the reasoning thus far has assumed the ability of isolated macrosystems to return into equilibrium from any nonequilibrium state. Formally one can apply the microcanonical distribution either to a single particle in a box or to an ensemble of point masses isolated in it. However, neither of these systems is appropriate for thermodynamic treatment. Once in a state different from equilibrium, such a system will remain that state, because point particles do not collide with one another and their speeds are conserved in elastic collisions with the walls. Thus, an originally non-Maxwellian distribution will remain so in all future times.

However, in real systems the Maxwell distribution of velocities and the Boltzmann distribution of positions are always restored. Transformation of the initial distribution into an equilibrium one is called a *relaxation process*, and it may proceed in velocity and coordinate spaces with different *relaxation times*. Why and how does relaxation occur? The answer to this question is to be found in the context of physical kinetics.

Relaxation

The answer first came through Krylov's stirring hypothesis. In outline, this is as follows. It is supposed that an isolated macrosystem is able, in a fairly short time, to get from almost any state accessible to it to any other element of the phase volume. In physical kinetics this hypothesis is as fundamentally important as the microcanonical distribution in the equilibrium theory.

Let us consider as an example the rapid expansion of gas into a larger volume. The point representing the system in Γ-space starts from the region initially accessible to the system and, following an intricate trajectory, crosses the expanded phase space in all directions. At first it cuts it roughly into a few pieces, which are in their turn divided later into smaller fractions and so on. Owing to this, in a rather short time approximately equal shares of the trajectory are found in equal volumes of the phase space. Further, the system point continues to move in such a way as to effectively continue this partitioning process. So, the longer the time elapsed from the start, the more truly one can say that all equal elements of phase space contain the same fraction of the phase trajectory. The more uniform is the "stirring" of the trajectory in the accessible phase space, the less the deviation from equilibrium. If most trajectories emerging from the small fraction of the phase space specified by the initial conditions are well stirred, then systems starting from adjacent or nearby points may be found in quite different places. As the starting point is not known with certainty, after a lapse of the relaxation time, the expanded system can be found in any element of the new phase space with the same (microcanonical) probability.

Not all the systems are capable of "stirring." Roughly speaking, the system has this property if it consists of a great number of particles bound to each other by a definite interaction. Obviously, neither a free particle in a box, nor an ensemble of point masses meet these requirements. When considering an ideal gas, the interparticle interaction ensuring stirring has not even been mentioned. However, this should be taken into account as soon as the question about the attainment of equilibrium arises. The efficiency of this interaction affects the relaxation time. In gases the interaction results in scattering of molecules during collisions. They makes the trajectory of each molecule extremely intricate, to say nothing of the gas on the whole. That is why the required property of stirring is inherent in real gases.

Not all interaction and not every system may result in sufficiently effective stirring. For gases, the existence of this property determined exactly by such a collisional mechanism was theoretically proved by Krylov and subsequently verified by computer simulations. By computer one can follow the trajectory of a many-particle system, on condition that it obeys Newton's laws, and collisions are elastic. It turns out that two or three collisions of a particle are enough for the original distribution in velocities to go over the Maxwell one, and for the entropy to attain its equilibrium, that is, maximum, value.

However, there are systems which are not capable of stirring at all. In this case, equilibrium is attained exclusively by external action. The interaction with the environment, which is in equilibrium itself, gradually brings the system under discussion into equilibrium state with the same temperature. This is how, for example, the equilibrium distribution over vibrational and rotational states is attained in the gas phase. Translational degrees of freedom act as the environment. The relaxation of intramolecular motion proceeds at the expense of the kinetic energy of colliding particles. Also the equilibrium properties of extremely rarefied gases are established due to inelastic collisions with the vessel's walls, provided that the walls are thermostated.

Velocity-Dependent Collision Rate

Is the rate of collisions in the gas phase high enough to ensure fairly rapid attainment of equilibrium? The answer is given by an elementary estimate of the number of collisions suffered by a rigid molecule of radius r_1, moving through a gas of hard spheres of radius r_2. The idea can easily be understood if we imagine that all particles are immobile except the molecule under consideration. It follows its own path, colliding from time to time with other particles. In a second the molecule covers a distance equal to its speed, and experiences as many collisions as there are hard spheres on the way.

The last statement needs clarification. In the case of a real, smooth interaction between particles, it is not so easy to determine which molecules collided, and which did not. The advantage of the hard sphere model is that here the problem is solved geometrically. Molecules collide, if they touch, otherwise, collision does not take place. Enclosing the center of the moving molecule in a sphere of the radius $r_1 + r_2 = R$, we can consider other molecules as the point masses. Collision will occur only with those of them whose distance from the trajectory of the moving molecule's center is not larger than R. In other words, in one second of motion, only particles whose centers are confined within a cylinder of cross-section $\sigma = \pi R^2$ and length v will collide with the moving molecule. The volume of the cylinder is σv, the density of particles in the unit volume is n, therefore, the number of collisions per unit time

$$\nu = n\sigma v . \tag{1.15.1}$$

The above estimate would be satisfactory if the velocity of the moving particle were considerably greater than the velocities of particles which are targets. For example, such is the case for the electrons moving in a gas whose thermal velocity is greater than that of the molecules because of the difference in masses. When molecules of comparable masses collide, to say nothing of identical particles, the calculation must be adjusted.

It turns out that even in this case the calculation can be reduced to the previous one. We need only apply the procedure which is extensively used in kinetics: to divide the ensemble of colliding particles into subensembles. Moving to a coordinate system associated with the chosen molecule which moves with velocity $\mathbf{v}$, we can classify other molecules by their relative velocity $\mathbf{w}$. This classification is determined by the Maxwell distribution of the velocities $\mathbf{v}'$ of the target molecules which may be represented as a distribution of the relative velocities $\mathbf{w} = \mathbf{v}' - \mathbf{v}$:

$$dW(\mathbf{v}') = \frac{1}{Z} e^{-m v'^2 / 2kT} d\mathbf{v}' = \frac{1}{Z} e^{-m(\mathbf{v}+\mathbf{w})^2 / 2kT} d\mathbf{w} = dW_\mathbf{v}(\mathbf{w}) . \tag{1.15.2}$$

Here m is the mass of target particles, different from M, which denotes the mass of the chosen particle.

Now if we select the subensemble of particles where all molecules have the given relative velocity $\mathbf{w}$ (Fig. 1.40), all arguments leading to (1.15.1) may be repeated once again. Thus the rate of collisions with the given subensemble is

$$d\nu_\mathbf{v} = \sigma w n \, dW_\mathbf{v}(\mathbf{w}) . \tag{1.15.3}$$

The total number of collisions that the molecule with the velocity v experiences with all subensembles is as follows

$$\nu_\mathbf{v} = \int_\mathbf{w} d\nu_\mathbf{v}(\mathbf{w}) = \frac{n\sigma}{\alpha} \left[\frac{1}{2v} \Phi(\sqrt{\alpha} v) + \alpha v \Phi(\sqrt{\alpha} v) + \sqrt{\frac{\alpha}{\pi}} e^{-\alpha v^2} \right] , \tag{1.15.4}$$

where the probability integral $\Phi(x) = (2/\sqrt{\pi}) \int_0^x e^{-z^2} dz$, and $\alpha = m/2kT$. With $\sqrt{\alpha} v \gg 1$, when the speed of the chosen particle is essentially greater than the mean speed of molecule-targets, $\Phi(\sqrt{\alpha} v) \approx 1$ and the rate of collisions

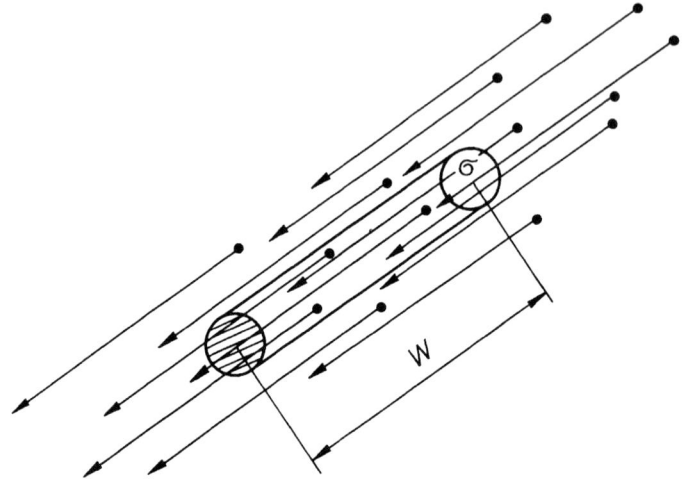

Figure 1.40 The target molecule (hatched) in a flux of particles moving with velocity w. Those of them contained in the cylinder with base σ and height w will collide within a second.

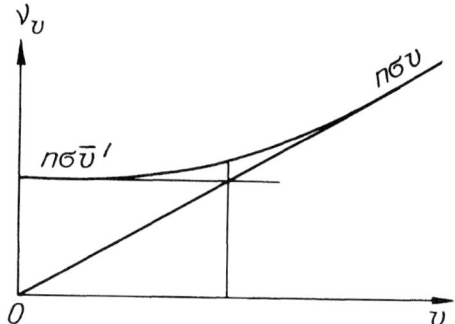

Figure 1.41 The increase of collision rate with velocity of a molecule moving through a gas.

$\nu_v \to n\sigma v$, just as in (1.15.1). However, unlike the case of immobile targets, at $v \to 0$ the number of collisions does not go to zero, but proves to be equal to $2\sigma n/\sqrt{\pi\alpha} = n\sigma\bar{v}'$. It is greater, the more intense the heat motion of surrounding molecules impacting the immobile particle (Fig. 1.41).

The Mean Collision Rate

Now let us make allowance for the fact that after each collision the velocity of the molecule changes. The fraction of time the molecule spends in the state with the given velocity v is determined by the Maxwell distribution $dW(\mathbf{v})$. The average number of collisions it experiences moving from time to time with different velocities is

$$\nu = \int \nu_v dW(\mathbf{v}) = \sigma n \int_v \int_{v'} w \, dW(\mathbf{v}') dW(\mathbf{v}) \, . \tag{1.15.5}$$

Converting to the variables $\mathbf{w}$ and $\mathbf{v}_0 = (M\mathbf{v} + m\mathbf{v}')/(M + m)$, we obtain

$$dW(\mathbf{v}) \cdot dW(\mathbf{v}') = dW(\mathbf{w}) \cdot dW(\mathbf{v}_0) \, ,$$

where $dW(\mathbf{w}) = (\mu/(2\pi kT))^{3/2} e^{-\mu w^2/2kT} d\mathbf{w}$ is the unconditional distribution of relative velocities, and $dW(\mathbf{v}_0)$ that of the velocity of the center of gravity of the colliding molecules ($\mu = mM/(m + M)$ is the reduced mass).

Upon integration over $\mathbf{v}_0$ in (1.15.5), we find

$$\nu = \sigma n \int w \, dW(\mathbf{w}) = \sigma n \bar{w} \tag{1.15.6}$$

or

$$\nu = n\sigma \sqrt{\frac{8kT}{\pi\mu}} = \sqrt{\frac{m+M}{M}} n\sigma \bar{v}' = \sqrt{\frac{m+M}{m}} n\sigma \bar{v} \, , \tag{1.15.7}$$

where $\bar{v} = \sqrt{8kT/\pi M}$, $\bar{v}' = \sqrt{8kT/\pi m}$.

Thus the total number of collisions is obtained from (1.15.1) by substituting w for v, and the dependence $\nu(\bar{v})$ is qualitatively similar to the dependence $\nu_v(v)$ given in Fig. 1.41, provided that $\bar{v}$ is considered as a function of M at fixed temperature. When $M = m$, (1.15.7) yields

$$\nu = \sqrt{2} n\sigma \bar{v} \, . \tag{1.15.8}$$

As is seen the motion of similar partners increases the number of collisions by a factor of 1.4 compared to the case where they are immobile.

Now we can estimate the order of the molecules' free path time under normal conditions: $n = 10^{19}$ cm^{-3}, $\bar{v} = 10^5$ cm/sec; $r_1 = r_2 = 2 \cdot 10^{-8}$ cm. In this case the cross-section of collisions is equal to $50 \cdot 10^{-16}$ cm^2, and

$$\nu = \sqrt{2} \cdot 50 \cdot 10^{-16} \cdot 10^{19} \cdot 10^5 \approx 7 \cdot 10^9 \text{ Hz} \, . \tag{1.15.9}$$

Consequently, if the Maxwell distribution of velocities is violated somehow, it is recovered within nanoseconds.

Experimental Verification

When a new physical idea or concept is advanced, it is always desirable to have direct experimental verification of its existence. In an experiment in which molecular beam was passed through a gas of rather low density, the collisions were visualized through the scattering of the beam molecules and the free path distribution was easily measured.

At some time each molecule experienced a first collision. As a result it changed its direction of motion, and left the beam. So the number of molecules in the beam decreased gradually with distance from the source (Fig. 1.42). This number could be measured by erecting a screen in the beam's path. It was established that the number of molecules which settled on the screen decreased exponentially depending on the distance between the screen and the source.

This exponential decrease was foreseen even by Clausius. The number of molecules involved in collisions and thrown out of the beam between x and $x + dx$ must be proportional to dx and to the number of molecules $N(x)$ which safely reached the borderline, that is, avoided collisions in the interval $(0, x)$. Therefore

$$-dN = a \cdot N(x) dx. \qquad (1.15.10)$$

In the gas phase, the coefficient a cannot depend on x, for there is no reason to believe that in any interval dx collisions occur more often than in any other.

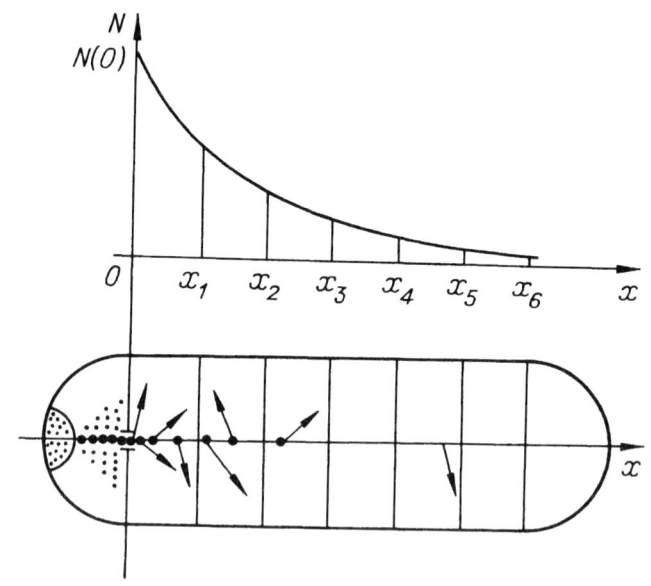

Figure 1.42 The scattering of a molecular jet in a gas, showing the resulting number density at section x.

This is a homogeneous flux of events, but their density in space a may depend on the velocity of the molecules in the beam, as in (1.15.4).

Solving (1.15.10), we obtain the number of particles which covered the distance x without a single collision

$$N(x) = N(0)e^{-ax}. \tag{1.15.11}$$

The thickness of the layer of molecules settled on the screen must be proportional to this value. This dependence was verified experimentally both qualitatively and quantitatively; moreover, plotting $\ln N$ against x, the value a was easily measured by the straight line slope.

Free Path

By definition the mean free path is

$$\ell = \int_0^\infty x \, dW(x)$$

where $dW(x)$ is the probability of colliding in the interval dx after passing the distance x without a collision. This probability is defined by the ratio of the number of molecules involved in the first collision in the interval dx to the total number of molecules $N(0)$ which "went to the starting line," that is,

$$dW(x) = -\frac{dN}{N(0)} = a\frac{N(x)}{N(0)}dx = ae^{-ax}dx.$$

Thus, the mean length of the molecules' free path in the gas

$$\ell = a \int_0^\infty x e^{-ax} dx = \frac{1}{a}, \tag{1.15.12}$$

is equal to inverse a and the distribution of free paths may be represented as follows:

$$dW(x) = \exp(-x/\ell)\frac{dx}{\ell}. \tag{1.15.13}$$

The free path length ℓ is related to the free path time $\tau_0 = \nu_v^{-1}$:

$$\ell = v\tau_0. \tag{1.15.14}$$

98   GASES

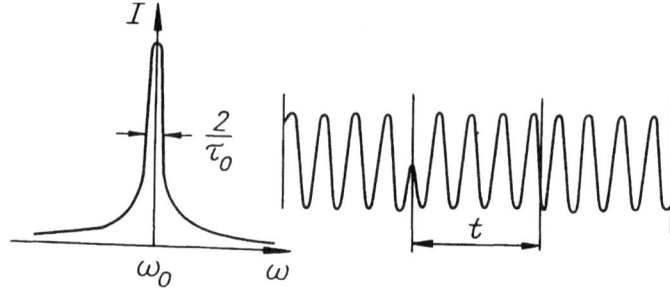

Figure 1.43 The spectrum of a Lorentzian wave: monochromatic radiation with its phase broken at each collision.

In view of $x = vt$, substitution of (1.15.14) into (1.15.13) yields the distribution in the free path times of

$$dW(t) = \frac{1}{\tau_0} e^{-t/\tau_0} dt . \tag{1.15.15}$$

This may be also subjected to experimental (spectroscopic) test. According to the Lorentz model, each collision results in an abrupt change in the phase of the electromagnetic wave frequency ω_0 radiated by the excited atom. The duration of a monochromatic piece of wave (Fig. 1.43) is determined by the probability of spending the time without collisions: $W_0 = \exp(-t/\tau_0)$. It can be shown that in this case the radiation spectrum shape is defined as

$$g(\omega) = \frac{Re}{\pi} \int_0^\infty \exp\left[-\frac{t}{\tau_0} + i(\omega_0 - \omega) t\right] dt = \frac{\tau_0/\pi}{(\omega - \omega_0)^2 \tau_0^2 + 1} . \tag{1.15.16}$$

Such a profile is called Lorentzian or impact. One can find the free path time of gas molecules from its width.

As τ_0 depends on v, the free path length $\ell(v) = v/\nu_v(v)$ increases monotonically with increasing velocity, tending to its upper limit $1/n\sigma$ which is reached at $v \gg \bar{v}'$. The average length of the free path may be defined as

$$\overline{\left(\frac{v}{\nu_v}\right)} \approx \frac{\bar{v}}{\nu} = \frac{1}{n\sigma} \sqrt{\frac{m}{m+M}} = \lambda . \tag{1.15.17}$$

It depends on the gas density only, and does not exceed the value $1/n\sigma$. Two mutually related parameters ν and λ are of fundamental importance in the construction of physical kinetics in the gas phase. At ordinary densities $n \sim 10^{19}$ cm^{-3} and $\sigma = 5 \cdot 10^{-15}$ cm^2 $\lambda = 2 \cdot 10^{-5}$ cm. However, λ may be essentially extended by exhausting the gas from the vessel: at $p = 10^{-4}$ atm

$\lambda \sim 0.2$ cm, while at $p = 10^{-5}$ atm $\lambda \sim 2$ cm. Under more intense evacuation of the vessel, λ exceeds the vessel size, and the gas turns into the so-called physical vacuum or ultrararefied gas. We shall see further that the latter is distinctly different in all its properties from a dense, ordinary gas.

Poissonian Statistics

It is of interest to ask what is the probability that only one collision occurs in the interval $(0, t)$ at time $t_1 \leq t$. Evidently this is a coincidence of two independent events:

- that the molecule moving through the gas experiences its first (after $t = 0$) collision in time element dt_1 and
- that this collision will be the first and the last, i.e., there will be no more collisions between t_1 and t.

The probability of such a realization of molecule's trajectory is given by the product of the corresponding probabilities: to have a first collision at the given time

$$dW = \frac{1}{\tau_0} e^{-t_1/\tau_0} dt_1$$

and to have no collisions between t_1 and t

$$W_0 = e^{-(t-t_1)/\tau_0}. \tag{1.15.18}$$

The result does not depend on t_1 if we assume for simplicity that τ_0 is v-independent ($M \gg m$):

$$dW_1 = dW \cdot W_0 = e^{-t/\tau_0} \frac{dt_1}{\tau_0}. \tag{1.15.19}$$

The probability that only one collision happened anywhere between 0 and t is the following integral

$$W_1 = \int_0^t dW_1 = e^{-t/\tau_0} \frac{t}{\tau_0}. \tag{1.15.20}$$

A simple generalization shows that the probability for two successive collisions to happen at times t_1 and t_2 is

$$dW_2 = e^{-t/\tau_0} \frac{dt_1}{\tau_0} \frac{dt_2}{\tau_0}, \tag{1.15.21}$$

while the probability of having just two collisions anywhere in the interval $(0, t)$ is

$$W_2 = \frac{e^{-t/\tau_0}}{\tau_0^2} \int_0^t dt_2 \int_0^{t_2} dt_1 == e^{-t/\tau_0} \frac{t^2}{2\tau_0^2}. \qquad (1.15.22)$$

For an arbitrary number of collisions n we obtain in the same way

$$W_n = \frac{1}{n!} \left(\frac{t}{\tau_0}\right)^n e^{-t/\tau_0}. \qquad (1.15.23)$$

This is the normalized Poissonian distribution of collisions: $\sum_{n=0}^{\infty} W_n = 1$. The average number of collisions in time t is evidently

$$\bar{n} = \sum_{n=0}^{\infty} n W_n = t/\tau_0, \qquad (1.15.24)$$

but the distribution about the mean is determined by the Poisson distribution (1.15.23).

Velocity Relaxation

Collisions of hard spheres are instantaneous, while those of real molecules are not. However, they are still considered as instantaneous in the so-called *impact approximation* as their duration in rare gases is much less than the free path time τ_0. Temporal change of the velocity in the impact approximation is a so-called *purely discontinuous random process*. While constant between impacts, the velocity abruptly changes at the moments of collisions as is shown in Fig. 1.44 for an arbitrary projection of velocity. The successive values of the projection v_i form a Markovian chain because any subsequent value depends only on a previous one. This dependence is given by the probability $F(v, v')dv$ to obtain the value v after collision if the velocity before it was v'.

Many different realizations of a random process lead to the same value v at time t when initially (at time 0) it was v_0. The chain between these values may be shorter or longer depending on how many collisions happen in the interval $(0, t)$. The conditional probability of the event $\varphi(v, t; v_0, 0)$ must sum up all possibilities as follows:

$$\varphi(v, t; v_0, 0) = \delta(v - v_0) W_0 + F(v, v_0) W_1 + \int F(v, v_1) dv_1 F(v_1, v_0) W_2 + \ldots$$
$$= [\delta(v - v_0) + \Phi] e^{-t/\tau_0}$$

$$(1.15.25)$$

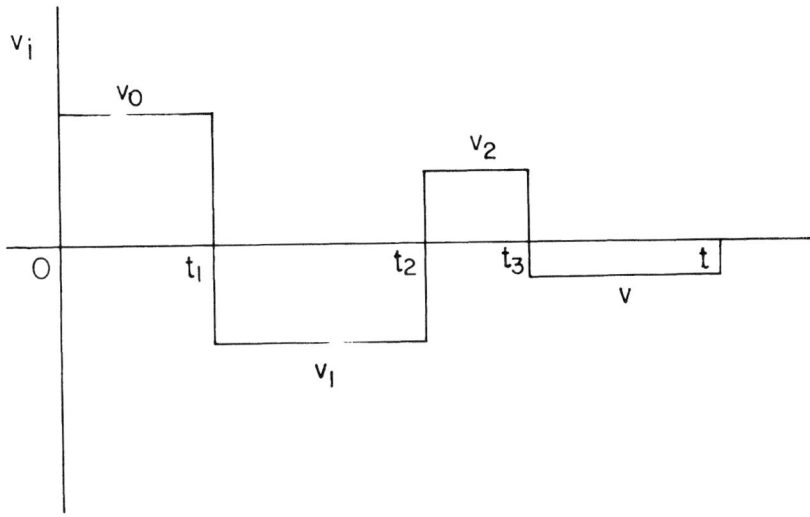

Figure 1.44 Time variation of a velocity component in three collisions: realization of a Poissonian process.

where

$$\Phi = F(v, v_0)\frac{t}{\tau_0} + \int F(v, v')dv' F(v', v_0)\frac{t^2}{2!\tau_0^2}$$

$$+ \int F(v, v')dv' \int F(v', v_1)dv_1 F(v_1, v_0)\frac{t^3}{3!\tau_0^3} \cdots \quad (1.15.26)$$

satisfies the integral equation

$$\Phi(v, t; v_0, 0) = \frac{1}{\tau_0} F(v, v_0) + \frac{1}{\tau_0} \int F(v, v')\Phi(v', t; v_0, 0)dv' \quad (1.15.27)$$

with initial condition $\Phi(0) = 0$. Differentiating (1.15.25) taking into account (1.15.27) we find the Kolmogorov–Feller equation for the conditional probability:

$$\dot{\varphi}(v, t; v_0, t) = -\frac{1}{\tau_0}\left[\varphi(v, t; v_0, 0) - \int F(v, v')\varphi(v', t; v_0, 0)dv'\right], \quad (1.15.28)$$

that must be solved with initial condition $\varphi(0) = \delta(v - v_0)$.

Starting from any initial value v_0 the velocity distribution must relax in time to the equilibrium Maxwellian distribution:

$$\varphi(v,\infty;v_0,0) = f(v). \tag{1.15.29}$$

Therefore $\dot{\varphi} \to 0$ at $t \to \infty$ and we find from Eqs. (1.15.28) and (1.15.29) the condition of stationarity of the equilibrium velocity distribution:

$$f(v) = \int F(v,v')f(v')dv'. \tag{1.15.30}$$

This may be considered as a direct consequence of the principle of *detailed balance* ensuring the equality of the rates of forward and back transitions between any two elements of the phase space:

$$F(v',v)f(v) = F(v,v')f(v') \tag{1.15.31}$$

The identity of the two statements is proved by integration of the last, taking into account the normalization condition: $\int F(v',v)dv' = 1$. This puts a limitation on the choice of $F(v,v')$.

The most popular model for this kernel was proposed by Keilson and Storer:

$$F(v,v') = F(v - \gamma v'). \tag{1.15.32}$$

Its particular shape is uniquely determined by Eq. (1.15.30) if $f(v)$ is known. For the one-dimensional Maxwellian distribution (1.2.13) it is

$$F(v - \gamma v') = \sqrt{\frac{\alpha}{\pi(1-\gamma^2)}} \exp\left(-\frac{\alpha(v-\gamma v')^2}{1-\gamma^2}\right), \tag{1.15.33}$$

where γ is a real numerical parameter with $|\gamma| \leq 1$. Its physical sense becomes clear after an estimate of the average velocity after collision:

$$<v> = \int vF(v - \gamma v')dv = \int_{-\infty}^{+\infty} zF(z)dz + \gamma v' \int_{-\infty}^{+\infty} F(z)dz = \gamma v'. \tag{1.15.34}$$

Collisions are *strong* at $\gamma = 0$ as each of them completely restores the equilibrium: after collision

$$F = f(v) \tag{1.15.35}$$

whatever is the velocity before it. Collisions are *weak* at $\gamma \to 1$ as the velocity decreases just a little (γ times) after a single collision and relaxes to 0 as a result of a long sequence of them (see Fig. 1.45).

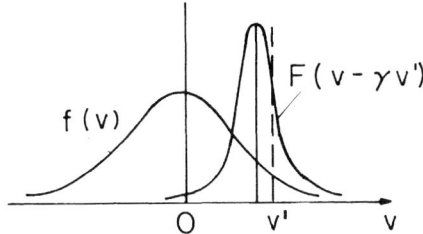

Figure 1.45 The Keilson–Storer kernel $F(v)$ for weak collisions in comparison with the one-dimensional Maxwellian distribution $f(v)$.

To find the relaxation law one has to multiply the Kolmogorov–Feller equation (1.15.28) by v and integrate, taking into account (1.15.34):

$$\dot{\bar{v}} = -\frac{\bar{v}}{\tau_0} + \frac{\gamma\bar{v}}{\tau_0} = -\Gamma\bar{v}, \qquad (1.15.36)$$

where

$$\Gamma = \frac{1-\gamma}{\tau_0} = n\sigma_v v \qquad (1.15.37)$$

is the relaxation rate of velocity. The effective cross-section of the process σ_v may be less than geometrical σ. The solution of Eq. (1.15.36) leads to exponential velocity relaxation

$$\bar{v}/v_0 = \exp(-\Gamma t), \qquad (1.15.38)$$

that is identical to $W_0 = \exp(-t/\tau_0)$ only in the limit of strong collisions.

Strong and Weak Collisions

The efficiency of hard sphere collisions depends on the masses of the colliding objects. If the target particle is so heavy that it may be considered as practically immobile, then the result of the collision is the elastic scattering of the light molecule center from a sphere of radius R. The scattering angle depends on the point of contact which is different for different molecules in a flux of point masses flying towards the sphere with the same initial velocity v'. Since the angle of incidence α is equal to the angle of reflection, the scattering angle $\Theta = 2\alpha$ (Fig. 1.46). This angle is the same for all points inside a differential cross-section $d\sigma$ which is an area increment of the circle of radius $R\sin\alpha$:

$$d\sigma = d(\pi R^2 \sin^2 \alpha) = 2\pi R^2 \sin\alpha \cos\alpha \, d\alpha. \qquad (1.15.39)$$

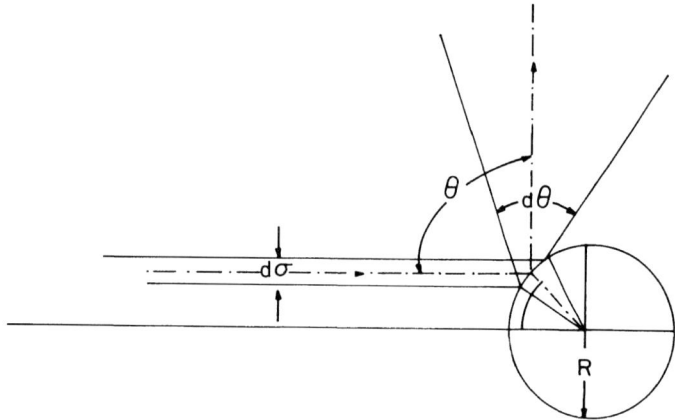

Figure 1.46 Elastic collision of a light molecule with a heavy one at an angle α with radius-vector of a contact point (the scattering angle $\Theta = 2\alpha$).

The solid angle of scattering is

$$d\Omega = 2\pi \sin\Theta\, d\Theta = 8\pi \sin\alpha \cos\alpha\, d\alpha. \qquad (1.15.40)$$

Hence, the differential cross-section

$$d\sigma = \frac{R^2}{4} d\Omega \qquad (1.15.41)$$

does not depend on α and the full cross-section $\sigma = (R^2/4) \int d\Omega = \pi R^2$ is purely geometrical.

The average projection of the velocity of the scattered molecule on the initial direction of motion is

$$<v> = -\int v' \cos\Theta\, d\sigma/\sigma.$$

From this we obtain, taking into account (1.15.39),

$$<v> = -v' \int_0^{\pi/2} \cos 2\alpha \sin 2\alpha\, d\alpha = 0. \qquad (1.15.42)$$

Thus the light particle (for example, an electron) is scattered uniformly in all directions, unlike the energy which is conserved in this approximation because $|\mathbf{v}|$ remains the same after collision. Therefore the real cross-section of electron energy relaxation though nonzero is much less than σ.

On the other hand, the velocity of a heavy molecule moving in a light gas is not significantly changed in a single collision and it requires many collisions to relax. Since both the velocity and the energy are slowly changing, the collisions may be considered as weak. The limit $\gamma \to 1$, when

$$M \gg m \quad \text{and} \quad \Gamma \ll \frac{1}{\tau_0}, \tag{1.15.43}$$

is appropriate for Brownian particles.

To give a proof of the strong collision limit is more problematic. Simultaneous relaxation to equilibrium of both the direction and magnitude of the velocity are required in each collision. In order to meet these requirements the colliding particles must be of comparable or equal masses:

$$\text{if} \quad M \approx m \quad \text{then} \quad \Gamma \approx \frac{1}{\tau_0}, \tag{1.15.44}$$

and hence $\gamma \approx 0$. Although approximate, the strong collision model is very useful for simple estimates and is widely used in kinetics.

Collisional relaxation of angular momentum is very similar to that of translational velocity and even the cross-sections are comparable. Vibrational relaxation, especially in diatomic molecules, is much slower and the rates of energy and phase relaxation are rather different. The slowest is spin relaxation in gases (governed by very weak magnetic interaction) whose cross-section may be many orders of magnitude less than σ.

1.16 TRANSFER PHENOMENA

Local Equilibrium

We can now proceed to the solution of the basic kinetic problem: the description of irreversible processes. Unfortunately, it is not so easy to apply general ideas of stirring to particular problems. These ideas serve to strengthen our belief that one or two strong collisions are quite sufficient for the complete recovery of the equilibrium velocity distribution at any point of the system. In practice, this assurance is embodied in the concept of *local equilibrium* which is of fundamental importance in considering the majority of irreversible, non-equilibrium processes. It is supposed that energy-momentum exchange between particles proceeds in such a way that after the first strong collision each molecule acquires equilibrium properties typical for the point of space where the collision occurred. The domain of validity of this hypothesis is bounded by slow irreversible processes, when the notion of local temperature, density, and so on, may be introduced at each point of the space. Primarily, these are transfer phenomena—heat conduction, diffusion, viscosity—which, as is evident in everyday experience, are fairly slow.

For example, if heat flows from a hot glass to a cool one through a plane gas layer in the double window frame, the "local temperature" at any point between the glasses can be easily measured, but no temperature can be assigned to the gas on the whole, which is in a nonequilibrium state. Thus one can speak about "local equilibrium" at its own temperature in each microregion with linear extent of the order λ. The flux of particles from other regions continually disturbs this local equilibrium, but it is recovered again and again in a time of the order $\tau_c = 1/\nu$. The spatial size of the microregions is determined by the fact that molecules retain their properties solely during the free path, and change them abruptly upon collision. That is why, despite the large velocities of molecules $\bar{v} \sim 10^5$ cm/s, the energy transfer proceeds gradually, from microregion to microregion, rather than immediately from one wall to another. Moving from hot regions to cool ones, a molecule releases its excess energy upon collision after each free path. It retains just the energy typical for the microregion where the collision occurred. On the other hand, molecules moving in the opposite direction heat up progressively, so they are not treated as alien inclusions in any region, and equilibrate with the "natives" immediately after the first collision.

This qualitative picture—the gradual adaptation of molecules to the properties of those space points (and at that time) where (and when) they experience collisions—forms the basis of the kinetic theory of quasistatic transfer phenomena. However, the calculations may be more or less rigorous, from quite primitive to rather complicated, depending on how accurately the different details of a phenomenon are taken into account. In order to elucidate the essence of the method, let us begin from the most elementary approach.

Heat Conduction

Consider heat transfer through some section perpendicular to the energy flux. If diffusion is absent the fluxes of particles j_+ and j_-, passing through the section from the right and from the left are equal to one another: $j_+ = j_- = j_0$. However, the energy transferred by particles moving in opposite directions is not the same. The particles moving from the left are "more hot," while those coming from the right are "more cool," if $dT/dx < 0$ (Fig. 1.47).

What are their energies? Here some simplifications are necessary. Suppose that all particles suffer their last collision before reaching the section at point x exactly at the distance λ from it. Then the particles, moving through the section from the left, will transfer the mean heat energy $\bar{\epsilon}(x - \lambda)$ characteristic for the point $x - \lambda$, while those coming from the right will approach the same section with the energy $\bar{\epsilon}(x + \lambda)$. The dependence of mean energy on x is conditioned by the fact that $\bar{\epsilon} = C_V T/N_0$, while $T = T(x)$. The resultant heat (energy) flux through the section is expressed as the difference of the two opposite fluxes:

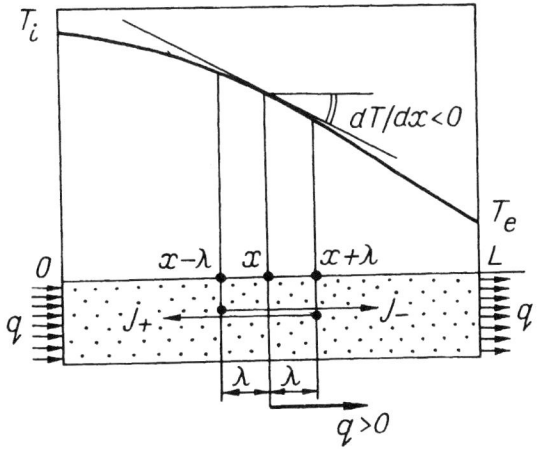

Figure 1.47 Heat transfer through a window with parallel glass panes at 0 and at L, one neighboring the high interior temperature T_i and the other the low exterior temperature T_e. The figure shows the variation of temperature in the intervening space, and denotes the flux of molecules in the direction of positive temperature gradient by j_+ and in the direction of negative gradient by j_-.

$$q = \bar{\epsilon}(x-\lambda)j_- - \bar{\epsilon}(x+\lambda)j_+$$
$$= \frac{C_V}{N_0} j_0 [T(x-\lambda) - T(x+\lambda)] = -\frac{2C_V j_0 \lambda}{N_0} \cdot \frac{dT}{dx}. \qquad (1.16.1)$$

The above expansion of $T(x-\lambda)$ and $T(x+\lambda)$ as a power series with an accuracy to the first-order terms is justified by the assumption that λ is small compared to the macroscopic scale of T variation.

Strictly speaking, it is incorrect to assign the same energy to all particles in the flux, since the kinetic energy depends on the velocity of motion and cannot be averaged independently of j_0. However, this error is partly justified by the fact that the flux itself is estimated by the Joule method, wherein the same velocity $\bar{v}$ is assigned to all particles. Besides, taking that all particles move solely in three perpendicular directions (along coordinate axes), with the same number of particles moving back and forth in each direction, one can easily see that $j_0 = \frac{1}{6}\bar{v}n$, so

$$q = -\frac{1}{3} \frac{C_V \bar{v} \lambda n}{N_0} \frac{dT}{dx} = -\kappa \frac{dT}{dx}. \qquad (1.16.2)$$

This is the "Fourier law." It claims that the heat flux is proportional to the temperature gradient and opposite to it in sign. Of course, the verification of this empirical dependence by direct kinetic calculation is important, but even more essential is a disclosure of the microscopic sense of thermal conductivity κ and its dependence on different gas parameters:

$$\kappa = \frac{1}{3} \cdot \frac{C_V \bar{v} \lambda n}{N_0} = \frac{1}{3} \cdot \frac{C_V}{M} \cdot \frac{m\bar{v}}{\sigma}, \qquad (1.16.3)$$

where M is a molar mass. In particular, though surprisingly at first sight, κ is independent of the gas density, as $\lambda \approx 1/n\sigma$.

Viscosity

Internal friction, or viscosity, is revealed when a gas is in motion: flows through a tube, or around some surface. Layers adjacent to solid surfaces are retarded by friction against the surface, while layers away from it rub against each other. As a result, the flow is gradually slowed down as it approaches the lower (immobile) surface (Fig. 1.48), and the relative velocity of the layer adjacent to an upper (moving surface) is also zero.

However, how does it happen that the gas layers experiencing internal friction though the molecules do not interact with each other most of time? Again, this is due to collisions. The heat velocity of molecules is much greater than the macroscopic speed of any gas layer $u(x)$. Being orientated in different directions, thermal motion freely carries a molecule from a slow layer to one moving more rapidly, and vice versa. Upon collision in the new layer, the molecule changes its macroscopic speed to become indistinguishable from its new neighbors, that is, the molecule is either accelerated or retarded, giving up the difference in momentum to an other. The total momentum carried by the "emigrants" from one layer to another per unit time is equal to the friction force acting in the section separating the layers (see Fig. 1.48).

Thus the calculation should proceed similarly to the previous one. The only difference is that in this case it is not energy which is transferred, but the macroscopic momentum $mu(x)$ where $u(x)$ is the velocity of the gas flux perpendicular to the x axis. Now C_V/N_0 is replaced by m, and $T(x)$ by $u(x)$; all the rest remains unchanged. So, by analogy with (1.16.2), we can immediately write

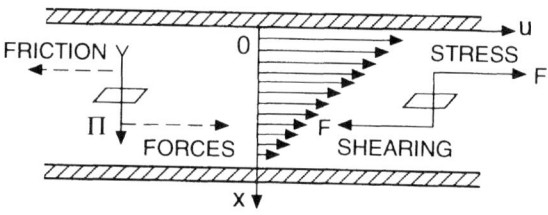

Figure 1.48 Velocity distribution in a laminar stream between two parallel plates the upper of which moves to the right.

$$F = -\frac{1}{3} m\bar{v}\lambda n \frac{du}{dx} = -\eta \frac{\partial u}{\partial x} . \qquad (1.16.4)$$

The viscosity

$$\eta = \frac{1}{3} nm\bar{v}\lambda = \frac{1}{3} \frac{m\bar{v}}{\sigma} \qquad (1.16.5)$$

is independent of the gas density as well as thermal conductivity (Maxwell's law).

Comparing formulae (1.16.5) and (1.16.3) gives an important relation between the kinetic coefficients, which is independent of the microscopic parameters:

$$\frac{\kappa M}{\eta C_V} = 1 . \qquad (1.16.6)$$

Experiments confirm that this ratio is really constant and universal. Such a relation between such apparently different phenomena could seem paradoxical, if they had not been considered from a common standpoint, which is a concept of local equilibrium.

In fact, the numerical value $\kappa M/\eta C_V$ varies from one gas to another, although it remains of order of magnitude unity. This is no surprise. In such rough calculations there is an inevitable error in estimating the numerical parameter. However, the information on the nature of the phenomena and semiquantitative estimates of the kinetic coefficients is so valuable that we have to reconcile ourselves to this inaccuracy of the method. Further we shall consider possible causes of the trouble and methods of eliminating them.

Diffusion

It is well known that diffusion is the penetration of one substance into another due to the thermal motion of molecules. If we deal with a gas, an example is the spread of perfume from a person who has just visited a hairdresser's or barber's shop. However, the gas mixing process may be accelerated by convectional flows due to wind or even breathing, that is, solely mechanical mixing which has nothing to do with diffusion. Thus pure diffusion can only be observed under experimental conditions, for example in the experiment carried out by Loschmidt.

In this experiment, two tubes filled with different gases at one and the same pressure are brought in contact. Then the partition separating them is taken away, and the gases begin mixing. Checking the gas composition some time later, one can see that the longer the time period since the start of the experiment, the more homogeneous the mixture. It is to be noted that this is a lengthy process (hours), so over a short time interval it may be treated as quasistationary, that is, with the concentration gradient approximately constant. Thus this process can be considered from the same standpoint as the above phenomena. For simplicity, let us concentrate on self-diffusion, which is observable in the case when molecules are different isotopes of the same gas distinguishable by some appropriate technique.

How can we calculate the flux of particles through a perpendicular section, if the gas is nonuniform in density? In other words, what should we mean by n in the Joule estimate $\frac{1}{6}n\bar{v} = j_\pm$? At which point should $n(x)$ be taken? According to the hypothesis of local equilibrium, the density of particles moving in columns towards the section is specified by the points where they started. After each free path λ, the columns are rearranged, adjusting to the density of particles at the points where they experienced collision. The last path before the section begins a distance λ from it; therefore, the density of particles in the flux passing through the section is $n(x - \lambda)$, if they move from left to right, and $n(x + \lambda)$ in the opposite case.

Thus $j_- = (\bar{v}/6)n(x - \lambda)$, and $j_+ = (\bar{v}/6)n(x + \lambda)$. Because of the difference between j_- and j_+ the resulting diffusion flux is

$$j = j_- - j_+ = \frac{1}{6}\bar{v}[n(x - \lambda) - n(x + \lambda)] = -\frac{1}{3}\bar{v}\lambda\frac{dn}{dx} = -D\frac{dn}{dx}, \quad (1.16.7)$$

where

$$D = \frac{1}{3}\bar{v}\lambda \approx \frac{\bar{v}}{3n\sigma} \quad (1.16.8)$$

is the diffusion coefficient.

Comparing (1.16.8) with (1.16.6) and (1.16.3), we find two more relations connecting the transfer coefficients

$$\frac{\eta}{\rho D} = 1, \quad (1.16.9)$$

$$\frac{\rho D C_V}{\kappa M} = 1, \quad (1.16.10)$$

where $\rho = nm$. A more rigorous theory adjusts the above relations for a gas of hard spheres:

$$\frac{\kappa M}{\eta C_V} = \frac{5}{2}, \quad \frac{\eta}{\rho D} = \frac{5}{6}, \quad \frac{\rho D C_V}{\kappa M} = \frac{12}{25}. \quad (1.16.11)$$

In view of these numerical corrections, we obtain

$$\kappa = \frac{\alpha}{\sigma}\sqrt{\frac{T}{M}}\frac{\text{cal}}{\text{cm} \cdot \text{deg} \cdot \text{sec}}, \quad \eta = \frac{\beta}{\sigma}\sqrt{TM}\frac{\text{g}}{\text{cm} \cdot \text{sec}}, \quad D = \frac{\gamma}{n k \sigma}\sqrt{\frac{T}{M}}\frac{\text{cm}^2}{\text{sec}},$$

where $\alpha = 6{,}2 \cdot 10^{-4}$; $\beta = 8{,}4 \cdot 10^{-5}$; $\gamma = 8{,}3 \cdot 10^{-3}$.

Refinement of Calculations

The method employed above simplifies rather brutally the picture of the phenomenon for the sake of brevity and simplicity of description. Nevertheless, the dependence of the kinetic coefficients on any varying parameters seems to be reasonable. To be sure of this, let us consider heat and matter transfer once again, abandoning, as far as possible, the most rough assumptions.

Let molecules move in all directions, with any velocity, and experience their last collision at arbitrary distance from near the section under consideration. From the entire flux of molecules passing through this section, we first separate those moving towards it with a speed v inside a solid angle $d\Omega$ which is inclined at angle θ to the x axis (Fig. 1.49). From these molecules we select those which suffered their last collision at a distance R from the section. How many such molecules pass through the chosen unit section per unit time? The flux is equal to $v_x dn = v \cos\theta \, n \, dW$; however, the probability dW that we are dealing with molecules starting at a distance R with the speed v is defined by the product of probabilities $dW(\mathbf{v}) dW(R)$. Here $dW(R)$ is the distribution in free path lengths (1.15.13), and $dW(\mathbf{v}) = g(v) d\mathbf{v}$ is the ordinary Maxwell distribution with the density

$$g(v) = \left(\frac{2\pi kT}{m}\right)^{-3/2} \exp\left(-\frac{mv^2}{2kT}\right), \quad (1.16.12)$$

where $T(x)$ is a function of the point where the collision occurred.

So the magnitude of the flux depends on the temperature at the point where the particle starts from and on the density there $n(x)$. The fluxes of particles

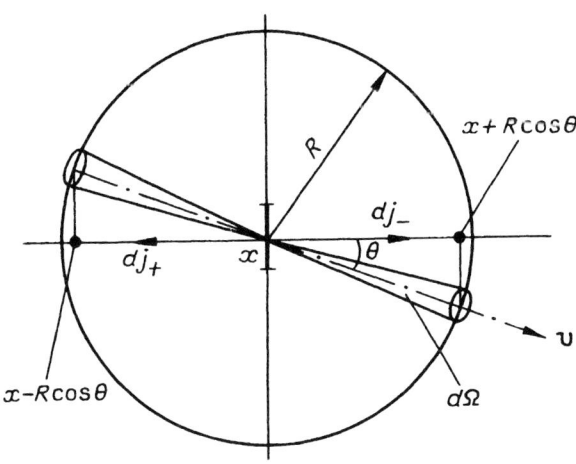

Figure 1.49 Opposite fluxes of particles moving with the same speed away from the points of last collision equidistant from section x.

from two opposite points lying on a sphere with radius R (see Fig. 1.49) differ only in the magnitude of the multiplier ng:

$$dj_- = v\cos\theta\,(ng)_-\,dv\,dW(R), \tag{1.16.13a}$$

$$dj_+ = v\cos\theta\,(ng)_+\,dv\,dW(R) \tag{1.16.13b}$$

The subscripts $+$ and $-$ mean that the value of ng is taken, respectively, at the points $x + R\cos\theta$ and $x - R\cos\theta$, where the last collisions before the "finishing line" took place. It is of interest that $dj_- \neq dj_+$ even at $n = \text{const}$, provided that there is temperature inhomogeneity. By virtue of the difference between g_- and g_+, the flux of high energy particles moves from hot points to cold ones, while low energy particles move from cold to hot points. The resulting flux of particles with the given velocity is as follows

$$dj = dj_- - dj_+ = v\cos\theta\left[(ng)_- - (ng)_+\right]dv\,dW(R). \tag{1.16.14}$$

The partial flux dj can differ from zero even when the total is zero (i.e. $j = \int dj = 0$). This ensures heat transfer from hot points to cold ones in the absence of diffusion.

If the starting points are offset by λ from the section and this distance may be considered small compared to the macroscopic scale, we can use the truncated expansion

$$(ng)_- - (ng)_+ = -2\frac{d(ng)}{dx}R\cos\theta. \tag{1.16.15}$$

Substituting (1.16.15) into (1.16.14) gives

$$dj = -2v\frac{d(ng)}{dx}v^2 dv\cos^2\theta\,d\Omega\,R\,dW(R). \tag{1.16.16}$$

This expression is easily integrated with respect to angles and free path lengths

$$dj = -2v\frac{d(ng)}{dx}4\pi v^2\,dv\,\overline{\cos^2\theta}\,\ell = -\frac{\tau_0}{3}v^2\frac{d(ng)}{dx}d\mathbf{v}, \tag{1.16.17}$$

since, according to (1.15.14), $\ell = \int R\,dW(R) = v\tau_0$, and $\overline{\cos^2\theta} = \int \cos^2\theta\,d\Omega/4\pi = \tfrac{1}{6}$.

Using a constant $\tau_c = 1/v$ instead of $\tau_0(v)$ and integrating over velocities in (1.16.17), we obtain

$$j = -\frac{1}{3}\frac{d}{dx}\left[n\int \tau_0(v)v^2 g(v)\,d\mathbf{v}\right] = -\frac{\tau_c}{3}\frac{d\left(\overline{nv^2}\right)}{dx} = -\frac{\tau_c k}{m}\frac{d(nT)}{dx}. \tag{1.16.18}$$

Differentiating, we find

$$j = -D\frac{dn}{dx} - B\frac{dT}{dx},\qquad(1.16.19)$$

where

$$D = \frac{kT\tau_c}{m} = \frac{\overline{v^2}\tau_c}{3},\qquad B = \frac{k\tau_c n}{m}.\qquad(1.16.20)$$

The second term in (1.16.19) describes the phenomenon known as *thermodiffusion*. It differs from ordinary diffusion in that the particles move from hot points to cold ones due to the higher velocity of motion rather than higher density.

Allowing another dependence $\tau_0(v)$ in (1.16.18) will result only in numerical changes to the conclusion. For example, in the case of the diffusion of light particles in a heavy gas $\tau_0 = [n\sigma v]^{-1} = \lambda/v$, and on averaging over v, we get $D = \frac{1}{3}\bar{v}\lambda = \bar{v}^2\tau_c/3$ (i.e., $D \propto \bar{v}^2$, not $\overline{v^2}$), just as in (1.16.8). Within the framework of the hard sphere model both results are reasonable, but in reality the dependence $\tau_0(v)$ should be found allowing for the actual intermolecular interaction, taking into account both repulsion and attraction.

Let us estimate the flux of heat, that is, thermal energy transferred by moving molecules. In the simplest case of monoatomic gas only kinetic energy is considered,* so at $\tau_0(v) = \tau_c$, we find using (1.16.17):

$$q = \int \frac{mv^2}{2} dj = -\frac{m\tau_c}{6}\frac{d(\overline{nv^4})}{dx} = -\frac{5\tau_c k^2}{2m}\frac{d(nT^2)}{dx}.$$

Differentiating, we have

$$q = -\chi\frac{dn}{dx} - \kappa\frac{dT}{dx},\qquad(1.16.21)$$

where

$$\chi = \frac{5}{2}\frac{(kT)^2}{m}\tau_c,\qquad \kappa = \frac{5k^2 T\tau_c n}{m}.\qquad(1.16.22)$$

It is seen that heat transfer is stimulated both by the temperature gradient and the concentration gradient, since energy is transferred together with the matter carrying it. This fact makes difficult a comparison between the result obtained

---

*For polyatomic molecules $q = \int[(mv^2/2) + i(kT/2)]\,dj$ where $i =$ total number of excited degrees of freedom except translational ones.

and the simple estimate of thermal conductivity given above. The latter was found under the additional condition $n = $ const, while Eq. (1.16.21) describes the more general and usual situation. For example, it arises when heat exits a house through a double window frame, the inner glass of which is at room temperature T_i, while the outer one at the outside temperature T_e (see Fig. 1.47). The pressure at all points of the air layer between the glasses is constant (normal). In the ideal gas approximation

$$p = nkT = \text{const} \quad \text{or} \quad \frac{1}{n}\frac{dn}{dx} = -\frac{1}{T}\frac{dT}{dx}, \qquad (1.16.23)$$

that is, the concentration gradient is proportional to the temperature gradient although opposite in sign. Eliminating it from (1.16.21) with the use of (1.16.23), we obtain the Fourier law for a stationary flux

$$q = -\left[\kappa - \chi\frac{B}{D}\right]\frac{dT}{dx} = -\mathcal{H}\frac{dT}{dx}, \qquad (1.16.24)$$

where $\mathcal{H}$ is the thermal conductivity in the ordinary sense. Substituting formulae (1.16.20) and (1.16.22) into (1.16.24), we can see that $\mathcal{H}$ differs from κ in (1.16.22) just by a factor of 2:

$$\mathcal{H} = \kappa - \chi\frac{B}{D} = \frac{5k^2 T \tau_c n}{2m} = \frac{5\pi k \bar{v}}{16\sqrt{2}\sigma}. \qquad (1.16.25)$$

The change in corpuscular density from the hot to the cold glass does not essentially affect the heat conductivity of the gas. Comparison of its magnitude with the approximate estimate (1.16.3) where $M = N_0 m$, $C_V = \frac{3}{2}kN_0$ shows that they differ only by a numerical factor. In view of the uncertainty in $\bar{v}$ in (1.16.3), this is of no fundamental significance. The most important features of the process, the dependence of $\mathcal{H}$ on temperature and its independence of density, are quite satisfactorily represented by the simplest calculation.

Heat Transfer

To demonstrate the significance of the above results let us apply them to the calculation of the heat flux through the double window frame. As the temperatures inside and outside the house vary rather slowly, the process may be regarded as being stationary, but the temperature-dependence of the thermal conductivity must be essentially taken into account:

$$\mathcal{H} = AT^{1/2}, \quad A = \frac{5\sqrt{\pi}}{8\sigma\sqrt{m}}k^{3/2}. \qquad (1.16.26)$$

This expression is used in Eq. (1.16.24) which can now be treated as a differential equation for the $T(x)$-dependence:

$$-\sqrt{T}\,\frac{dT}{dx} = \frac{q}{A}, \qquad (1.16.27)$$

where q is a constant to be defined later. Solving (1.16.27), we find

$$T^{3/2} = C - \frac{3}{2}\frac{q}{A}x. \qquad (1.16.28)$$

The boundary conditions are given by the temperatures of the internal and external glasses:

$$T(0) = T_i; \qquad T(L) = T_e. \qquad (1.16.29)$$

The first condition serves to determine the integration constant in (1.16.28): $C = T_i^{3/2}$. Then, using the second condition, one can calculate the flux itself

$$q = \frac{2}{3} A\,\frac{T_i^{3/2} - T_e^{3/2}}{L}. \qquad (1.16.30)$$

In this final expression the heat flux is expressed directly in terms specified by the experimental conditions: T_i, T_e, L. If $\mathcal{H}(T)$ were other than (1.16.26) this expression could be different.

The independence of thermal conductivity on density deserves special consideration. This fact is really paradoxical and points to a limited applicability of the results obtained. Indeed, how it can be explained that the decrease in the number of particles involved in energy transfer does not affect heat transfer? In normal gas this is attributed to the fact that the increasing length of the free path compensates for the decrease in molecular density. However, one can exhaust a gas from a vessel until vacuum is reached. Obviously, in this case the heat conduction will go to zero due to the absence of heat carriers. However, the mere existence of such an alternative cannot be inferred from the results obtained: $\mathcal{H}$ = const whatever the degree of the gas rarefaction.

Ultrararefied Gas

In view of the above reasoning, it is necessary to discuss the limits of application of the whole approach, based on the concept of local equilibrium. It was supposed that each point within a microregion of the scale λ may be considered as almost at equilibrium, with its own local characteristics of state: temperature and density. However, is this always true? Remember that $\lambda = 1/n\sigma = kT/p\sigma$; therefore, by decreasing the pressure we can make the path length a macroscopic quantity which exceeds considerably the vessel's size. In this situation,

116 GASES

the molecules will collide against the walls more often than with other molecules. From a kinetic standpoint, such a ultrararefied gas (physical vacuum) behaves quite differently to dense gases.

It should be noted that collisions with the walls are not necessarily elastic. Moreover, the molecules may be adsorbed onto a wall, leveling the energy with it. If the wall is warmer than the gas, the molecules take away some energy, and give it up in the opposite case. The heat exchange between gas and environment proceeds owing to adsorption–desorption processes. In order to heat a gas, we need only heat up the walls of the vessel.

The external heat exchange promotes attainment and maintenance of equilibrium in ultrararefied gas. The "stirring" property of such a system is so depressed that equilibrium is more rapidly established by the interaction with the exterior medium. This fact has an essential effect on all transfer phenomena. The concept of local equilibrium becomes meaningless. Since molecules move freely in the vessel, heat is transferred directly from one wall to another, rather than in the relay fashion (from molecule to molecule), as before.

Let us assume that from j molecules colliding against the wall per second, αj adhere to it. Under stationary conditions the same number of molecules evaporate each second taking with them the mean energy corresponding to the temperature of the surface. Also, there are $(1 - \alpha)j$ particles in the reflected flux whose energy remains unchanged due to elastic collisions with the wall. In such a model there are two groups of particles between the walls moving in opposite directions (Fig. 1.50). The total flux and density are correspondingly

$$j = j_1^+ + j_2^+ = j_1^- + j_2^- \quad \text{and} \quad n = n_1^+ + n_2^+ + n_1^- + n_2^- . \tag{1.16.31}$$

Here the signs denote the direction of motion, and the indices label which wall the molecules are in equilibrium with. Obviously, the heat flux is

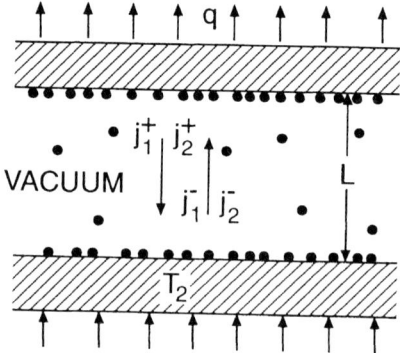

Figure 1.50 Opposite fluxes in an ultrararefied gas carrying out the heat transfer.

$$q = \frac{C_V}{N_0}\left[T_2\left(j_2^- - j_2^+\right) + T_1\left(j_1^- - j_1^+\right)\right], \qquad (1.16.32)$$

In order to calculate q, one has to determine the relative magnitude of all fluxes. It is clear that the molecules with the "foreign" temperature only appear in the reflected flux $j_1^- = (1-\alpha)j_1^+$, while those with the native one are either reflected or desorbed: $j_2^- = (1-\alpha)j_2^+ + \alpha j$. Hence, it follows from the balance considerations that

$$\begin{cases} j_2^- = j_2^+ + \alpha j_1^+ ; \\ j_1^- = (1-\alpha)j_1^+ ; \end{cases} \qquad \begin{cases} j_2^+ = (1-\alpha)j_2^- ; \\ j_1^+ = j_1^- + \alpha j_2^- . \end{cases}$$

The system on the left refers to the lower wall (Fig. 1.50), while that on the right to the upper one. Thus

$$(1-\alpha)j_2^- = j_2^+ = j_1^- = (1-\alpha)j_1^+ . \qquad (1.16.33)$$

Each of the fluxes consists solely of particles moving "forth" (they return in another flux). So, unlike (1.4.12), we have

$$j_i^\pm = \tfrac{1}{2} n_i^\pm \bar{v}_i, \qquad i = 1, 2, \qquad (1.16.34)$$

Generally speaking, all $n_i^\pm$ are different.

The difference in density between the opposing fluxes can be readily illustrated by the following experiment. Joining two vessels filled with rarefied gas at temperatures T_1 and T_2 by a short tube, one can see that eventually the fluxes in the tube in both directions become equal to one another: $j_1^+ = j_2^-$. In view of (1.16.34), this is only possible at $n_2/n_1 = v_1/\bar{v}_2$, since the gas density is related to its pressure, so $p_2/p_1 = T_2\bar{v}_1/T_1\bar{v}_2 = (T_2/T_1)^{1/2}$, that is, the pressure in the joined vessels together ceases to be the same, which is just what was observed.

Using (1.16.33) and (1.16.34) in (1.16.31), we get

$$j = (2-\alpha)j_2^- = (2-\alpha)j_1^+ = \frac{n}{2}\frac{\bar{v}_1 \cdot \bar{v}_2}{\bar{v}_1 + \bar{v}_2}. \qquad (1.16.35)$$

On substitution of (1.16.33) into (1.16.32) and in view of the above result, we can easily obtain

$$q = \frac{C_V \alpha}{N_0(2-\alpha)} \cdot \frac{n}{2}\frac{\bar{v}_1 \cdot \bar{v}_2}{\bar{v}_1 + \bar{v}_2}(T_2 - T_1). \qquad (1.16.36)$$

118 GASES

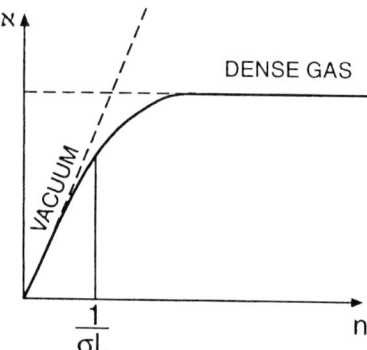

Figure 1.51 Density-dependence of the effective thermal conductivity κ.

The quantity

$$\mathcal{H} = \frac{C_V \alpha}{N_0 (2 - \alpha)} \cdot \frac{nL}{2} \cdot \frac{\bar{v}_1 \cdot \bar{v}_2}{\bar{v}_1 + \bar{v}_2} \qquad (1.16.37)$$

has the meaning and dimension of thermal conductivity and is linear in the density of the remaining molecules.

This result holds when $\lambda \gg L$. With increasing of density the sign of the inequality is reversed, and local equilibrium is restored. Only then the previous estimate of $\mathcal{H}$ (1.16.25) becomes valid. Thus the estimate independent of the gas density holds in the limited density range $n \gg 1/\sigma L$ (Fig. 1.51).

Current in Gases

Although at room temperature and atmospheric pressure any gas is a dielectric, there is always a small equilibrium concentration of charges brought about by heat ionization of molecules. When the electric conductivity of a gas becomes observable, the concentration of ions in the unit volume is still much less than the density of neutral particles. Note that this concentration can exceed the equilibrium one, if it is created by external sources: penetrating radiation or emission of electrons from a cathode. Whatever their origin, the charges which find themselves in the spark-gap respond to the applied field and drift in the direction of the force acting upon them. Moving with velocity w, they transfer electricity between electrodes, thus making the gas a conductor.

When electric conduction is unipolar, that is, when all carriers have the same sign, the current density

$$i = ewn, \qquad (1.16.38)$$

where n is the concentration of ions, and e is the absolute magnitude of their charge. If the drift speed w is directly proportional to the field

$$w = uE, \qquad (1.16.39)$$

then Ohm's law holds

$$i = \sigma E. \qquad (1.16.40)$$

The conductivity is

$$\sigma = eun, \qquad (1.16.41)$$

where u is the mobility of the current carriers. To prove Ohm's law, it is sufficient to assume that ions completely change the direction of their motion in each collision. Because of this, the mean velocity after collision is equal to zero as in Eq. (1.15.42). In other words, on the average, each collision makes the ion stop, and its acceleration in the next free path starts afresh. That is why charge drift is uniform, despite quadratic increase of the free path with time typical for uniformly accelerated motion (Fig. 1.52).

Indeed, the free path in the direction of the field averaged over initial velocity directions,

$$\bar{x} = \bar{v}_x t + \frac{eE}{2m} t^2 = \frac{eE}{2m} t^2,$$

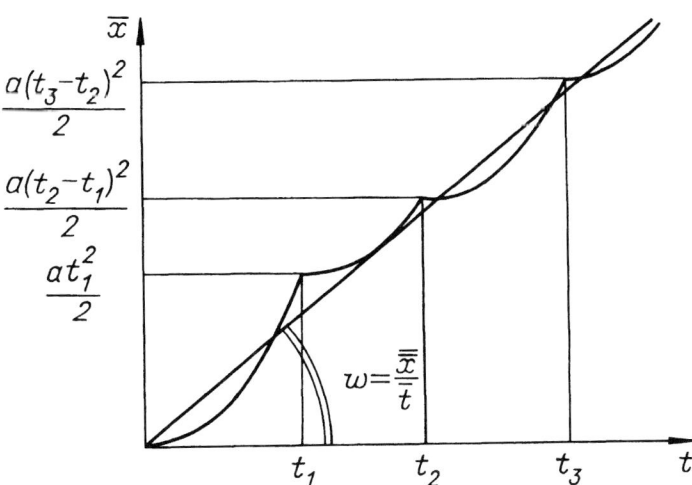

Figure 1.52 Averaged ion drift (straight line) accomplished by a sequence of free paths in an external electric field (polygonal line).

should also be averaged over the distribution of free path times (1.15.15):

$$\overline{\overline{\dot{x}}} = \frac{eE}{2m} \int_0^\infty t^2 dW(t) = \frac{eE\tau_0^2}{m}.$$

Now it is clear that the drift speed is proportional to the field

$$w = \int \frac{\overline{\overline{\dot{x}}}}{\tau_0} dW(v) = u E, \qquad (1.16.42)$$

and the mobility of carriers is equal to

$$u \approx \frac{e\tau_c}{m} = \frac{e\lambda}{m\sqrt{8kT/\pi\mu}}. \qquad (1.16.43)$$

Naturally, the lighter the charge carriers and the lower the density of the gas they pass through, the greater the mobility.

These qualitative conclusions were verified experimentally by Rutherford, although the measured mobility was found to be less by an order of magnitude than the expected result

$$u = \frac{e\tau_c}{m} = \frac{5 \cdot 10^{-10} \cdot 3 \cdot 10^{-9}}{300 \cdot 10^{-22}} \approx 50 \text{ cm}^2/(\text{V} \cdot \text{sec}) \qquad (1.16.44)$$

This anomaly is partly due to an inappropriate estimation of the ion mass. As a rule, owing to their charge, the gas ions attract some neutral particles and form associates which move as a unit. Such ions have greater mass, and therefore are less mobile. Another source of error is the rough nature of the model of hard spheres. The charged particle is the source of a long-range Coulomb field. This affects the ion's trajectories far beyond the limits of their own diameter, that is, in fact, collisions occur at greater distances than one can infer on simple geometric grounds. To put it another way, the effective cross-section of an ion exceeds the sizes of the neutral molecules. Consequently, the frequency of collisions is greater, and the interval between them τ_c is less than was assumed in (1.16.44). These peculiarities of charged particles gave a stimulus for the refinement of electric conduction theory which proved to be quite satisfactory for ions of any substance and both signs.

The single but very important exception from the rule is the case where charge is carried by free electrons. This mechanism of electric conduction was taken to explain the extremely high mobility of negative carriers in some experiments (up to 10^4 cm^2/V · sec). In inert gases and strong fields the electrons adhere weakly to molecules and become free, undergoing only elastic collisions. In this case the charge mobility becomes much greater, because the electron's mass is less than that of a negative ion by a factor of 10^5. On closer examination of formula (1.16.43), it is seen that the mobility increases by a factor of $\sqrt{M/m} \approx 200$, and this was verified experimentally.

However, this is not all. When free electrons take the role of carriers, the character of their movement through the gas changes. Due to the sharp difference in mass, the energy exchange between electrons and molecules resulting from collisions is hampered. The share of energy transferred from light to heavy particles in collisions is determined by the ratio of their masses. This can easily be seen if one tries to play cherry stones against billiard balls. The energy $\Delta \epsilon$ lost by an electron upon collision accounts for just a small fraction $q = \Delta \epsilon / \epsilon = m/M = 10^{-5}$ of the total energy ϵ it carries. Thus the local equilibrium principle which formed the basis of the previous consideration is violated. Collisions are unable to correct the electron's energy in such a way as to make it correspond to the local temperature of the points where the collisions occur. On the contrary, moving through the gas along the field, the electron accumulates gradually the energy imparted to along the free path. Only an insignificant share of this energy is converted into heat at collisions.

Although unable to take away all the electron's energy, collisions essentially affect the direction of its motion. Owing to this, the work executed by the field accelerates the motion of electrons uniformly in all directions, thus establishing a distribution similar to the Maxwell one. However, the *electron temperature* T_e appearing in this distribution is greater than the local temperature T of the gas. It is determined by the mean electron energy $\bar{\epsilon} = \frac{3}{2} k T_e$.

As long as the energy of electrons is fairly small the electron temperature increases linearly approaching the anode:

$$\frac{3}{2} k T_e = \frac{\overline{mv^2}}{2} \approx eEx. \quad (1.16.45)$$

However, this rise of temperature cannot continue indefinitely, since upon each collision a qth portion of the electron's energy is lost. At higher ϵ, it is no longer a small value compared with the additional energy acquired in free path. Eventually, ϵ ceases to increase, and all the energy taken from the field is converted to heat transferred to the surroundings.

Accumulation of energy by a drifting electron is described by the equation

$$\frac{d\epsilon}{dt} = eEw - \frac{q}{\tau_c}[\epsilon - \epsilon_0],$$

where $\epsilon_0 = \frac{3}{2} kT$. If the second term is neglected, then $\epsilon = eEwt = eEx$, as in (1.16.45). On the other hand, in the stationary regime $d\epsilon/dt = 0$, while

$$\epsilon = \epsilon_0 + \frac{eEw\tau_c}{q} \approx \frac{eEw\tau_c}{q},$$

if electrons become hot enough. Using (1.16.42) and (1.16.43), we obtain

$$\epsilon = \frac{mw^2}{q} = \frac{\overline{mv^2}}{2},$$

or

$$w = \left(\frac{q}{2}\overline{v^2}\right)^{1/2}.$$

As is seen, the drift speed is less than the root mean square velocity by a factor of $\sqrt{q/2}$. Revealing the meaning of w with the help of Eqs. (1.16.42) and (1.16.43), we find

$$\frac{e\lambda}{m\bar{v}} E = \sqrt{\frac{q}{2}\overline{v^2}},$$

and

$$\frac{3}{2}kT_e = \frac{\overline{mv^2}}{2} \approx \frac{e\lambda E}{\sqrt{q}}. \quad (1.16.46)$$

Thus the kinetic energy of an electron in a stationary drift is $q^{-1/2}$ times as great as the work executed by the field in one free path. The electron temperature T_e exceeds considerably the temperature of the environment, and it is T_e that should be substituted for T in formula (1.16.43) to estimate the electrons' mobility

$$u = \left(e\lambda\sqrt{q}/mE\right)^{1/2}. \quad (1.16.47)$$

So it turns out that, due to the dependence of the electron temperature on the field, the direct proportionality between w and E is replaced by the relation

$$w = uE = \left(e\lambda E\sqrt{q}/m\right)^{1/2}. \quad (1.16.48)$$

Thus the notion of mobility becomes less significant, as this quantity is no longer a constant. Under normal conditions, for a field of 10 V/cm, $u \sim 10^3 \text{cm}^2/\text{V} \cdot \text{sec}$, $T_e \sim 1000$ K, and the kinetic energy of electrons is about 0.1 eV.

Formulae (1.16.45) and (1.16.46) describe two opposite limits. The former accounts for the electron acceleration interval where losses are very low, while the latter the stationary regime where the gain and expenditure of energy compensate one another. On its way from the cathode to anode, the electron passes from one region to another. Obviously, the boundary is the point where, according to both approximate estimates, the electron temperature is the same: $(e\lambda E)/\sqrt{q} = eEx_0$. Thus

$$x_0 = \frac{\lambda}{\sqrt{q}} \qquad (1.16.49)$$

is approximately one hundred times greater than the free path length. At normal pressure, it is still a small value: $x_0 \sim 10^{-2}$ cm. So, under normal conditions, the electron drift is stationary over almost the entire discharge gap, except in a very narrow layer. However, at pressures a thousand times less than atmospheric, the acceleration interval extends to macroscopic sizes. If the electrodes are separated by a distance of the order of several centimeters, the stationary regime is not reached at all. As is seen from formula (1.16.45), in this case the electron temperature is measured merely by the distance covered. Thus any *a priori* prescribed energy may be imparted to the electrons by installing a grid near the cathode within the acceleration interval. The energy of the electrons which reach the grid will be specified by the difference in potential between the grid and the cathode.

So, the violation of local equilibrium results in the coexistence of two gases in the spark-gap: molecular and electron, each having its own temperature. The difference in temperature is maintained and controlled by the field: the source of energy accumulation. Thus one can easily increase the field to such a degree that the energy of heat motion of electrons will be sufficient for excitation, and ultimately for ionization of neutral particles of the matter. Atoms and molecules excited upon collisions with electrons can radiate the acquired energy in the visible range. In principle, the operation of gas-filled tubes which remain relatively cool, despite the daylight they create, is based just on this phenomenon. At high electron temperatures, ionization of molecules also becomes possible. The chain ionization results in an avalanche increase of current carriers. Positive ions also contribute to gas conductivity. Among other things, they bombard the cathode knocking additional electrons out of it. Thus at a definite potential difference, the "breakdown" of a gas may take place—a sudden conversion into a highly ionized conducting matter (*plasma*).

2

RADIATION

2.1 PHENOMENOLOGY

From the corpuscular standpoint, electromagnetic radiation is similar to a gas in a number of the phenomenological properties. Like a gas, radiation cannot be kept within some limited volume without placing it into a closed container. Also, radiation exerts pressure on a wall in its path. The analogy with an ideal gas suggests itself: if light is treated as an ensemble of photons which move freely and rectilinearly, collide against the container's walls and rebound, then the above properties may be qualitatively accounted for.

However, here we reach the end of the analogy. Unlike a molecular gas, light is able to penetrate into any wall and be absorbed into it. By Lambert's first law, which was established experimentally, in passing through matter, light of frequency ν becomes progressively less intense, as it penetrates deeper into the substance. Still, at any finite distance, the intensity remains different from zero

$$I(x) = I(0) \exp(-\epsilon x). \qquad (2.1.1)$$

Here $I(0)$ is the intensity in the section $x = 0$, $I(x)$ is that at x, and $\epsilon(\nu)$ is the so-called extinction coefficient, which depends essentially on the frequency of the penetrating light.

If a light flux with intensity I_0 falls on a wall, a fraction ηI_0 is reflected from it. The remaining light penetrates inside (Fig. 2.1), and $\alpha(\nu)$, the absorption coefficient, is connected with η by the simple relation $\alpha = 1 - \eta \leq 1$. Some fraction of the remaining portion passes through the wall and escapes, while the remaining light is absorbed by the wall. Thus any attempt to study the

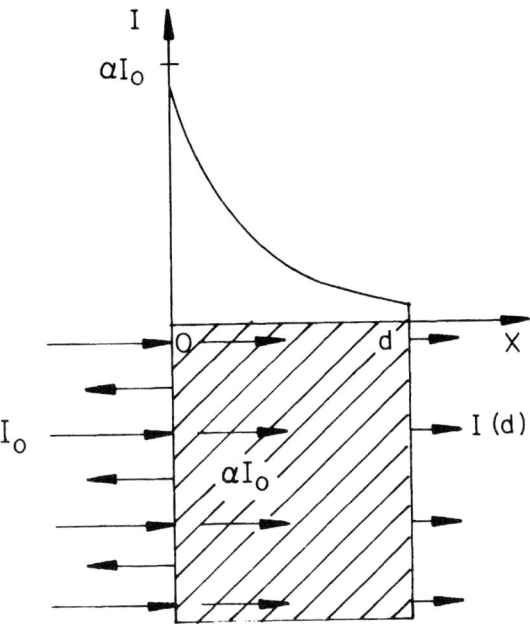

Figure 2.1 Light reflection and transmission through a homogeneous medium (shaded). Plotted is the light intensity in the medium as a function of x.

thermodynamic properties of radiation in a transparent container will be a failure: the light will leave the vessel, and only slightly heated walls will remind one of its existence. In order to confine radiation to a closed volume, it is necessary to place it into a thick-walled container of black material, or, alternatively, into a thin-walled transparent vessel immersed in a practically unbounded medium. In both cases, we can be sure that radiation does not pass beyond the limits of the surroundings of the container. However, there is absorption within the medium. As absorption can never be avoided, one can get the impression that nothing can confine radiation to the container. This is partly true: sooner or later it will be absorbed by the vessel's walls and converted into heat. However, at any temperature different from zero, the medium surrounding the vessel radiates light which, eventually, finds itself inside the container. Two alternative processes—absorption and emission—create the conditions for the attainment of zero balance: the amount of light which penetrates into the medium is equal to that returning to the container owing to thermal radiation. Under equilibrium conditions, the balance must be detailed: the absorbed energy is compensated by an equal amount of radiated energy in each spectral component separately. Therefore, the condition for detailed balance at frequency ν is of the form

$$E_0(\nu) = \alpha(\nu)I_0(\nu) . \qquad (2.1.2)$$

Here $E_0(\nu)$ is the *emissive power* of the walls, that is, the total energy radiated in all directions from a unit area in a unit frequency range per unit time, and $I_0(\nu)$ is the light energy flux within the same frequency range which falls on a unit area of the inner surface of the container.

According to Lambert's second law, the amount of light energy radiated into the solid angle $d\Omega$ from the surface element dS of a heated body in time interval dt and in the frequency range from ν to $\nu + d\nu$ is as follows

$$dE = E_0(\nu)d\nu d\Omega dt dS \frac{\cos \vartheta}{\pi}, \qquad (2.1.3)$$

where ϑ is the angle of the light ray with the normal at dS, and the line spectrum $E_0(\nu)$ is the characteristic property of the substance. The physical nature of the matter also affects the frequency dependence of the absorption coefficient $\alpha(\nu)$ that determines its transparency in various spectral ranges. It could have been assumed that $I_0(\nu)$ is also a specific property of the container. However, this is not the case: $I_0(\nu)$ is the same independent of the container. This conclusion can be verified experimentally by making a small hole in the vessel and studying the equilibrium radiation emitting from it. This is also supported by theoretical arguments presented by Kirchhoff. These arguments are of fundamental importance and deserves special consideration.

2.2 KIRCHHOFF'S THEOREM

Let us first consider a somewhat idealized case, where radiation in a confined in the so-called "black box"— a vessel with absolutely absorbing walls ($\alpha = 1$). This situation is simple only in the sense that there is no reflected light inside the black container. In this case, the radiation is always between the surface area which emitted it and the opposite side where it should be absorbed. Thus the energy density inside the black box depends solely on the emissive power of its walls. It can be determined by taking into account that the light emitted in some direction in a time dt is in the element of the cone with the side length cdt and the base area $ds = r^2 d\Omega$ perpendicular to the beam

$$du = \frac{dE}{dscdt} . \qquad (2.2.1)$$

The energy density thus defined is still differentially small, since it is just the fraction which penetrated into the volume element from the area dS of the radiating surface of the container (Fig. 2.2a). In view of (2.1.3), it is equal to

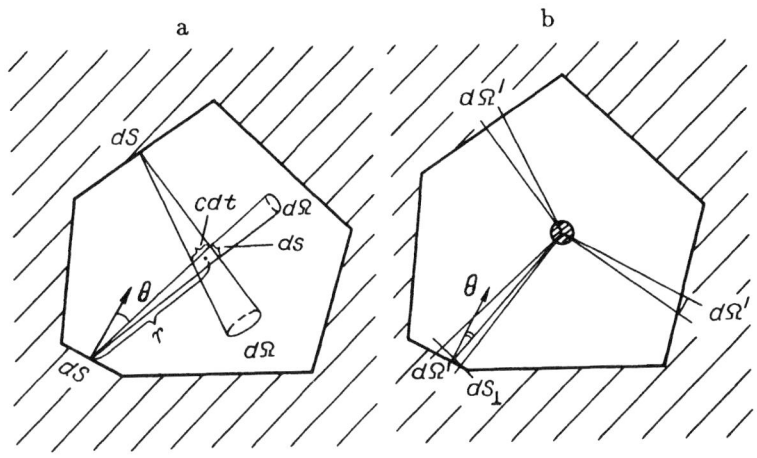

Figure 2.2 (a) Radiation of surface elements dS of a black hole and (b) their view from inside.

$$du = \frac{dE}{r^2 d\Omega c dt} = E_0(\nu)d\nu \frac{dS\cos\vartheta}{\pi c\, r^2}. \qquad (2.2.2)$$

Discerning the radiating surface element from the volume element under consideration, we see that projection of dS on the plane perpendicular to the direction of observation, has the area $dS_\perp = dS\cos\theta$. The solid angle subtended from this surface element is $d\Omega' = dS_\perp/r^2 = (dS\cos\theta)/r^2$ (Fig. 2.2b). So, the total energy density in the interval $d\nu$ may be readily obtained from (2.2.2) by integrating over the angles

$$du(\nu) = \frac{E_0(\nu)d\nu}{\pi c}\int d\Omega' = \frac{4 F_0(\nu)}{c} d\nu = \rho(\nu)d\nu. \qquad (2.2.3)$$

Here $\rho(\nu)$ is the spectral density of the radiation energy in the unit volume of the container:

$$\rho(\nu) = \frac{4 E_0(\nu)}{c}. \qquad (2.2.4)$$

By definition of the black body, $\alpha = 1$, so its emissive power $E_0(\nu) = I_0(\nu)$. Eliminating this from (2.2.4), we find the radiation flux from a black body:

$$I_0(\nu) = \tfrac{1}{4} c \rho(\nu). \qquad (2.2.5)$$

This result is easy to understand. As usual, the energy flux is proportional to the product of its density and the propagation velocity. Only the numerical factor $\frac{1}{4}$ could not have been foreseen without the above calculations.

Substituting Eq. (2.2.5) into the energy balance for a real body (2.1.2), we can present it in the following form

$$\frac{E_0(\nu)}{\alpha(\nu)} = \tfrac{1}{4} c\rho(\nu) . \tag{2.2.6}$$

It has been proven already that at $\alpha = 1$, $\rho(\nu)$ is the equilibrium energy density in a black box. Will this density as well as $I_0(\nu)$ be the same in any other box (different in color)? A positive answer to this question is given by the famous Kirchhoff theorem.

The theorem is proven by *reductio ad absurdum*. Let us assume that in different boxes thermostated at the same temperature, a different density of equilibrium radiation is established, depending on the material of the boxes. In this case, joining the boxes by a window transparent to radiation, we create the conditions for the exchange of radiation energy. If the radiation density in one box is greater than that in another, the energy flux from it should dominate over that flowing in the opposite direction. As a result, the heat equal to the difference in the fluxes will pass from one box to another, thus decreasing the energy density in the first box and increasing it in the second one. Simultaneously, this will break the equilibrium between the radiation and the walls maintained by the magnitude of ρ in each box. In the first box emission will dominate over absorption, while in the second the situation will be the opposite. Thus the walls of the first box will cool down, and those of the second box will heat up. As a result, radiation contact between the boxes will bring about a spontaneous heat transfer from a cool body to a hot one which is strictly forbidden by the second law of thermodynamics. The impossibility of this process ensures that the equilibrium density of radiation is the same in any medium.

However, the following subtle point should be taken into account. If the light intensity $I_1(\nu)$ from the first box were greater than the opposite flux only for certain frequencies, while for other frequencies the reverse were true, then the total energy flux through the window would be equal to zero, when $\int I_1(\nu)d\nu = \int I_2(\nu)d\nu$. In this case, the above reasoning would seem not very convincing. Still, this possibility is easily ruled out. Place in the window a light filter with transmission characterized by the carrying capacity $\xi(\nu)$. As the filter is transparent only to certain frequencies, the condition for the balance is more complicated: $\int \xi(\nu)I_1(\nu)d\nu = \int \xi(\nu)I_2(\nu)d\nu$. Since we can install any light filter we like, this condition must be fulfilled at any $\xi(\nu)$, and this is possible with $I_1(\nu) = I_2(\nu) = I_0(\nu)$ only.

Thus the ratio between the emissive power and the absorption coefficient must be the same for all media. This statement becomes clearer if we take into account that for all real bodies both $E_0(\nu)$ and $\alpha(\nu)$ are saw-like curves with

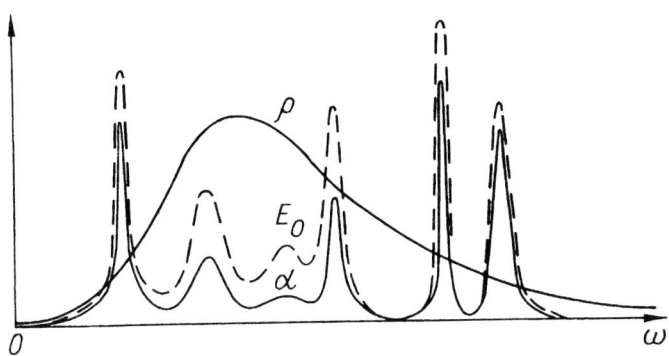

Figure 2.3 Absorption (α) and emission (E_0) spectra in comparison with the density of equilibrium radiation ρ as a function of frequency.

sharp peaks corresponding to the resonance frequencies of the substance. On the contrary, $\rho(\nu)$ is a smooth curve which is the same for all media and varies only with temperature. Relation (2.2.6) holds only in the case where the absorption spectrum $\alpha(\nu)$ faithfully copies the emission spectrum $E_0(\nu)$: maximum to maximum, minimum to minimum, so their ratio $\rho(\nu)$ varies monotonically (Fig. 2.3).

As experience shows, we run into this regularity at every step. What seemed to be a dark spot on a metal surface will prove to be the brightest area upon heating. At room temperature an ampoule containing a gas absorbs exactly the frequencies it emits when the gas is heated. The continuous spectrum of the Sun contains black (Fraunhofer) lines resulting from the absorption of these frequencies by hydrogen atoms in the corona, and at the same frequencies bright lines are observed in the spectrum of a hydrogen gas burner. These and some other illustrations of Kirchhoff's law substantiate the fundamental nature of relation (2.2.6) where the equilibrium distribution of the radiation density $\rho(\nu)$ is of primary importance. Given this function, we need know only one of the substance characteristics, either $\alpha(\nu)$ or $E_0(\nu)$, as the other is defined by relation (2.2.6).

Specific features of radiation as a thermodynamic system are also adequately represented by this relation. Whereas the gas density is uniquely determined by the number of molecules which remain inside the vessel, radiation is not conserved in principle. At some time radiation can flow into the container, or out of it. However, owing to the balanced exchange with the medium, in the time average $\rho = $ const. Besides, both the frequency dependence and the magnitude of ρ are completely determined by the medium temperature. In other words, unlike gases, radiation density can be neither increased nor decreased by varying the volume it occupies. Once equilibrium has been attained, it always satisfies the same universal function $\rho(\nu)$.

Owing to the ability of radiation to permeate through any wall, the radiation density inside the equilibrium medium which, speaking metaphorically, is a sponge saturated with radiation, is the same as in the container. We considered a vacuum cavity in a body as a radiation container just for the reason that this allows one to separate the radiation from the matter. In fact, it makes no difference whether radiation is "squeezed" out of the sponge into the container, or considered inside the medium emitting it. The latter is even more convenient. For example, imagine a vessel with ideal mirrors as the walls and a certain amount of gas inside. The radiation should be in equilibrium at the temperature of the gas, that is, the equilibrium distribution $\rho(\nu)$ is to be set at each point inside the vessel.

2.3 ULTRAVIOLET CATASTROPHE

The above arguments show that knowing the function $\rho(\nu)$ is of primary importance. This information was obtained experimentally late in the nineteenth century through direct spectral investigation of equilibrium radiation. However, the theoretical interpretation of the results presented some fundamental difficulties. It turns out that a statistical treatment of equilibrium radiation leads to a paradoxical conclusion: the integral (over all frequencies) of the energy density is infinite. If this were the case, the equilibrium between radiation and substance would be impossible at any temperature. If radiation had infinite heat capacity, it would "pump" the energy out of any body, thus cooling it to absolute zero. Since in reality this is not the situation, it is obvious that the paradox is due to an error in the theoretical estimate of $\rho(\nu)$.

Heat Capacity of Radiation

In order to find the error, let us estimate the spectral density of equilibrium radiation. To do this, we need only know

- the number of electromagnetic vibrations dG for the interval $d\nu$ per unit volume,
- the mean heat energy $\overline{E}(\nu)$ of each vibration in equilibrium.

The value of the first quantity is simple to determine. An isolated vibration is a monochromatic wave of a definite polarization and direction of motion. In a closed volume such vibrations are infinite in number, but still countable. Each is defined by its frequency ν or wave length λ or wave number $k = 1/\lambda$, and

$$\nu = ck. \tag{2.3.1}$$

In order to specify the direction of the wave motion, vibration is characterized by a vector **k** of magnitude $k = \nu/c$ which is perpendicular to the wave propagation front. Different values of the vector **k** mark possible states of the wave motion and allow one to classify them in magnitude and direction, just as the states of the free particle are distinguished by specifying its momentum **p**. By analogy, we can claim that the density of vibrations in the interval dk is proportional to $4\pi k^2 dk$, just as the density of free particle states is proportional to $4\pi p^2 dp$. Physical reasoning similar to that given in Chapter 3 (Section 3.2) makes it possible to convert this proportionality into an exact equality

$$dG = 2 \cdot 4\pi k^2 dk . \qquad (2.3.2)$$

The multiplier 2 accounts for the two independent vibrations of the same frequency with orthogonal light polarization. According to (2.3.1) and (2.3.2), the total number of different vibrations in a unit volume and in the interval $d\nu$ is as follows

$$dG = \frac{2 \cdot 4\pi \nu^2 d\nu}{c^3} . \qquad (2.3.2a)$$

The equilibrium energy density in the interval $d\nu$ is merely the sum of energies of all vibrations per unit volume

$$du = \bar{E}(\nu) dG = \rho(\nu) d(\nu) . \qquad (2.3.3)$$

Thus the calculation of

$$\rho(\nu) = \bar{E}(\nu) \frac{8\pi \nu^2}{c^3} \qquad (2.3.3a)$$

completely reduces to solving the second problem: what is the mean energy $E(\nu)$ of one electromagnetic vibration in equilibrium?

Answering this question from the classical standpoint, we cannot avoid the general equipartition law: under equilibrium conditions, the average energy of any harmonic vibration, including that of light, is equal to kT. Assuming $\bar{E} = kT$ in (2.3.3), we get the distribution

$$du = \frac{8\pi \nu^2 kT}{c^3} d\nu , \qquad (2.3.4)$$

known as the Rayleigh law. From this distribution it immediately follows that

$$u = \int du = \frac{8\pi kT}{c^3} \int_0^\infty \nu^2 d\nu = \infty .$$

Field Quantization

The use of the word "catastrophe" is indeed justified when one realizes that the foundations of the theory were essentially affected and radical changes needed to be made to modify it. Searching for the correct spectral distribution, Planck was lucky to find an analytical formula with a single parameter h (the Planck constant) which described adequately the frequency and temperature dependence of radiation energy density

$$\rho(\nu) = \frac{8\pi h\nu^3/c^3}{\exp\left(\dfrac{h\nu}{kT}\right) - 1}. \tag{2.3.5}$$

On integrating (2.3.5), we obtain the important result

$$u = \int_0^\infty \rho(\nu)\,d\nu = \frac{8\pi(kT)^4}{(ch)^3}\int_0^\infty \frac{x^3\,dx}{e^x - 1} = \frac{8\pi^5 k^4}{15 c^3 h^3}T^4, \tag{2.3.6}$$

which agrees with the thermodynamic calculation of $u(T)$ given in (5.5.17) and the Stefan–Boltzmann law for radiation energy flux

$$q = \int_0^\infty I_0(\nu)\,d\nu = \frac{1}{4}cu = \sigma T^4, \tag{2.3.7}$$

where $\sigma = (2\pi^5 k^4)/(15 c^2 h^3) = 5.67 \cdot 10^{-5}$ erg/cm²·deg⁴·sec.

However, comparing (2.3.3a) with the Planck law (2.3.5), we must reach the conclusion that

$$\overline{E} = \frac{h\nu}{\exp\left(\dfrac{h\nu}{kT}\right) - 1}. \tag{2.3.8}$$

This conclusion is inconsistent with the equipartition law, which is a fundamental conclusion of the classical theory.

As was explained in detail in Section 1.12, Planck made far-reaching conclusions from his lucky guess. Trying to substantiate formula (2.3.8), he interpreted it as the average of discrete energies of light vibration

$$E = N\epsilon, \tag{2.3.9a}$$

where

$$\epsilon = h\nu .\qquad(2.3.9b)$$

The next step forward in this direction was made by Einstein who saw in the quantum postulate (2.3.9) the possibility of rehabilitating the corpuscular description of light known as the photon model of radiation.

Photon Model

If we take it that electromagnetic quanta are light particles with the energy ϵ, then $N = E/\epsilon$ is the number of such particles in a wave with energy E. Fig. 2.4 gives a good idea of the sense of this change in terminology. The vertical lines are energy axes of the corresponding vibrations divided by quantization into segments (quanta) of equal size which increase linearly with increasing frequency. From the wave point of view, the Nth graduation mark on the energy scale corresponds to the Nth excitation level of the light oscillator. From the corpuscular standpoint, N is the number of photons moving with the energy ϵ. Thus either wave (ν and E) or corpuscular (ϵ and N) variables can be plotted on each axis, as abscissa and ordinate. These are alternative models. The wave energy E and photon energy ϵ have nothing in common: the first is ordinate,

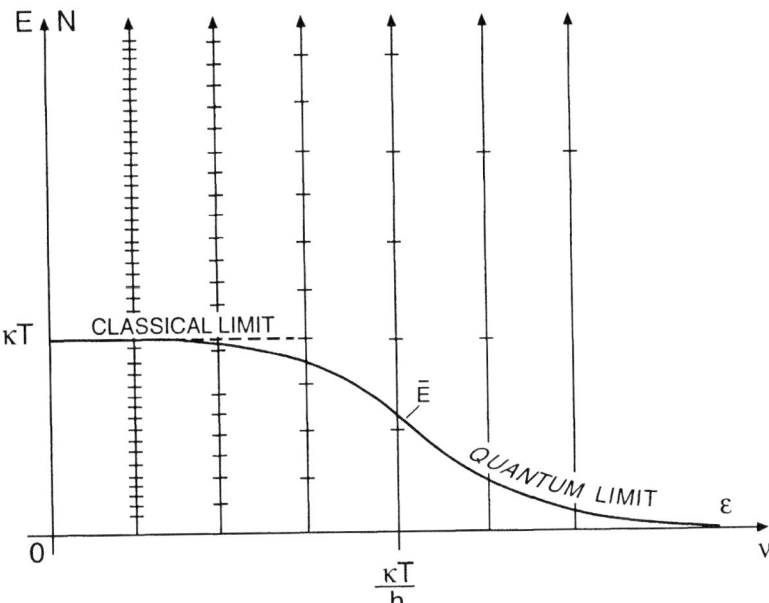

Figure 2.4 Description of radiation in corpuscular (N and ϵ) and wave (E and ν) variables. Dashes show the quantized energy levels at different frequencies, and the curve $\bar{E}$ is the mean energy at a given temperature.

the second abscissa, and there is no relation between them. It would make no sense to try to imagine quantized waves as sinusoids cut into slices. With this change of description, the image of a wave should just give way to another image, a flux of point particles (photons) moving with the energy ϵ.

The statistical description of radiation in the alternative models differs in a similar way. From the wave point of view the range of low-frequency vibrations is exclusively classical, since $\overline{E} = kT$ (see Fig. 2.4). Quantum anomalies are observable only in the ultraviolet wing of the spectrum where $\overline{E} \to 0$. On the other hand, from the corpuscular standpoint, the exponential decrease in N in the ultraviolet region agrees with the classical distribution of particles with energy. At the same time the distribution of low-energy photons is essentially different from the Maxwell distribution: their number increases hyperbolically rather than exponentially when $\epsilon \ll kT$ (Fig. 2.5). Indeed,

$$\overline{N} = \sum_{N=0}^{\infty} N \frac{\exp\left(-\dfrac{N\epsilon}{kT}\right)}{\sum_{N=0}^{\infty} \exp\left(-\dfrac{N\epsilon}{kT}\right)} = \frac{1}{\exp\left(\dfrac{\epsilon}{kT}\right) - 1}, \quad (2.3.10)$$

so at $\epsilon \gg kT$ $\overline{N} = \exp(-\epsilon/kT) \ll 1$, while at $\epsilon \ll kT$ $\overline{N} = kT/\epsilon \gg 1$ and tends to infinity with $\epsilon \to \infty$ which is by no means characteristic of particles with classical behavior. Thus, photons are essentially quantum particles. Their distribution over the entire energy range $(0 \leq \epsilon \leq \infty)$ never reduces to the classical one. In this sense, the photon gas is "degenerate," that is, anomalous, exclusively quantum at any temperature.

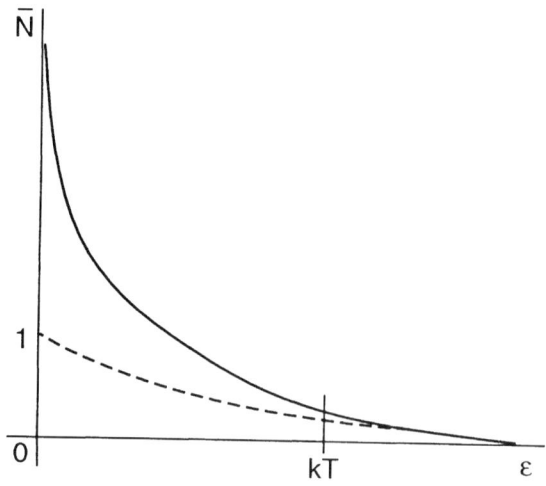

Figure 2.5 The photon distribution $\overline{N}(\epsilon)$ and its high-energy asymptotics (dashed line).

The choice of the model (either wave or corpuscular) is at one's convenience. Neither is universal: quantization is not inherent in classical wave motion, while interference and diffraction do not appear in the corpuscular picture. However, using either model aids in interpretation of the results and does not lead to incorrect estimates, provided that quantum peculiarities and correspondence rules are taken into account. The corpuscular model is preferable in describing equilibrium radiation, for it makes it possible to reduce the system to a gas model. Since equilibrium radiation consists of a great variety of vibrations differing in frequency, polarization and phase, the possibility of interference and other wave phenomena is reduced to zero. As for radiation passed through a monochromator which can be split and then made to interfere with itself or induce resonant processes, the situation is quite different. In this case, the wave model is often preferable to the photon one due to radiation coherence. However, in studies of equilibrium radiation, we can continue to use corpuscular terminology.

2.4 PHOTON MODEL

The Basic Features

The statistical description of a photon gas is defined by the distribution of particles in the phase space divided into cells of volume h^3:

$$dN = \overline{N}(\epsilon) \cdot 2 \cdot \frac{dp_x dp_y dp_z dx dy dz}{h^3} = \frac{2}{e^{\epsilon/kT} - 1} \frac{d\mathbf{p}d\mathbf{r}}{h^3} . \qquad (2.4.1)$$

Along with (2.3.10), the above expression includes the double degeneracy of each state of motion due to the possibility of different polarizations of light. The potential energy of photons in the Earth's gravitational field varies only slightly at reasonable distances and was neglected in Eq. (2.4.1) As the coordinate part of this distribution is of no interest, it would be natural to turn from (2.4.1) to the volume density of photon gas $dn = dN/dxdydz$ which is obviously equal to

$$dn(\mathbf{p}) = \frac{2}{\exp\left(\frac{\epsilon}{kT}\right) - 1} \cdot \frac{d\mathbf{p}}{h^3} , \qquad (2.4.2)$$

On integrating over the angles, we have

$$dn(p) = \frac{8\pi p^2 dp/h^3}{\exp\left(\frac{\epsilon}{kT}\right) - 1} . \qquad (2.4.3)$$

Now, for complete definiteness, we need only know the dependence $\epsilon(p)$, the basic mechanical characteristic of an ideal gas.

To determine it, let us compare the last result with the photon density following from the Planck law (2.3.5)

$$dn = \frac{du}{\epsilon} = \frac{\rho(\nu)d\nu}{h\nu} = \frac{8\pi\nu^2 d\nu/c^3}{\exp\left(\frac{h\nu}{kT}\right) - 1}. \qquad (2.4.4)$$

These formulae differ only in that Eq. (2.4.4) is expressed in terms of wave variables, while Eq. (2.4.3) is expressed in terms of corpuscular variables. Identity is obtained on condition that

$$p = h\frac{\nu}{c} = hk, \qquad \mathbf{p} = h\mathbf{k}. \qquad (2.4.5)$$

The above equalities, which set up a correspondence between corpuscular and wave variables, close the corpuscular model. Eqs. (2.4.5) and (2.3.9b) yield

$$\epsilon = pc. \qquad (2.4.6)$$

Eliminating one of the two variables (either ϵ or p) from the distribution (2.4.3) we can get either the distribution of momenta

$$dn(p) = \frac{8\pi p^2 dp/h^3}{\exp\left(\frac{pc}{kT}\right) - 1}, \qquad (2.4.7)$$

or energies

$$dn(\epsilon) = \frac{8\pi\epsilon^2 d\epsilon}{\left[\exp\left(\frac{\epsilon}{kT}\right) - 1\right] c^3 h^3}. \qquad (2.4.8)$$

Thus relation (2.4.6) makes the above distributions uniquely determined. Now the mean energy, pressure, flux of particles and so on can be calculated within the framework of the purely corpuscular model using the usual recipes.

The relation between energy and momentum is the main individual characteristic of a gas. This relation specifies the mechanics that gas particles obey. The constant appearing in the $\epsilon(p)$ dependence is the characteristic invariant of a gas. In the Maxwell gas it is the particle mass m; in a photon gas it is the velocity of light c.

Comparison of the Models

Direct comparison of the above ideal gases clearly shows the distinctions in their mechanical and statistical properties:

$m = \text{const}$	$c = \text{const}$
$\epsilon = \dfrac{p^2}{2m}$	$\epsilon = pc$
$\bar{N} = \dfrac{N_0}{Z} e^{-\epsilon/kT}$	$\bar{N} = 1/(e^{\epsilon/kT} - 1)$
$N_0 = \text{const}$	—
$p = mv$	$p = mc$

The fact that in a photon gas not the mass but the velocity is conserved leads to some interesting peculiarities which make the photon gas qualitatively different from the Maxwell one. In particular, the accidental similarity of distributions over momenta and velocities disappears as it is a direct consequence of the constancy of mass. Due to constancy of c, the distribution of photons over velocities is a distribution only of directions of motion

$$dW(\mathbf{c}) = \frac{d\Omega}{4\pi} = \frac{\sin\theta \, d\theta \, d\varphi}{4\pi} . \qquad (2.4.9)$$

This is therefore by no means similar to (1.4.1), while between distributions of molecules and photons over momenta there is, at least, a structural resemblance.

Moreover, if we calculate from (2.4.9) the probability of the z-projection of the velocity, we obtain

$$dW(c_z) = \frac{dc_z}{2c} . \qquad (2.4.10)$$

This is not surprising, since a quantity which does not affect the energy may be expected to be homogeneously distributed over the entire domain: from $-c$ to $+c$. By virtue of the space symmetry, c_y and c_z have the same distributions. From this it immediately follows that $dW(\mathbf{c}) \neq dW(c_x)dW(c_y)dW(c_z)$, in contrast to the Maxwell gas relation (1.2.5). Different values of the velocity projections cannot be considered as independent events. For example, if $c_x = c$, then, obviously, $c_y = c_z = 0$.

The peculiarity of photons as field quanta also lies in the fact that their total number is not fixed. This is seen from the above table: N_0 is not conserved for photons. As for the photon mass, it is defined by a relation formally similar to a classical one but having quite a different meaning due to $c = \text{const}$. It is interesting to compare this information with that obtainable from another, quite independent, source: the theory of relativity. Such a comparison reveals no contradictions; moreover, it points to excellent agreement. In relativistic mechanics, relation (2.4.6) is a natural consequence of the fact that the mass of a photon at rest is equal to zero. Besides, substituting $p = mc$ into (2.4.6), we have

$$\epsilon = mc^2, \qquad (2.4.11)$$

in perfect agreement with the general inference of the theory of relativity.

It can easily be seen that in the region of visible light ($\nu = 10^{15}$ Hz) the photon mass $m = h\nu/c^2 \approx 10^{-47}\nu \sim 10^{-32}$ g. Even hard X-rays ($\nu \sim 10^{19}$ Hz) consist of particles with a mass 10 times less than the electron mass. Only γ-radiation quanta ($\nu \sim 10^{21}$ Hz) have mass comparable with or even greater than that of an electron.

Photon Weight

Not only the inertial mass, but also the weight of a particle is associated with the concept of mass. In this aspect, a photon does not differ from other particles, and in a homogeneous gravitational field its equation of motion has the form

$$\dot{p} = F = mg = pg/c. \qquad (2.4.12)$$

When a photon is falling its mass does not remain constant, but increases exponentially just as momentum:

$$p = p(0) \exp(gt/c) = p(0) \exp(gx/c^2). \qquad (2.4.13)$$

As $gx \ll c^2$, expanding the exponent in (2.4.13) into a power series we obtain in the lowest approximation:

$$\Delta p = p(x) - p(0) = \Delta m \cdot c = mgx/c. \qquad (2.4.14)$$

The increase in the energy of photon after falling a distance x is

$$\Delta \epsilon = c\Delta p = mgx. \qquad (2.4.15)$$

This is merely the difference of potentials over the distance covered.

The last result was brilliantly verified by the direct weighing of a photon. A source of exclusively monochromatic γ-radiation obtained by the Mössbauer method was located in such a way that the photons traveled a distance H of the order of one meter along the vertical before they entered the frequency analyzer. Each photon in free fall acquired the additional energy $\Delta\epsilon = mgH$ under the action of the gravity force. Owing to the extreme precision of the Mössbauer method, this furnished an opportunity to note the frequency shift $\Delta \nu = mgH/h$. If a photon moved downwards, the light entering the analyzer had shorter wavelength than that emitted. In the opposite case, the reverse situation was observed. When a photon followed a horizontal path, the light frequency remained unchanged. Thus a photon, just as other particles, experiences a gravitational force the magnitude of which is expressed in terms of its mass in an ordinary way. It would be much more complicated to calculate the gravitational frequency shift within the wave model of light.

2.5 PHOTON GAS PROPERTIES

From the corpuscular point of view, the walls of a container filled with radiation experience the same "bombarding" by photons as the walls of a vessel with a gas. The photon flux striking the walls can easily be calculated using (2.4.10): the result is

$$j = n \int c_x dW(c_x) = \tfrac{1}{4} nc, \qquad (2.5.1)$$

This is exactly the same as the classical flux if we relate the velocity of light to the mean absolute velocity $\bar{v}$ of Maxwell's particles. Moreover, since the distribution in velocities and that in polarizations and energies are independent, the photon density in this formula may be either total or partial:

$$dj = \tfrac{1}{4} cdn. \qquad (2.5.2)$$

Multiplying it by ϵ, we can see that the radiation energy flux

$$dq = \epsilon dj = \tfrac{1}{4} c\, du = \tfrac{1}{4} c\rho(\nu)d\nu = I(\nu)\, d\nu \qquad (2.5.3)$$

also agrees with the expression (2.2.5) we obtained previously.

Pressure

Though a fraction of the photon flux striking the wall is absorbed, we can nevertheless consider that all photons are elastically reflected from the wall, transferring momentum $2p_x$ to it. Of course, this is only possible due to detailed balance, whereby for each photon absorbed by the wall there is identical photon emitted by it. The absorbed photon gives up momentum p_x to the wall, while that emitted takes the momentum $(-p_x)$. So, treating both acts as one process, we can take it that all photons are reflected elastically, as from a mirror wall. In this case, the general recipe for the calculation of pressure remains the same

$$p = \int 2p_x dj. \tag{2.5.4}$$

However, as it is necessary to perform averaging over the angles, one should not use result (2.5.2) which has already been averaged over these variables. Calculations should be continued as follows

$$p = \int 2p_x c_x dn = \int 2pc \cos^2 \vartheta \, dn = \int 2\epsilon \cos^2 \vartheta \, dn(\epsilon) \frac{d\Omega}{4\pi}. \tag{2.5.5}$$

The averaging over energies and angles is separated, and integration over the semispace yields

$$p = \overline{2\cos^2 \vartheta} \cdot \int_0^\infty \epsilon \, dn = \tfrac{1}{3} u. \tag{2.5.6}$$

As in the classical case, the photon gas pressure is proportional to the volume density of energy, but the numerical coefficient is different from that in Eq. (1.1.8).

This result is also obtained in the electromagnetic theory of light, and is independent of the spectral composition. It was verified by extremely precise measurements of the light pressure carried out by P. N. Lebedev in the late nineteenth century. These days, owing to the advent of powerful light sources, this effect is much easier to observe. However, equilibrium radiation pressure is still difficult to measure due to its extreme smallness at room temperatures.

Indeed, substituting the equilibrium density of energy from (2.3.6) into (2.5.6) gives

$$p = rT^4, \tag{2.5.7}$$

where

$$r = \frac{8\pi^5 k^4}{45 c^3 h^3} = \frac{4\sigma}{3c} = 2.5 \cdot 10^{-15} \frac{\text{dyn}}{\text{cm}^2 \cdot \text{K}^4}. \tag{2.5.8}$$

At 300 K the pressure $p = 2.5 \cdot 10^{-15} \cdot 3^4 \cdot 10^8 = 2 \cdot 10^{-5}$ dyn/cm², thus its existence may be verified only indirectly; for example, by measuring the radiant energy flux proportional to it. According to (2.3.7), the magnitude of flux emitting by any body at room temperature is

$$q = \sigma T^4 = 5.7 \cdot 10^{-5} \cdot 3^4 \cdot 10^8 = 5 \cdot 10^5 \, \text{erg/cm}^2 \cdot \text{sec}.$$

This is readily noted, even a long way off.

Equation of State

Formula (2.5.7) is actually a photon gas equation of state similar to the ideal molecular gas equation $p = nkT$. A comparison between them shows considerable distinctions determined by the different statistics and the fact that the total number of photons is not conserved. This is due to this fact the number of photons does not appear in (2.5.7) as an independent parameter. If we calculate it by integrating (2.4.7) or (2.4.8), we see that

$$n = \int dn = \frac{8\pi(kT)^3}{c^3 h^3} \int_0^\infty \frac{x^2 dx}{e^x - 1} = gT^3 \qquad (2.5.9)$$

is defined by temperature as uniquely as the pressure itself. That is why the pressure is not function of two variables, but only one, temperature. Accordingly, the diagram of states given in Fig. 2.6 shows a single curve $p(T)$ instead of a set of such curves, as in the case of a molecular gas.

A more rapid increase in pressure with temperature is due to a sharp increase in the number of photons per unit volume. As is seen from (2.3.6) and (2.5.9), their average heat energy is as follows:

$$\bar{\epsilon} = \frac{u}{n} = qkT. \qquad (2.5.10)$$

It is not surprising that the numerical coefficient

$$q = \int_0^\infty \frac{x^3 dx}{e^x - 1} \bigg/ \int_0^\infty \frac{x^2 dx}{e^x - 1} \sim 1$$

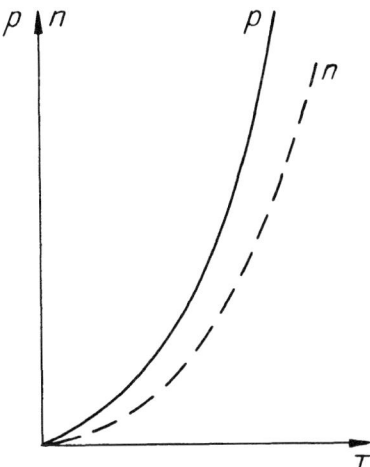

Figure 2.6 The temperature dependence of pressure p and the number density n of a photon gas.

differs from 3/2. The equipartition law in its original form is not applicable to quantum particles. More importantly, the direct proportion $\bar{\epsilon} \sim kT$ is valid for radiation as well as for a molecular gas, confirming the ordinary sense of absolute temperature.

Since n is uniquely determined by temperature, the photon gas state remains unchanged under an isothermal change of its volume. This is no surprise, since radiation resides both in the container and in the equilibrium medium surrounding it. In increasing the container's size, we just remove the atoms of the substance, but do not increase the volume accessible to radiation. In other words, it is impossible to expand or compress radiation isothermally.

2.6 EINSTEIN COEFFICIENTS

Absorption and Emission

In developing the idea of the photon model, one can easily imagine an atom as a target which is bombarded by light particles. Absorption is the result of a direct hit which leads to the excitation of the atom from state 1 to a higher energy state 2 at the expense of the energy of the photon absorbed. Such an event is possible when the target is not a point particle, that is, the atom must have an effective cross-section σ_{12}. The length of a photon free path between the act of emission and that of absorption is $\lambda = 1/n\sigma_{12}$. As it passes through the medium consisting of unexcited atoms, the photon flux gets thinner according to the law which is the corpuscular analogue of (2.1.1),

$$N = N(0)\exp\left(-\frac{x}{\lambda}\right)$$

where $\lambda = 1/\epsilon$.

When a photon strikes an excited atom it can be absorbed by it, initiating a resonance transition to an even higher energy level if any there are appropriate. Alternatively, it can knock out a new photon — an exact copy of the impacting one. Then the new photon will move away together with the perpetrator of the accident, while the atom drops back to a lower state due to the loss of the quantum $h\nu = E_2 - E_1$. This process of *induced radiation* predicted by Einstein makes it possible to extract a powerful light flux from an *active medium* consisting mostly of excited atoms.

Since it is difficult to say beforehand whether an atom releases its energy as readily as it acquires it, we will for now ascribe to induced radiation a cross-section σ_{21} different from σ_{12}. In passing through the substance, the flux of light resonant with the $1 \leftrightarrow 2$ transition either loses photons or acquires them, depending on what is greater: $\sigma_{12}N_1$ or $\sigma_{21}N_2$ (N_2 and N_1 are the numbers of atoms in excited and ground states).

Apart from the processes mentioned above, there is also a *spontaneous emission* of photons which manifests itself in thermal radiation of heated bodies. There is a certain probability for excited atoms to decay per unit time A_{21}. According to the kinetic equation

$$\dot{N}_2 = -A_{21}N_2,$$

they decay exponentially emitting light.

Excitation and Deactivation

Each atom collides not only with photons but also with the other atoms constituting the substance. As a result of interparticle interactions, the atom moves back and forth between levels 1 and 2, either giving energy to the partner or taking it away. The flux of energy transfering between the atoms is the same in both directions, $p_{12}N_1^r = p_{21}N_2^r$, but equilibrium population of excited state $N_2^r = N_0 \exp[-(E_2 - E_1)/kT]$ is less than that of the groundstate ($N_1^r = N_0$). Therefore the probability p_{12} to acquire energy (per unit time) is less than the probability p_{21} to release it (see Fig. 2.7):

$$\frac{p_{12}}{p_{21}} = \exp\left(-\frac{E_2 - E_1}{kT}\right). \qquad (2.6.1)$$

Rates of Induced Transitions

Following Einstein we should *a priori* characterize the photoinduced transitions by the corresponding rates which are actually the rates of collisions between the atom and photons. For either excitation or deactivation the rate

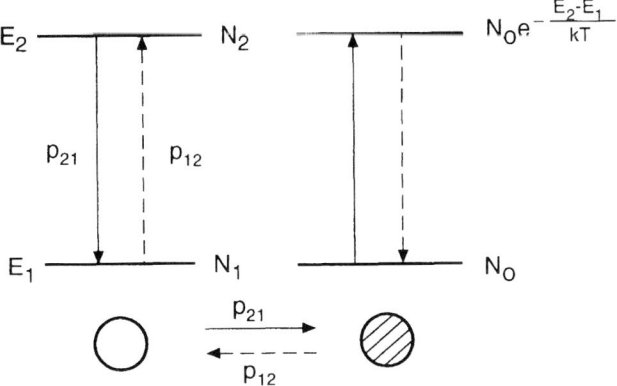

Figure 2.7 Resonant energy exchange between two-level system (left) and the equilibrium medium at fixed temperature T (right). p_{12} and p_{21} are the rates of activation and deactivation correspondingly.

$$W = \int \sigma c \, dn = \int \frac{\sigma(\nu)c}{h\nu} \rho(\nu) d\nu \qquad (2.6.2)$$

is proportional to the corresponding cross-section σ. Transitions occur only in the case where the light frequency is in the vicinity of the resonant frequency of the atom, because σ is a sharply peaked function of the frequency similar to $\alpha(\nu)$ or $E_0(\nu)$. For an isolated spectral line the dependence $\sigma(\nu)$ is similar to a δ-function, at least compared to $\rho(\nu)$, which is a fairly smooth curve. Thus

$$W = \int B(\nu)\delta(\nu - \nu_0) \rho(\nu) d\nu = B(\nu_0)\rho(\nu_0), \qquad (2.6.3)$$

and the uncertainty of the effective cross-section σ is transferred to B. Assuming that induced absorption and emission proceed at different rates, we should also distinguish

$$W_{12} = B_{12}(\nu_0)\rho(\nu_0) \quad \text{and} \quad W_{21} = B_{21}(\nu_0)\rho(\nu_0). \qquad (2.6.3a)$$

The constants B_{ik} characterizing the atomic transitions are called the Einstein coefficients. Since the equilibrium is always detailed, we can confine the discussion to two states of the atomic spectrum which are resonant to the frequency ν_0. We can also easily write the basic kinetic equation for them proceeding from elementary balance considerations. The number of particles leaving one level should be equal to that added to another:

$$\frac{dN_2}{dt} = -\frac{dN_1}{dt} = (W_{12}N_1 - W_{21}N_2 - A_{21}N_2) + (p_{12}N_1 - p_{21}N_2). \qquad (2.6.4)$$

The components in the first set of parentheses describe the interaction of the atom with photons, and those in the second that with the equilibrium surroundings.

Einstein's Relations

Under detailed equilibrium conditions, not only does $\dot{N}_2 = -\dot{N}_1 = 0$, but also the expressions in each bracket are equal to zero. This means that the energy exchange should be balanced both with the medium

$$p_{12}N_1 = p_{21}N_2, \qquad (2.6.5a)$$

and with the light

$$W_{12}N_1 - W_{21}N_2 - A_{21}N_2 = 0. \qquad (2.6.5b)$$

In other words, the number of the photons absorbed should be equal to the number of those emitted. Otherwise the light energy would be converted into heat, or vice versa. However, under equilibrium conditions, this should not take place. From (2.6.5a) and (2.6.1) it immediately follows that

$$\frac{N_2}{N_1} = \frac{p_{12}}{p_{21}} = \exp\left(-\frac{E_2 - E_1}{kT}\right). \tag{2.6.6}$$

On the other hand, (2.6.5b) and (2.6.6) yield quite an unusual relation

$$W_{12} = (W_{21} + A_{21}) \exp\left(-\frac{E_2 - E_1}{kT}\right), \tag{2.6.7}$$

which connects the rates of induced and spontaneous radiation to one another. Substituting (2.6.3a) into (2.6.7) and taking into consideration that $h\nu_0 = E_2 - E_1$, we derive

$$\rho(\nu_0) = \frac{A_{12}}{B_{12} \exp\left(\frac{h\nu_0}{kT}\right) - B_{21}}. \tag{2.6.8}$$

Up to now our discussion has been based on extensive application of the detailed equilibrium principle. However, we also have knowledge of the actual spectral density of equilibrium radiation defined by the Planck law (2.3.5). It is reasonable to require formula (2.3.5) to coincide with (2.6.8):

$$\frac{8\pi h\nu_0^3/c^3}{\exp\left(\frac{h\nu_0}{kT}\right) - 1} = \frac{A_{21}}{B_{12}\left[\exp\left(\frac{h\nu_0}{kT}\right) - \frac{B_{21}}{B_{12}}\right]}. \tag{2.6.9}$$

It is notable, that the only way to transform (2.6.9) into an identity is to admit that the Einstein coefficients are connected by the relations established Einstein himself,

$$B_{12} = B_{21}, \tag{2.6.10}$$

$$\frac{A_{21}}{B_{21}} = \frac{8\pi h\nu_0^3}{c^3}. \tag{2.6.11}$$

If there were no induced radiation (unknown before Einstein's discovery), the results would be inconsistent. Indeed, for $B_{21} = 0$ we would have

$$\frac{A_{21}}{B_{12}} \exp\left(-\frac{h\nu_0}{kT}\right)$$

in the right-hand side of (2.6.9), opposed to the left-hand side. In such a situation Einstein had nothing to do but postulate the existence of a new effect.

Not only did Einstein introduce this phenomenon, but he also established the equality of the rates of induced emission and absorption, stated in Eq. (2.6.10). This allows us not to distinguish between the rates of light-induced transitions between two levels, therefore omitting indices of B. Of course, if we consider several spectral lines, the indices should be retained to distinguish B_{12} from B_{13} or B_{23}. This is also true for A_{ik}.

Partial Rates

By analogy with (2.6.3), the stationary light flux propagating in a solid angle $d\Omega$ and having a definite polarization α generates resonant transitions between two levels with the rate

$$dW_\alpha = b_\alpha(\nu_0)\rho(\nu_0)\frac{d\Omega}{8\pi}. \qquad (2.6.12)$$

Such a stochastic description of atomic relaxation is feasible if the radiation intensity is not too high, and the light spectrum is sufficiently wide. If these conditions are violated, the stochastic approach to light-induced relaxation becomes meaningless. Such is the case with powerful monochromatic sources, when the wave nature of light manifests itself in its coherent interaction with an atom. If so, the process is no longer stochastic, and obeys equations which are much more complicated than (2.6.4). However, this is not true for natural sources of light, because their spectra are usually wide enough. They induce transitions with the partial rate (2.6.12) which is simply related to the integral ones given above, if one takes into account that the equilibrium spectral density is $\rho_\alpha(\nu_0, \Omega) = \rho(\nu_0)/8\pi$:

$$W = \int_\Omega \sum_{\alpha=1}^{2} dW_\alpha, \qquad B(\nu_0) = \int \sum_{\alpha=1}^{2} b_\alpha(\nu_0)\frac{d\Omega}{8\pi}. \qquad (2.6.13)$$

Similarly,

$$A = \int \sum_{\alpha=1}^{2} a_\alpha(\nu_0)\, d\Omega. \qquad (2.6.14)$$

Imposing detailed balance, we get similarly to (2.6.11):

$$\frac{a_\alpha(\nu_0)}{b_\alpha(\nu_0)} = \frac{h\nu_0^3}{c^3}. \qquad (2.6.15)$$

The difference of 8π is related to the fact that the coefficient $b_\alpha(\nu_0)$ defines the rates of induced absorption and emission of a plane wave of a definite polarization rather than isotropic and depolarized equilibrium radiation. The rate is

$$W_\alpha(\nu_0) = b_\alpha(\nu_0)\rho_\alpha(\nu_0) \ . \tag{2.6.16}$$

Note that there is no difference in the frequency, polarization and direction of motion of the emitted photons and those which induced the process.

The importance of relations (2.6.11) and (2.6.15) lies in the fact that they remove the need to calculate or measure separately both coefficients b and a. If a is determined from the emission spectrum, the intensity of absorption is easy to find by recalculation. The qualitative content of relation (2.6.15) is analogous to the Kirchhoff theorem.

It follows from relation (2.6.15) that the rate of spontaneous radiation increases with increasing frequency as ν^3 as compared with the rate of light adsorption. Due to this, high-energy excitations of the inner atomic shells decay rapidly, emitting X-ray quanta. Most bodies are permeable to X-rays due to their weak absorption. In the optical range spontaneous radiation still dominates over induced. Probably this is the only reason why induced radiation was not discovered long before Einstein. Daylight dispersed in a room may be seen due to reflection, but it does not induce noticeable radiation. Its intensity is too low for the induced rates to compete with spontaneous ones. Now let us pass to radio-frequency range (centimeter and kilometer waves). Here the situation is radically different. Hours and days pass between sequential events of spontaneous radiation of radio-frequency quanta. That is why in magnetic spectroscopy which makes use of these frequencies, the process of spontaneous radiation is usually neglected without any effect on the results. As for induced radiation, it should necessarily be taken into account along with absorption, since the states shown on Figs. 1.37 and 1.38 are almost equally populated at $\Delta U = h\nu_0 \ll kT$.

2.7 STATIONARY LIGHT TRANSFORMATION

Now we turn to stationary interaction between an atom and nonequilibrium radiation. In considering only transitions between two levels, we assume that the light under study does not involve frequencies significantly different from $\nu_{12} = (E_2 - E_1)/h$. To put it another way, the width of the radiation spectrum must be less than $\Delta\nu$, the distance to the resonance frequencies adjacent to ν_{12} in the frequency scale. Then this light will excite only the transition 1–2, without affecting other transitions (Fig. 2.8). (There was no need for this restriction on equilibrium radiation, since, by virtue of the detailed balance, we could consider any frequency separately).

Assuming $\dot{N}_2 = -\dot{N}_1 = 0$ in Eq. (2.6.4) and solving the resulting algebraic equation for N_2/N_1, we find, making use of (2.6.16),

$$q = \frac{N_2}{N_1} = \frac{b_\alpha \rho_\alpha + p_{12}}{b_\alpha \rho_\alpha + A + p_{21}} \ . \tag{2.7.1}$$

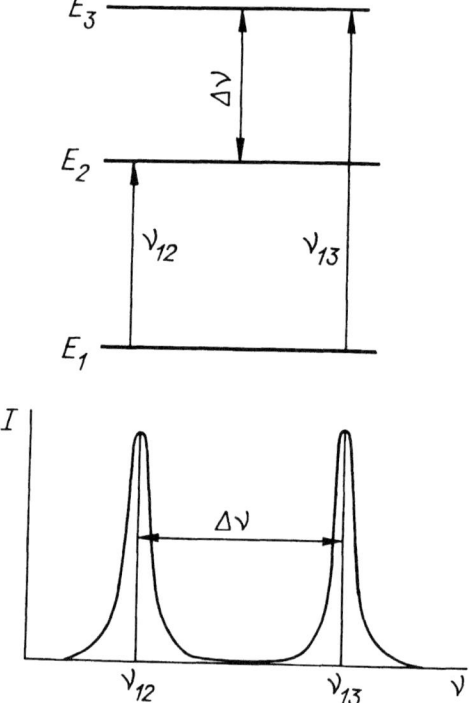

Figure 2.8 A three-level system and its spectrum $I(\nu)$.

A plane wave can induce the same relaxation in atoms as equilibrium radiation, if its spectral density obeys the obvious condition $b_\alpha \rho_\alpha = B\rho$. If and only if this is met we get from (2.7.1) $q = \exp(-h\nu_{12}/kT)$. However, in this case, the Boltzmann distribution over levels does not indicate that the light attains equilibrium with the matter. It is permanently transformed. Light of a definite direction and polarization is absorbed, and that of arbitrary polarization and direction is spontaneously emitted.

In general, the energy absorbed by the substance is

$$Q = h\nu_{12}(p_{21}N_2 - p_{12}N_1). \qquad (2.7.2)$$

This expression becomes zero for an equilibrium population of the levels,

$$\frac{N_2}{N_1} = \frac{p_{12}}{p_{21}} = \exp(-h\nu_{12}/kT) = q_r.$$

At the same time, it is clear that $Q > 0$ at $q = N_2/N_1 > q_r$ and $Q < 0$ at $q = N_2/N_1 < q_r$. Thus any deviation of q from $q_r = e^{-h\nu_{12}/kT}$ results either in radiation or absorption of light:

$$Q = h\nu_{12}p_{21}N_0 \frac{q - q_r}{1 + q} \quad (N_0 = N_1 + N_2). \quad (2.7.3)$$

Both situations are so often observed in everyday life and experiment that it is difficult to choose any one of a great number of examples. A torch, the Sun and smoldering coals shine transforming, heat energy into radiation. Even man, despite the low temperature of his body, is shining (in the ordinary sense of the word). However, one should be careful with this conclusion: in a cool laboratory any photometer sensitive in the infrared range detects the approach of a living being. On the other hand, somewhere in the tropics where the temperature is above 40°C, absorption of the light dispersed in space dominates over radiation, and only intensive evaporation from the skin stabilizes the temperature of body. This is especially true for optical and, particularly, ultraviolet rays of direct solar radiation. Absorption of such rays heats up the surface and, more importantly, induces a complex chain of photochemical reactions. You see the end result as a sunburn.

The point here is that deactivation of a molecule excited by light does not always lead to transformation of the acquired quantum to heat. The essence of photosynthesis taking place in green leaves is that the major fraction of energy taken from light is used for ionization and charge separation followed by oxidation and reduction. Finally the light energy is converted into chemical energy of the newly formed organic molecules. The accumulation of light energy in an organic substance is a very complex process, but owing only to this process the very existence of flora, and all the rest of life on the Earth, became possible.

3

CRYSTALS

3.1 PHENOMENOLOGY

While the fundamental property of gases is their ability to expand without limit, solids tend to hold their volume and even shape, regardless of the presence of external actions. Such contrasting properties point to essential differences in the molecular structure of these phases. Even ancient followers of atomic theory could infer from mere observation that solids are a collection of atoms firmly bonded to one other.

Indeed, atoms of a solid are always tightly bound together, while those of a gas practically do not interact with each other, spending most of the time in free path. However, strong interaction is not a specific feature of solids; it is the common property of all condensed phases, including liquids. Still, this property is particularly pronounced in crystals. Actually, in the ideal approximation we can neglect the heat motion of atoms, restricting ourselves to their static interaction. In this sense, ideal crystals are the exact reverse of ideal gases where one could confine the consideration to atomic motion, taking into account the kinetic energy, but ignoring the potential one. To clarify this dichotomy, we call attention to the fact that in a liquid heat motion cannot be neglected, even in the zero approximation, despite tight coupling of particles. Heat motion is as fundamental to the existence of liquids as it is to that of gases. Thus the potential energy of bonding in crystals is not just high, but very high; to be more exact, it is much greater than the kinetic energy of heat motion, while in liquids these quantities are comparable.

The necessity of taking the interaction force into account from the outset makes the problem much more complicated. We have already seen that even in a gas the potential energy can be taken into consideration only using a very rough interaction model. Specific features of the interaction between molecules should be carefully allowed for, otherwise it is impossible to obtain a real equation of state with reasonable accuracy.

The variety of atoms and molecules is in itself responsible for the existence of intercenter interaction types differing essentially in strength and in kind. The situation is even more complicated by the fact that in the course of crystal formation original elements often change their state: atoms turn into ions, molecules dissociate into atoms. Thus any attempt to develop a universal model describing the properties of ideal crystals will be a failure; moreover, it will be not justified. First, it is necessary to classify all crystals into groups. Then, to each group should be associated its own specific model describing static properties of solids (their bonding energy, structure, etc.). However, since the heat motion and kinetic effects in crystals are less sensitive to the types of bonding, we can restrict ourselves to a very rough but necessary classification.

Types of Crystals

The so-called "ionic crystals" are most simple in structure. Two different elements, a typical electron donor (for example, a basic metal atom) and a typical acceptor (any halogen) participate in their formation on more or less equal terms. When they come into contact, the electron moves from the donor to the acceptor, and both assume opposite charges. So the resulting bond is primarily of electrostatic origin. The strongest interaction is observed between the ion and its closest neighbors. The greater the distance, the less the interaction. The total energy is obtained by summing over all ions, which are arranged regularly $(+-+-+-)$. From the electrostatic standpoint, the closer the sites of this sequence to each other, the greater the gain in energy. However, if the proximity becomes too close, the repulsive forces generated when electron shells of ions come into contact begin to act. The compromise is reached at a definite distance a between the adjacent ions which is called the lattice period. At this distance attraction and repulsion balance one another, and the potential energy minimum is attained.

In another large class of compounds, metals and valence crystals, bonding is by valence electrons on a collective basis. For example, consider a crystal of pure sodium. Each of its atoms has one valence electron, thus there is no reason for the formation of a mosaic of opposite charges: all atoms of the crystal are equivalent. However, by virtue of this equivalence and the close neighborhood of atoms, the electron becomes the collective property of all positive ions of the lattice. At a rough approximation, all the valence electrons of metal may be treated as a negative liquid poured into the positive frame, thus cementing it. X-ray studies of structure show that the actual distribution of the electron density is such that interatomic spacings contain compactions of electron density which play the same role as negative ions in ionic crystals.

Valence crystals differ from metals in that they are formed of atoms with an even number of electrons, and each electron participates in the formation of chemical (covalent) bond. One electron of the coupled atoms is involved in each bond. In a single-valent bonding, this would cause a crystal to split into pairs of atoms, which obviously contradicts the high symmetry of crystals. However, if all adjacent atoms are held together by covalent bonds, and the same for their neighbors, then a high-strength crystal is formed. For example, this is how the lattice of a diamond—one of the crystal phases of carbon—is structured. Four valence electrons of carbon are bonded to four neighbors arranged in a tetrahedron.

The above types of bonding are typical for atomic crystals. Molecular compounds condense to give molecular crystals, either monoatomic (helium, argon) or such as solid nitrogen, oxygen, ice, sugar. As the polyatomic substances condense into crystals, molecules retain their specific features due to the strong intramolecular bond (as compared to the intermolecular, crystal one). As all bonds are saturated, the intermolecular interaction responsible for crystal phase formation is, as a rule, of a van der Waals nature. In organic crystals additional bonds are sometimes observed. This is because the hydrogen atoms contained in them participate in intramolecular bonding and become partly ionized. Their electrostatic interaction with negatively charged centers (most commonly oxygen), incorporated into the neighboring molecules, establishes the so-called "hydrogen bonding."

Strength

In terms of bond strength, valence crystals prove to be the most stiff. Next are metal, ion and molecular crystals. This classification is not absolutely strict: sometimes it is difficult to distinguish between covalent and ionic, van der Waals and hydrogen bonds. Moreover, electrons of inner shells of atoms are also involved in bond formation.

However, some properties are common to all crystals. Any extension or compression of a crystal leads to an increase of bonding energy and, therefore, gives rise to stresses which oppose applied external forces. This is responsible for the well-known mechanical properties of crystals: elasticity, hardness, and so on. As a rule, external forces are less than intermolecular ones, thus they result not in disruption of the structure, but only in deformations and shifts insignificant compared with the lattice period. However, a load comparable with the bonding forces can bring about a flow (macroscopic shifts of the adjacent layers), and then rupture. In fact breakage is usually observed at considerably lower loads than one would predict from the known bonding energy. This is due to stacking faults which always arise in real crystals: some atom is missing from its intended site, or another atom is wrongly embedded between sites, some atoms of the basic substance are replaced by impurity ones, and so on. These defects and their accumulations (dislocations) decrease essentially the strength of solids and play an important role in a great number of transfer processes.

Anisotropy

When considering mechanical properties of crystals, we run into one more typical peculiarity of the crystal phase—anisotropy. In crystals, loads are transmitted differently in different directions. That is why we distinguish tension deformation and shearing, length and width alteration (instead of the change in the volume). This means that generally neither pressure nor volume can be treated as universal characteristics of solids. Since different directions in crystals are not equivalent, it is necessary to specify how the load is divided: what force is exerted along the length, what strain along the width, and so on. Accordingly, the response to this force—lengthening, broadening—is different. The complex relation between the applied load and crystal deformation is the analog of gas isotherm. In order to describe spatial anisotropy of stresses the special tensor formalism is used. To avoid the necessity of applying it, we shall not consider the elastic response of solids in detail, and restrict ourselves to highly symmetrical, cubic lattices (Fig. 3.1). By considering compression or tension applied from all sides, one can avoid complications related to anisotropy, and study the heat motions and equation of state of solids by way of the simplest example.

For the sake of simplicity we will address primarily a simple cubic lattice. In this lattice each atom in the center of a cubic cell has six nearest neighbors at equal distance a. For comparison, the cell of a face-centered lattice is a dodecahedron formed by planes bisecting the distance between a molecule located at the origin and its twelve nearest neighbors (Fig. 3.2). This simple lattice may be represented as an ensemble of linear chains parallel to the Cartesian axes with atoms equally spaced, reducing the problem to a one-dimensional one. Although not necessary, this is a very useful simplification that we will often employ later.

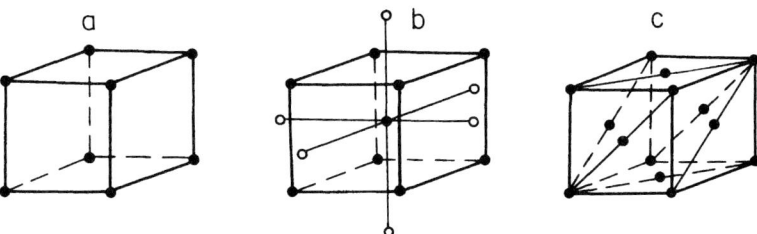

Figure 3.1 Cubic lattices: (a) simple, (b) body-centered and (c) face-centered.

154 CRYSTALS

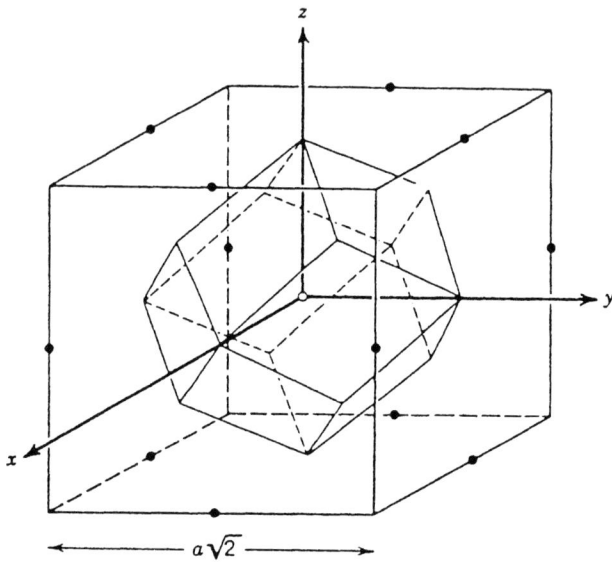

Figure 3.2 The dodecahedral cell of a face-centered lattice. (From R. J. Buehler, R. H. Wentorf, J. O. Hirschfelger, C. F. Curtiss, *J. Chem. Phys.* **19**, 61 (1951).)

3.2 CRYSTAL LATTICE MOTION

Interaction Forces

The atoms of a solid are so tightly bound together that the energy of heat motion does not affect noticeably their regular arrangement. It can induce only slight displacements from the equilibrium position insignificant compared with the lattice period a. The force resulting from the displacement counteracts the heat motion and tries to bring the atom back to its original position. It is one of the most important characteristics of crystals. Obviously, this restoring force is determined by the interaction potential of the particles of the crystal, so, generally speaking, it depends on its specific features. Nevertheless, a linear relation between the force and displacements (as long as they are small) is the general rule for all crystals.

The nature of this dependence can easily be clarified if we consider a linear sequence of atoms and assume that each of them interacts solely with the two nearest neighbors. This assumption simplifies the actual physical situation where more distant lattice sites are also involved in the intermolecular interaction. However, valence and molecular crystals are characterized by such a sharp decrease in interaction with distance, that the approximation taking only the nearest neighbors into account proves to be quite satisfactory.* Each atom of the sequence is bonded to its neighbors, which in turn are linked to their

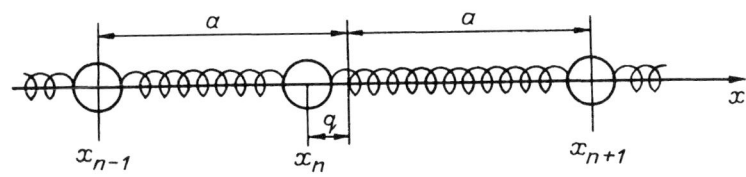

Figure 3.3 The model of an atomic chain with the lattice period a built from balls and springs.

neighbors, thus, none is free of even the farthest removed atom of the sequence (Fig. 3.3).

A qualitative notion of the potential in which the atoms of such a sequence move can be obtained by combining potential curves characterizing the interaction of the atom with its neighbors on the right and on the left: u_+ and u_-. Each curve is similar to that depicted in Fig. 1.12, and near the minimum at $r = a_0$ may be represented as

$$u(r) = u_0 + \frac{\beta_0}{2!}(r - a_0)^2 - \frac{g_0}{3!}(r - a)^3 + \ldots \quad (3.2.1a)$$

Their sum is the cell potential

$$\Phi(x_n) = u_- + u_+ = u(x_n - x_{n-1}) + u(x_{n+1} - x_n), \quad (3.2.1b)$$

symmetric about the displacement from the middle position between neighbors (Fig. 3.4). The restoring force exerted on the nth atom is

$$F_n = -\frac{d\Phi}{dx_n} = -f(x_n - x_{n-1}) + f(x_{n+1} - x_n) = f_- + f_+. \quad (3.2.2)$$

It is the sum of forces acting from the left and from the right. Each of them is the same derivative of the pair potential

$$f(r) = \frac{du}{dr} \quad (3.2.2a)$$

plotted in Fig. 3.5. Using expansion (3.2.1a), we can easily establish that near the lattice site,

---

*No cubic lattice is stable in the short-range approximation. Simple and body-centered lattices can be crumpled without any resistance, as there is a way of doing this without changing the distance between the nearest neighbors. The only exception is the face-centered lattice, that is, the closely packed structure which is typical for crysals bound by the Lennard–Jones short-range interaction (1.10.30).

156 CRYSTALS

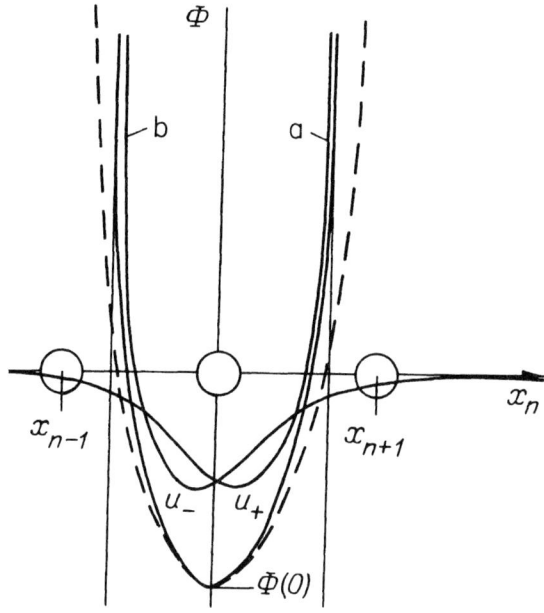

Figure 3.4 The cell potential in a linear atomic chain $\Phi(x_n)$ which is the sum of the interatomic interaction with nearest neighbors from the right (a) and from the left (b). The harmonic approximation of the potential is shown by the dashed line.

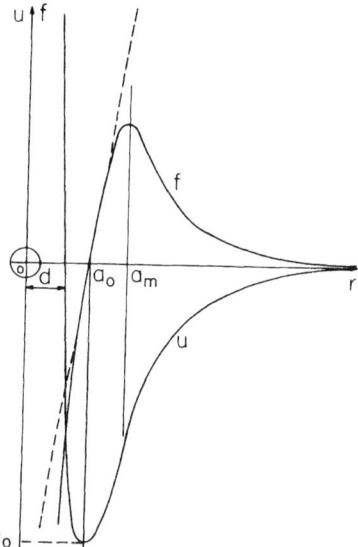

Figure 3.5 The intermolecular potential u, the force $f = du/dr$ and the harmonic approximation to it (dashed line) as functions of the particle separation r (d, a_0, a_m are coordinates of zero, minimum and inflection point of the potential).

$$f(r) \approx \beta_0 (r - a_0) - \frac{g_0}{2} (r - a_0)^2 \approx e\Delta a + \beta (r - a) - \frac{1}{2} g (r - a)^2, \quad (3.2.2b)$$

where $\Delta a = a - a_0$, and

$$e = \beta_0 - \tfrac{1}{2} g_0 \Delta a, \quad \beta = \beta_0 - g_0 \Delta a, \quad g = g_0. \quad (3.2.2c)$$

Generally speaking, $\Delta a \neq 0$, since the lattice can be either in a compressed or in a stretched state. The permanent force $f(a) = e\Delta$ is equal to zero only in the *tensionless state*, when $\Delta a = 0$, and the minima of $u_\pm$ coincide with the minimum Φ.

Even if the minima u_- and u_+ do not coincide, at small $\Delta a/a$ the sum of potentials has a parabolic shape (Fig. 3.4, dashed line). Neglecting quadratic terms in (3.2.2b), the forces acting on the atom from different sides are as follows

$$\begin{aligned} f_- &= -e\Delta a - \beta (x_n - x_{n-1} - a), \\ f_+ &= e\Delta a + \beta (x_{n+1} - x_n - a). \end{aligned} \quad (3.2.3)$$

When combined, they give the force

$$F_n = f_- + f_+ = \beta (x_{n+1} + x_{n-1} - 2x_n) = -2\beta q, \quad (3.2.3a)$$

where stresses caused by the lattice deformation ($\pm e\Delta a$) cancel each other, and the resultant stress is linear in the deflection of the nth atom from the central position between its neighbors: $q = x_n - (x_{n+1} - x_{n-1})/2$. In this sense, the force (3.2.3a) is analogous to (1.11.7) acting in diatomic molecule. The only difference is that it is twice as large owing to the combined contributions of both partners.

It is obvious that slight displacements of atoms in a solid should resemble harmonic vibrations around the equilibrium positions. At the same time, these vibrations are bound to be correlated, since all displacements of atoms in a crystal are related to each other. This is seen from the fact that the force (3.2.3a) depends both on the individual coordinates of the atom and on the coordinates of its neighbors. As isolated links of one chain, in the model under consideration, atoms are bonded solely to the nearest neighbors. However, oscillation of one atom immediately causes all other atoms to vibrate as well. It will be more correct to say that the entire crystal vibrates as a unit, not each atom separately.

Acoustic Vibrations

Mathematically, this is described in following way. The N Newton equations for N atoms of the sequence form a set of equations $m\ddot{x}_n = F_n$ where F_n is defined in (3.2.3a). Therefore,

$$m\ddot{x}_n = \beta(x_{n+1} + x_{n-1} - 2x_n). \tag{3.2.4}$$

It is natural to search for the solution of this set of equations in the form

$$x_n = \bar{x} + \Delta x_n = A \sin[2\pi \nu t - 2\pi kan + \delta] + \bar{x}_n, \tag{3.2.5}$$

where $\bar{x}_n = an$ is the coordinate of the nth site of the chain, and Δx_n is the displacement of the atom from this site. The parameters A and δ have the meaning of the amplitude and phase of the longitudinal mode, ν its frequency, and k the wave number. Direct substitution shows that this solution satisfies (3.2.4), provided that the frequency is related to the mode vector as

$$\nu = \pm \frac{1}{\pi} \sqrt{\frac{\beta}{m}} \sin(\pi ka). \tag{3.2.6}$$

Since the frequency cannot be negative, this relation means that the displacement waves can propagate both in the positive and negative directions, differing in the sign of k. Considering the dependence $\nu(k)$ (Fig. 3.6), one can also note that the wave numbers of all independent solutions of (3.2.5) lie within the interval $(-1/2a, 1/2a)$. Waves with vectors k falling outside this interval duplicate the solutions found: for each $k' = k + ia$, the solution is identical to that for k. So it is quite sufficient to restrict ourselves to the domain $-1/2a \leq k \leq +1/2a$. The frequency of vibrations increases monotonically with $|k|$ over the entire domain (Fig. 3.7). At $|k| = 1/2a$, it is at a maximum equal to

$$\nu_m = \frac{1}{\pi} \sqrt{\frac{\beta}{m}}. \tag{3.2.7}$$

Thus atoms bound in a chain can vibrate with various frequencies between zero and ν_m, while vibrations at higher frequencies are impossible. This is because the force acting on an atom changes sign too rapidly. Having no time to

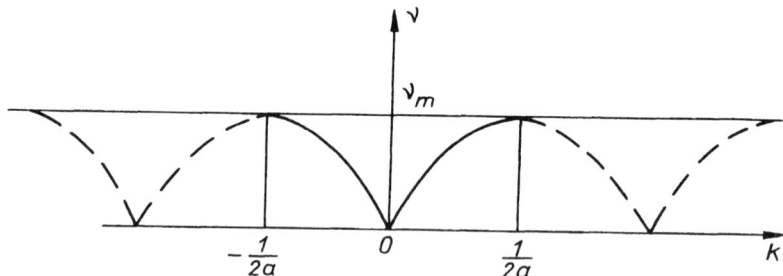

Figure 3.6 Frequency dispersion curve of linear chain vibrations for a lattice of period a.

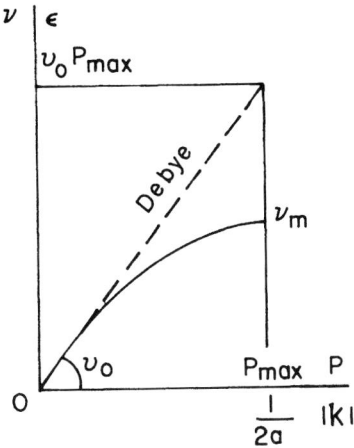

Figure 3.7 Acoustic branch of the dispersion curve of linear chain vibrations and its Debye approximation (dashed line) plotted in wave and corpuscular variables.

respond to its variations at high frequencies the atom senses only the action averaged in time, which is obviously equal to zero. This can be illustrated by an analogy. A man obeying the command "turn right"–"turn left"–"turn right" – "turn left," and so on, is unable to comply if orders are issued too rapidly.

The reverse is also true: electromagnetic vibrations of any frequency are possible due to the continuous structure of vacuum, in contrast to solids where vibrating particles are separated by finite spatial intervals. This is easily seen if we hold the linear density $\rho = m/a$ =const, and let m, and, therefore, a tend to zero. This hypothetical passage to smaller and nearer particles will eventually lead us to the ideal model of a continuous chain. Its mechanical properties were studied thoroughly long before the advent of the atomic concept. As is the case with a vacuum, the vibrational spectrum of such a chain is not limited: $\nu_m = \lim_{m \to 0} \sqrt{\beta/m\pi^2} = \infty$, and the range of linear dependence $\nu(k)$ extends to infinity

$$\nu = \nu_0 \cdot k, \quad 0 \leq k \leq \infty. \tag{3.2.8}$$

Here

$$\nu_0 = a\sqrt{\frac{\beta}{m}} = \pi a \nu_m \tag{3.2.9}$$

is nothing but the sound velocity replacing c in relation (2.3.1). This quantity remains finite with $a \to 0$, since the decrease in a increases the steepness of the potential $\beta \sim 1/a \to \infty$. However, taking into account that any chain is an atomic structure with a finite lattice period, the range validity of the continuous model should be bounded by inequalities $\nu \ll \nu_m$ or $|k| \ll 1/2a$ where the dependence $\nu(k)$ is actually close to linear. Beyond this range the sound velocity $v_0 = d\nu/dk$ decreases gradually, and goes to zero at $\nu = \nu_m$ (see Fig. 3.7).

Thus in this respect the analogy between the vibrations of the lattice and those of the vacuum is severely limited. Whereas the dependence (2.3.1) is absolutely rigorous, its crystal analogue (3.2.8) is valid only approximately and in a finite frequency range. This appears to be the most radical difference between the material medium as a carrier of mechanical vibrations and the vacuum as a carrier of electromagnetic modes. To run ahead a little, it should be noted that within the corpuscular model of the heat motion of solids introduced by analogy with light, this distinction takes the form of the nonlinear dependence $\epsilon(p)$ relating the energy of a phonon $\epsilon = h\nu$ to its momentum $p = hk$. In any case, this dependence is by no means universal. Each crystal, and even vibrations of different polarization, are associated with their own specific form of the above dependence.

The sound velocity in solids is $\sim 3 \cdot 10^5$ cm/sec and can be roughly estimated with the use of (3.2.8) assuming $\nu_m = v_0 k_m = v_0/2a = 3 \cdot 10^5/2 \cdot 10^{-8} \approx 10^{13}$ Hz. So the limiting frequency of mechanical vibrations in solids falls within the infrared range. This means that in studies of acoustic properties of crystals, atomic structure can well be ignored, since in the scale of Fig. 3.7 the whole vibrational range which is of practical interest, including the ultrasonic range, $0 < \nu < 10^9$ Hz, is extremely close to zero frequency where the deviations from (3.2.8) are negligible. However, in the high temperature range, heat properties of solids are strongly affected by the cutoff of the vibrational spectrum, and the continuous medium approximation is inapplicable.

Density of States

Since the estimation of the crystal heat capacity always involves a calculation of the density of states, it will be useful to elucidate here, without abandoning the wave model, how the atomic structure of a crystal affects the number of independent vibration of its lattice. Note that any modes of the family (3.2.5) are possible in the unbounded medium only. However, in any real crystal of finite extent of linear size L, special restrictions are imposed on the displacements of extreme atoms. For example, if the ends of a crystal are fixed (clamped on both sides), then $\Delta x(0) = \Delta x(L) = 0$. If ends are free, the boundary conditions are quite different, but, in any case, the number of allowable vibrations can be calculated. The frequencies assume values of a definite discrete series, and the intermediate values are impossible. It should be emphasized that here the discrete nature of k and ν has nothing to do with quantization. This classical phenomenon is a direct consequence of the atomic structure of matter. It is explained by the fact that the degrees of freedom of any system involving a discrete number of elements can be counted.

When the ends of the chain are fixed, only standing waves are possible and, moreover, only those which have nodes rather than crests of the wave at the ends of the chain. However, any standing wave is a superposition of two identical traveling waves propagating in opposite directions. Choosing vibrations (3.2.5) as such waves, we have

$$\Delta x_n = \tfrac{1}{2}[\Delta x_n(k) - \Delta x_n(-k)] = A\cos(2\pi\nu t + \delta)\sin(2\pi kan). \quad (3.2.10)$$

Of the above standing waves, we should select only those which satisfy the boundary conditions $\Delta x_0 = 0$ and $\Delta x_n = 0$. To meet the first condition, we need only choose the sign $(-)$ when combining traveling waves in (3.2.10), just as has been done. The second condition is less trivial. It can be satisfied only if $\sin(2\pi kNa) = 0$, that is, $2\pi kNa = \ell\pi$ or

$$k = \frac{\ell}{2Na} = \frac{\ell}{2L}, \quad (3.2.11)$$

where ℓ is an integer. Since for standing waves k is positive and varies within the range $0 \le k \le 1/2a$, ℓ takes successively all values from zero to N: $\ell = 0, 1, 2, ..., N$. The first and the last values represent the quiescent state, while all the other ℓs number distinct independent vibrations of the system. Their total number $N - 1$ is equal to the number of atoms in the chain, minus the extreme, fixed ones. This is the total number of the degrees of freedom of a one-dimensional chain, one degree of freedom per atom. This number does not depend on whether vibrations are individual or coordinated. Rather than independent vibrations of different atoms, we observe the same number of different oscillations of the chain as a unit. Since, according to (3.2.11), all permitted states of motion are uniformly distributed along the k axis, the interval dk can include $dk/(1/2L)$ different standing waves.

If the ends of the chain are not fixed, the traveling waves are possible. The domain of k increases by two owing to the inclusion of negative k, but the number of independent vibrations remains the same, and equal to the number of atoms in the chain. Accordingly, the density of traveling waves in k-space is half as great as that of standing waves: there are $dk/(1/L)$ different modes in the interval dk.

In a three-dimensional crystal the number of different lattice vibrations is found by generalizing the last result which is valid for each of the axes. All three axes of k-space are divided into equal intervals of length $1/L$, so the permitted values form a so-called "inverse lattice": with period $1/L$. Consequently, the space element $d\mathbf{k}$ contains $dk_x dk_y dk_z / L^{-3} = L^3 d^3\mathbf{k} = 4\pi k^2 dk L^3$ sites of this lattice: all permitted vibrations of the three-dimensional structure. It is clear that the number of such vibrations per unit volume is simply $d\mathbf{k} = 4\pi k^2 dk$. Thus the density of states depends neither on the crystal size nor on the lattice period. The crystal density, given by

$$dG = 3 \cdot 4\pi k^2 dk = 3 d\mathbf{k}, \quad (3.2.12)$$

differs from the density of light waves propagating in a continuous medium only by a numerical multiplier which allows for the number of independent polarizations compatible with a given k. In a crystal it is equal to 3 instead of 2, since, along with the two vibrations transverse to k, the lattice is also capable of longitudinal oscillation with the same k. Since vibrations of this kind are not inherent in an electromagnetic field, they were not taken into account in (2.3.2).

Optical Vibrations

If crystals are made up of atoms of different kinds, the variety of vibrations is not restricted to those mentioned above. Oscillations of another type are associated with the displacements of different atoms with respect to one another, and resemble the vibrations of free diatomic molecules.

To gain an impression of this type of motion, let us consider a one-dimensional chain of alternating atoms of two types which reproduces the characteristic structure of ions crystals. All even atoms are described by (3.2.4), while odd ones by the same set of equations but with different mass M:

$$m\ddot{x}_{2n} = \beta \left[x_{2n+1} + x_{2n-1} - 2x_{2n} \right],$$

$$M\ddot{x}_{2n-1} = \beta \left[x_{2n} + x_{2n-2} - 2x_{2n-1} \right]. \quad (3.2.13)$$

Since the masses are different in the trial solution

$$\Delta x_{2n} = A \sin \left[2\pi \nu t + 2\pi k a \cdot 2n + \delta \right],$$

$$\Delta x_{2n-1} = B \sin \left[2\pi \nu t + 2\pi k a (2n - 1) + \delta \right] \quad (3.2.14)$$

the amplitudes A and B can also differ; however, k, ν and δ should be the same. As a lattice of the AB type is a single interacting lattice and not two independent lattices A and B embedded into one another, both types of particles are involved in coordinated oscillations. For the same reason, the amplitudes A and B are not independent, although different. Each wave has only one arbitrary parameter—its intensity—specified by one of the amplitudes. One amplitude is uniquely expressed in terms of the other one: the stronger the vibrations of A particles, the stronger those of B, if they are involved in one process.

This becomes evident upon substitution of (3.2.14) into (3.2.13). It is seen that satisfying the equalities

$$-4\pi^2 \nu^2 mA = 2\beta[B\cos(2\pi ka) - A],$$

$$-4\pi^2 \nu^2 MB = 2\beta[A\cos(2\pi ka) - B] \quad (3.2.15)$$

is the obligatory condition for the existence of a solution in the form (3.2.14). However, set (3.2.15) has nonzero solutions only if

$$\left| \begin{matrix} \beta - 2\pi^2\nu^2 m & -\beta\cos(2\pi ka) \\ -\beta\cos(2\pi ka) & \beta - 2\pi^2\nu^2 M \end{matrix} \right| = 0 \qquad (3.2.16)$$

or

$$\nu^2 = \frac{\beta}{4\pi^2}\left(\frac{1}{m} + \frac{1}{M}\right) \pm \frac{\beta}{4\pi^2}\sqrt{\left(\frac{1}{m} + \frac{1}{M}\right)^2 - \frac{4\sin^2(2\pi ka)}{mM}}. \qquad (3.2.17)$$

The appearance of two signs in this expression points to the existence of two types of vibrations. At small k, we get from (3.2.17)

$$\text{(a) } \nu = a\sqrt{\frac{2\beta}{m+M}} \cdot k \quad \text{and} \quad \text{(b) } \nu = \sqrt{\frac{\beta}{2\pi^2}\left(\frac{1}{m} + \frac{1}{M}\right)} \qquad (3.2.18)$$

and, accordingly, (3.2.15) yields

$$\text{(a) } A = B \quad \text{and} \quad \text{(b) } \frac{A}{B} = -\frac{M}{m}, \qquad (3.2.19)$$

where (a) corresponds to the lower branch of the vibrational spectrum (Fig. 3.8), while (b) to the upper one. The lower branch represents acoustic vibrations, similar to those considered above. As before, the frequency of these

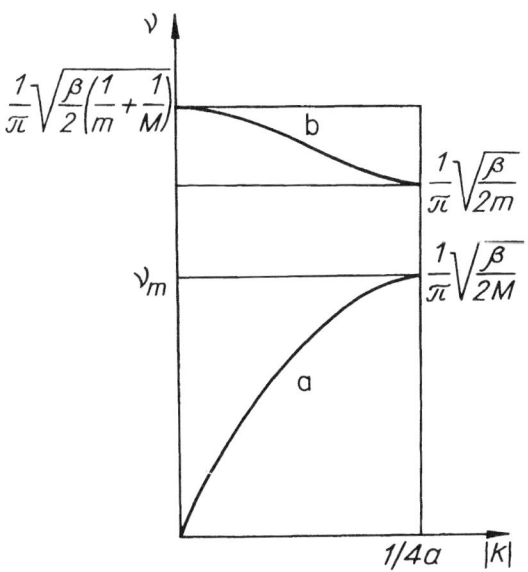

Figure 3.8 (a) Acoustic and (b) optical branches of the dispersion curve of linear chain vibrations.

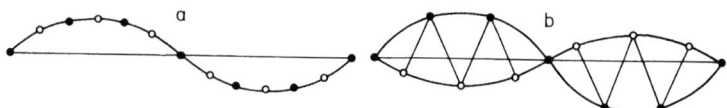

Figure 3.9 Pictorial representation of (a) acoustic and (b) optical vibrations of a two-component linear chain.

vibrations increases linearly with k at $\nu \ll \nu_m$, although their velocity $a\sqrt{2\beta/(m+M)}$ involves the average mass, taking account of the inertia of particles of both kinds. As the vibrational amplitudes are equal, we can conclude that this motion involves the crystal as a unit bending of the whole chain, not of its separate links. This fact is illustrated in Fig. 3.9a with a transverse wave as an example.

High-energy vibrations shown in Fig. 3.9b are of different nature. The fact that atoms are shifted out of phase and in inverse proportion to their masses indicates that they oscillate about the position of the common mass center. The bond length oscillates in time just as in a diatomic molecule of the same sort. Here, the specific features of the crystal structure manifest themselves only in that the deformation of bonds is synchronized, and proceeds simultaneously in all atoms of the chain. It is maximal in pairs near the wave crest, and is almost zero for those in the neighborhood of the node. The frequencies of such vibrations are within the infrared range, and this is why they are called *optical*.

As k increases, the energy gap between the acoustic and optical branches decreases, but does not disappear completely, since even at $k = k_{\max} = 1/4a$,

$$\text{(a)}\ \nu_m = \sqrt{\frac{\beta}{2\pi^2 M}}, \quad \text{and} \quad \text{(b)}\ \nu_m = \sqrt{\frac{\beta}{2\pi^2 m}} \quad (M > m). \qquad (3.2.20)$$

It is important that as k varies from zero to $k_{\max}$, the frequency of optical oscillations decreases only by a factor of $(1 + m/M)^{1/2}$ instead of changing by many orders, as in the case of the acoustic branch. Thus in a rough approximation the frequency dispersion determined by the crystal structure can be neglected. In this case, the specific features of the crystal as a unit disappear. Now the relative displacements of atoms can be treated both as coordinated and as independent, similar to vibrations of N diatomic molecules. In the acoustic frequency range the collective character of oscillations is the essential and unavoidable fact which determines both the mechanical and thermal properties of the lattice.

Molecular Vibrations

Apart from acoustic vibrations, molecular crystals can also exhibit intramolecular oscillations. It turns out that these vibrations also belong to the entire crystal rather than to separate molecules. Having arisen in one molecule, a vibration is transferred to its nearest neighbors, and successively to the whole crystal. The nature of this phenomenon is somewhat different from that of collective motions of the lattice discussed above. The transfer of vibration from one pendulum to another can serve as an illustration. If the pendulums are in resonance, a slight interaction is sufficient to transfer energy from one to another. It is exactly this resonance transfer that is responsible for the propagation of intramolecular excitations throughout the crystal.

Along with vibrations, a molecule as an entire unit or a separate group of its constituent atoms can also execute rotational motions. However, as a rule, close packing of particles in a crystal structure prevents free rotation due to steric difficulties. In this case, limited rotation (libration) is observed, in which a molecule or a group of atoms exhibits partial rotation in both directions within the limits of a backlash. Such librations are typical for molecular crystals and polymers.

3.3 PHONON GAS

Model

Acoustic vibrations of the lattice are the only type of thermal motion common to all crystals. These are longitudinal and transverse waves of different frequency and intensity which propagate with the velocity of sound in all directions, reflecting from the crystal boundaries. At low temperatures, when optical oscillations are frozen out, this is the only possible type of motion.

An analogy with radiation suggests itself. The fact that the material structure of a crystal and not a vacuum is the agent transferring acoustic vibrations affects only their spectrum, which is bounded above by the limiting frequency ν_m. Nevertheless, we can consider harmonic vibrations of a crystal in the same way as any other oscillations, be they vibrations of molecules or the vacuum. Oscillations of the crystal lattice, just as any other, should be quantized. Since this procedure has been performed more than once, now we can just agree that they may be treated as quantum oscillators with all their peculiarities, including zero energy.

Without going into details, we call the readers' attention to the fact that, according to (1.13.7), each quantized vibration with frequency ν has the energy

$$E = \left(N + \frac{1}{2}\right) h\nu = N\epsilon + \frac{\epsilon}{2}. \tag{3.3.1}$$

It can be replaced by an equivalent number of particles N with energy equal to $\epsilon = h\nu$. As the vibration spectrum is wide ($0 \leq \nu \leq \nu_m$), these quasiparticles, called *phonons* (sound quanta), can differ significantly in energy: $0 \leq \epsilon \leq \epsilon_{max} = h\nu_m$. Thus we have a picture characteristic for a gas: the volume limited by the crystal surface contains phonons distributed in their energies ϵ.

This fact in itself is not the main reason for using the terminology and methods of a gas model. This approach becomes feasible if each phonon is associated, apart from energy, with some momentum defined as $\mathbf{p} = \hbar\mathbf{k}$ by analogy with the photon gas. As has already been established in the analysis of the one-dimensional chain, k varies within the range $-1/2a \leq k \leq 1/2a$. Since this is true for any of the crystal axes, any projection of the phonon momentum is defined in the interval $(-h/2a, h/2a)$, and its absolute value $p = (p_x^2 + p_y^2 + p_z^2)^{1/2}$ does not exceed $h\sqrt{3}/2a$.

The basic mechanical characteristic of the phonon gas, just as for any other gas, is the relation between the energy and the momentum of its particles: $\epsilon(p)$. Using the one-dimensional chain as an example, we can see that this dependence replicating $\nu(k)$ is by no means linear. According to (3.2.6),

$$\epsilon = \epsilon_{\max} \sin\left(\frac{\pi a p}{h}\right).$$

In the three-dimensional case it is even more complicated and can prove to be different for vibrations of different polarizations and directions of motion. However, at small p the relation between ϵ and p is always linear, as in (3.2.8) and (3.2.18a)

$$\epsilon = v_0 p, \quad \text{at} \quad p \ll p_{\max}, \qquad (3.3.2)$$

where v_0 is the velocity of sound. The use of phonon model leads to a quite unexpected result: thermal motion in a solid—a tightly bound collection of atoms—is qualitatively reduced to a gas phenomenon. A crystal can be treated as a reservoir for a gas of phonons having any momenta and energies within allowable limits and moving in different directions with the velocity of sound. As was seen in the preceding section, the number of different states of phonon motion per unit volume is determined by formula (3.2.12). If written in corpuscular terms, it takes the form

$$dG = 3\frac{d\mathbf{p}}{h^3} = \frac{3p^2 dp d\Omega}{h^3} = 3\frac{4\pi p^2 dp}{h^3}. \qquad (3.3.3)$$

Thus the density of states in the interval $(p, p+dp)$ in a phonon gas is three times greater than in the Maxwell one and $\frac{3}{2}$ times greater than in a photon gas. The reason is that the state of a classical particle is uniquely determined by specifying the momentum, and a photon in a similar state can exist in two modifications differing in polarization. As for phonons, they are of three types: two kinds of transverse phonons, just like photons, and longitudinal ones.

From the statistical standpoint, phonons do not differ from photons. In equilibrium conditions, the average number of phonons in the state with the given energy N is determined by formula (2.3.10), as for any oscillator. So the density of phonons with the energy $\overline{N}$ is

$$d\tilde{n} = \overline{N}dG = \frac{3}{\exp(\epsilon/kT) - 1} \cdot \frac{d\mathbf{p}}{h^3} = \frac{12\pi}{\exp(\epsilon/kT) - 1} \cdot \frac{p^2 dp}{h^3}. \quad (3.3.4)$$

This differs from the corresponding distribution of photons (2.4.3) only by the statistical weight $\frac{3}{2}$.

Debye Approximation

Despite this similarity, the distribution (3.3.4) is more difficult to use than its photon analog. The problem is that the relation between the energy and momentum, necessary for any particular calculation, is linear only when $\epsilon \ll \epsilon_{max}$. At ϵ close to ϵ_{max}, this dependence is nonlinear even in the case of a one-dimensional chain. In a real three-dimensional crystal with its own specific crystal structure and anisotropy, it becomes even more complicated. The easiest way to allow for the main peculiarity of a crystal—its finite frequency spectrum—without going into details, is to assume that the linear relation

$$\epsilon = v_0 p \quad (3.3.5)$$

remains valid up to the highest energies, while phonons with $\epsilon > v_0 p_{max}$ and, accordingly, with $p > p_{max}$, do not exist. This is the Debye approximation (see Fig. 3.7). Of course, it is not completely consistent with experiment; however, it provides qualitative and even semiquantitative agreement.

In the Debye approximation, the main feature distinguishing a phonon gas from a photon one is the limited range of its energy spectrum and momentum variation. All other properties and characteristics of the model are the same, except the difference in the velocities of phonons and photons ($v_0 \ll c$). Besides, phonons are polarized differently: both perpendicular and parallel to the direction of motion.

For three-dimensional lattice, as well as for linear, the maximum values of energy and momentum are determined by the volume density of various vibrations, that is finite and equal to the number of degrees of freedom (i.e., all possible phonon states) per unit volume of crystal. As each atom can oscillate in only three mutually perpendicular directions, for a particle density n, the number of degrees of freedom per cubic centimeter is $3n$. According to (3.3.3), this is obtained as

$$\int dG = \frac{3}{h^3} \int_0^{p_{max}} 4\pi p^2 dp = \frac{4\pi}{h^3} p_{max}^3 = 3n. \quad (3.3.6)$$

Thus in the Debye approximation

$$\epsilon_{max} = v_0 p_{max} = v_0 h \left(\frac{3n}{4\pi}\right)^{1/3}. \quad (3.3.7)$$

Though not very accurate, this estimate adequately represents the situation, and even reproduces the order of magnitude of ϵ_{max}. It is seen that phonons with arbitrarily large energies can exist only at $n \to \infty$, in the limit where a crystal becomes a continuous medium. As for real crystals, their density is not greater than 10^{23} cm^{-3}, so at $v_0 = 3 \cdot 10^5$ cm/sec the energy of acoustic phonons does not exceed $\epsilon_{max} = 5 \cdot 10^{-14}$ erg ≈ 0.03 eV.

The existence of an energy limit determined by the discrete nature of crystals is responsible for a qualitative difference in the behavior of photon and phonon gases. While at $kT \ll \epsilon_{max}$ they differ only in velocity and polarization, for $kT \approx \epsilon_{max}$ and higher, the phonon gas becomes quite different from the photon one which retains its exclusively quantum behavior at all temperatures.

It is also obvious that phonons, unlike photons, are not material particles in the ordinary sense. This is seen at least from the fact that any attempt to define a mass for the phonon by relation $p = mv_0$ leads to a meaningless result even in the Debye approximation: $\epsilon = pv_0 = mv_0^2$, instead of the well-known Einstein formula $\epsilon = mc^2$. Further, the motion of the phonon both in the horizontal and the vertical direction does not change the position of the center of mass of the oscillating sequence of atoms. In other words, unlike for a photon, the energy of a phonon moving in a gravitation field is unaffected. Thus the analogy with photons is limited, at least because of the fact that a phonon cannot be associated with a mass as a measure of energy and gravitation. Consequently, the phonon can be considered as a particle only by convention, in the framework of the model describing thermal motion in a crystal by analogy with a gas.

The Lattice Heat Capacity

In the wave description, the energy of the crystal motion is the average total energy of all types of lattice vibrations, while in the particle description, the total energy is the zero energy plus the energy of all available phonons. The energy density is

$$u = \int_0^{\epsilon_{max}} \overline{E} dG = u_0 + \int_0^{\epsilon_{max}} \epsilon d\tilde{n}, \qquad (3.3.8)$$

where, according to (3.3.1), $\overline{E} = \epsilon/2 + \epsilon \overline{N}$. The phonons energy is the only heat contribution to $\overline{E}$. The zero energy $u_0 = u(0)$ is essentially of quantum origin. It is a measure of the mechanical motion always executed by oscillators which obey quantum mechanics.

In the Debye approximation (3.3.5), distributions (3.3.3) and (3.3.4) may be expressed as

$$dG = \frac{12\pi\epsilon^2 d\epsilon}{v_0^3 h^3}, \qquad (3.3.9)$$

$$d\tilde{n} = \overline{N}dG = \frac{12\pi\epsilon^2 d\epsilon}{[\exp(\epsilon/kT) - 1]v_0^3 h^3}. \qquad (3.3.10)$$

The zero energy defined in (3.3.8) is easily calculated using (3.3.9):

$$u_0 = \frac{1}{2}\int_0^{\epsilon_{max}} \epsilon\, dG = \frac{3\pi\epsilon_{max}^4}{2v_0^3 h^3} = \frac{9}{8}k\Theta n. \qquad (3.3.11)$$

The Debye temperature appearing in the above expression,

$$\Theta = \frac{\epsilon_{max}}{k} = \frac{v_0 h}{k}\left(\frac{3n}{4\pi}\right)^{1/3}, \qquad (3.3.12)$$

is the only parameter affecting the vibrational energy density

$$u = u_0 + \frac{12\pi}{h^3 v_0^3}\int_0^{\epsilon_{max}} \frac{\epsilon^3 d\epsilon}{e^{\epsilon/kT} - 1} = \frac{9}{8}k\Theta n + 9nkT\left(\frac{T}{\Theta}\right)^3 \int_0^{\Theta/T} \frac{x^3 dx}{e^x - 1}. \qquad (3.3.13)$$

The total Debye energy of the lattice is

$$\mathcal{E}_D = uV, \qquad (3.3.14)$$

while for one mole of the substance $nV = N_0$ and

$$\frac{\mathcal{E}_D}{R\Theta} = \frac{9}{8} + 9\left(\frac{T}{\Theta}\right)^4 \int_0^{\Theta/T} \frac{x^3 dx}{e^x - 1}. \qquad (3.3.15)$$

This quantity is a universal function of the ratio T/Θ. The same is valid for molar heat capacity of the lattice

$$C_V = \left(\frac{\partial \mathcal{E}_D}{\partial T}\right)_V = \left(\frac{\partial(\mathcal{E}_D/\Theta)}{\partial(T/\Theta)}\right)_V = C_V\left(\frac{T}{\Theta}\right), \qquad (3.3.16)$$

since $\Theta = \Theta(V)$, and differentiation is taken at constant volume.

The heat capacity of crystals, like that of freely moving molecules, behaves differently at low and high temperatures. The Debye temperature plays the role of a characteristic temperature near which vibrational motion is frozen out. In view of the typical values of the velocity of sound and the density of particles in the solid state, one can estimate Θ using (3.3.12): $\Theta \approx 360$ K. Although this is a very rough estimate giving just an idea of right scale, it is quite close for metals: for example, for aluminum, it is 400 K, and for copper 320 K. For ionic crystals these temperatures are usually lower: for NaCl 280 K, for KCl 230 K, and for KBr 180 K. However, the lowest Debye temperatures are usually found for molecular crystals and solid noble gases: for Ar 93.3 K, for Kr 71.7 K, for Xe

64.0 K. It should be noted that these data are not calculated but obtained from the comparison between theory and experiment. Although deviations from calculated values are always observed, they are rather insignificant and to be expected. The Debye theory as given here not only presents the vibrational spectrum of a crystal in a rough form, but also ignores the difference between longitudinal and transverse vibrations characterized by a different propagation rate. For the above reasons and due to the possibility of anisotropy, Θ is usually a function of temperature varying within the range of 10% and is considered to be a fitting parameter in (3.3.18).

Now let us compare the states of the crystal above and below the Debye temperature. At fairly low temperatures, integration in (3.3.13) can be extended to infinity without affecting the result:

$$u = u_0 + \frac{4\pi^5 k^4}{5 v_0^3 h^3} T^4 = \frac{9}{8} k\Theta n \left[1 + \frac{8\pi^4}{15} \left(\frac{T}{\Theta}\right)^4 \right] \quad \text{at } T \ll \Theta. \quad (3.3.17)$$

Since in this case the existence of the energy limit is of no importance, the phonon component in this formula differs from its photon analog only in having v_0 instead of c and by the multiplier $3/2$ allowing for the number of polarizations. Clearly also that everywhere within this quantum range the contribution of zero vibrations is comparable with the thermal contribution (Fig. 3.10). As for the molar heat capacity, according to (3.3.16), it increases by the "Debye T^3 law"

$$C_V = \frac{12\pi^4}{5} R \left(\frac{T}{\Theta}\right)^3, \quad (3.3.18)$$

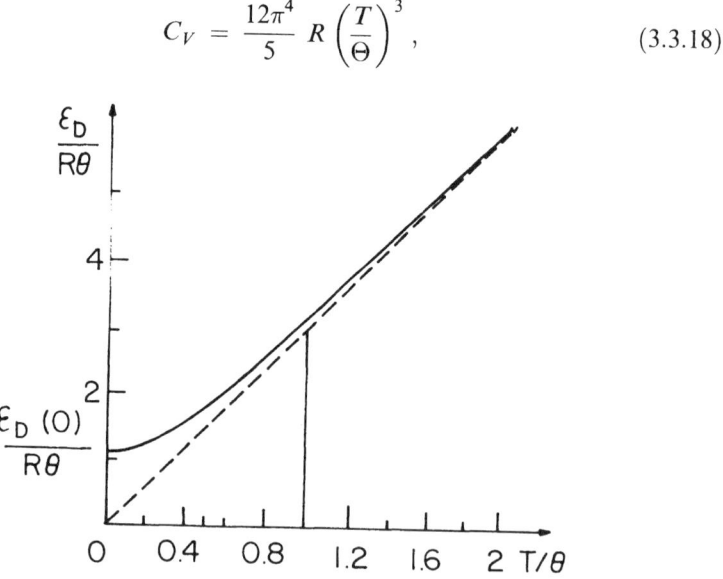

Figure 3.10 Mean energy of crystal vibrations in the Debye approximation $\mathcal{E}_D(T)$. The dashed line is the classical equipartition law of Dulong and Petit. Θ is the Debye temperature.

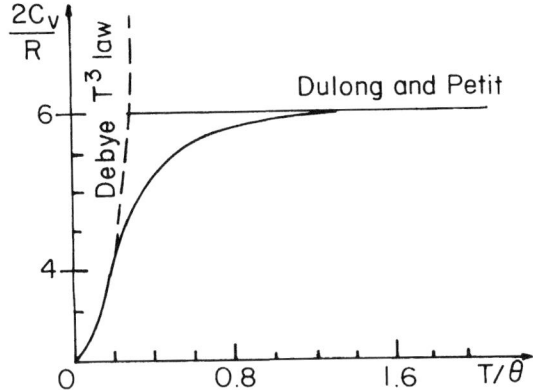

Figure 3.11 Phonon heat capacity for crystals, showing the asymptotic regimes of the Dulong–Petit and the Debye T^3 law.

that is, as rapidly as for a photon gas.

This rise is retarded abruptly as the Debye temperature is approached, and stops completely when Θ is considerably exceeded (Fig. 3.11). In the high temperature limit we can restrict ourselves to the first terms of the expansion following from Eq. (1.13.11):

$$\overline{E} = \frac{\epsilon}{2} \operatorname{cth}\left(\frac{\epsilon}{2kT}\right) = kT\left[1 + \frac{1}{12}\left(\frac{\epsilon}{kT}\right)^2 + \ldots\right].$$

Substitution into (3.3.8) yields

$$u = 3nkT\left[1 + \frac{1}{20}\left(\frac{\Theta}{T}\right)^2 + \ldots\right]. \tag{3.3.19}$$

As the temperature rises, all correction terms vanish, and we get the classical expression for the energy density following from the equipartition law: $\mathcal{E}_D = 3nkT$, with kT for each of three vibrational degrees of freedom of N atoms. In this limit

$$C_V = 3N_0 k = 3R \tag{3.3.20}$$

which is also consistent with the classical result of heat capacity theory. It is known as the Dulong–Petit law: "molar heat capacity of solids is equal to 6 cal/mole · K, independent of temperature."

No such thing occurs for a photon gas. Since for photons $\Theta = \epsilon_{\max}/k = \infty$, the first of the limiting cases considered here extends over the entire temperature range. Reaching the classical limit in the case of the lattice vibrations is a direct consequence of a limited extent of their spectrum. Seen as waves, the

172 CRYSTALS

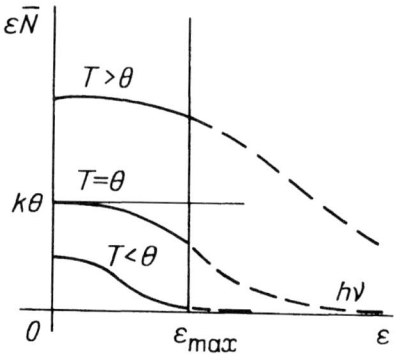

Figure 3.12 Mean energy of phonons within permitted energy interval ($0 < \epsilon < \epsilon_{max}$) at different temperatures (compare with Fig. 2.4).

lattice vibrations are of quantum nature only as long as a fraction of them is frozen out (Fig. 3.12). However, at $T \gg \Theta$, even the highest-frequency vibrations are excited and can be described by a classical theory. That is why solids obey the correspondence principle to the same extent as gases (see Section 1.13).

From the corpuscular point of view, the average energy of phonons in the low temperature range at $T \ll \Theta$ is of the order of kT, and the rapid increase of $u(T)$ in (3.3.17) is due to the increasing density of the phonon gas: $\tilde{n} \sim T^3$. This follows from the arguments leading to (2.5.9) which are valid in the given case as well. At higher temperatures ($T \gg \Theta$) the average energy of phonons tends to the limit $k\Theta$, while the phonon density continues to increase, though linearly: $\tilde{n} \sim 3nT/\Theta$, and $u \approx \bar{\epsilon}\tilde{n} = 3kTn$.

A different situation arises with the average energy of thermal motion of atoms. In contrast to photons, the number of atoms remains unchanged at any temperature: $n = $const. Their mean square velocity (as well as the velocity distribution) is correctly described by classical statistics only at temperatures above the Debye one. Below this temperature, in the quantum range, the average energy of thermal motion of atoms is considerably less than the classical value, and its temperature dependence is more sharp.

Correlation of Vibrations

If a crystal consists of atoms of two types, there is one optical oscillation for each acoustic vibration. Accordingly, the heat energy density of a two-component crystal is the sum of expression (3.3.13) and the following component due to the optical branch

$$u = 3n \cdot \bar{E} = 3n \left[\frac{h\nu_0}{\exp(h\nu_0/kT) - 1} + \frac{h\nu_0}{2} \right]. \tag{3.3.21}$$

Since the frequency of optical oscillations is of the same order of magnitude at any k, in (3.3.21) it is taken to be ν_0. In such an approximation, the heat capacity associated with optical vibrations does not differ from the heat capacity of an equal number of diatomic molecules.

It is of interest to note that this is just how the heat capacity related to acoustic vibrations was originally estimated by Einstein. It was shown that $C_V \to 0$ at $T \to 0$, which improved the result of Dulong and Petit. However, the exponential decrease was inconsistent with the T^3 law, which was reproduced only after Debye made allowance for the continuous spectrum of acoustic vibrations. Beginning at the zero frequency, this spectrum always contains vibrations which can be excited at low temperatures.

Curiously, the classical result of Dulong and Petit can be obtained both from (3.3.13) and (3.3.21), and so is independent of whether the crystal lattice motion is considered as collective or consisting of individual oscillations of atoms in their potential wells. This indicates that, since at high temperatures an atom participates simultaneously in a large number of lattice vibrations, their random overlap makes its motion so chaotic that no correlation with the motion of its neighbors can be discerned.

The excess of individual energy disrupts the orderly pattern of collective motions; moreover, under suitable conditions, it can even throw an atom from its site. This is how thermal defects (dislocations and vacancies) arise. As long as their concentration is not high, the effect they have on the crystal heat capacity may be neglected. However, in the premelting range, the number of defects increases abruptly, and the expenditure of heat on their formation (partial rupture of bonds) become significant.

Heat Conduction of the Lattice

The existence of dislocations indicates that a crystal is not ideal. However, long before this can be discerned, the difference between the real crystal and its harmonic model becomes apparent. This distinction lies in the fact that the actual restoring force acting on the crystal atoms is nonlinear in the deflection from the equilibrium position. The degree of nonlinearity is specified by the magnitude of the next-order corrections in the potential expansion. Making allowance for even the first correction is bound to cause a change in the character of the lattice vibrations. It was suggested that this correction should be considered as a weak interaction binding together various harmonic vibrations involved in the lattice spectrum. Though such an interaction has an energy which is insignificant compared with the energy of vibrations themselves, it has the result that, once excited, a vibration does not always remain the same. As time passes, the excitation is transferred from oscillator to oscillator, changing the frequency and orientation of the wave vector.

Within the particle framework, the transfer of vibrational quanta among various modes of a lattice points to the existence of the phonon–phonon interaction. The change in their energy and direction of motion can be naturally interpreted as the result of their collisions. Thus, owing to anharmonic components of the potential, the phonon gas is no longer ideal but real. Now its particles can collide with each other, which affects their paths and retards heat transfer. If the collision rate, or the mean free time τ_0, is defined, this also specifies the free path length

$$\lambda = v_0 \tau_0, \qquad (3.3.22)$$

where v_0 is the velocity of sound, as usual. Now a theory of heat conduction can be developed by analogy with that already constructed, as we can proceed from the concept of local equilibrium at each point of space within the volume element of size λ. So we can immediately write formula (1.16.3) for the heat conduction coefficient, without going into details:

$$\kappa = \tfrac{1}{3} c_V v_0^2 \tau_0 = \tfrac{1}{3} c_V v_0 \lambda, \qquad (3.3.23)$$

where c_V is the specific heat capacity of a crystal.

The distinctive features of a phonon gas manifest themselves primarily in that collisions are of a somewhat unusual character: the number of particles involved in them is not constant. The decay of one phonon into two, or, vice versa, the coalescence of two phonons into a bigger one proceed most rapidly. Anharmonic interactions responsible for phonon collisions not only control heat transfer but also restore equilibrium in the phonon gas. This is the essential difference between the phonon gas and the photon one which cannot attain equilibrium by itself. In other words, photons behave as point masses, while phonons are similar to particles of finite cross-section.

3.4 EQUATION OF STATE

Balance of Forces

Although isobaric expansion of solids over a wide temperature range is quite small, it certainly occurs. On the other hand, it is almost impossible to bring about significant changes in the volume and even shape of hard crystals by usual means. That is why it is not reasonable to consider the size of a crystal as a measure of what really happens to it.

The actual situation is as follows: as the temperature rises, the intensity of thermal motion of particles in crystals, just as in gases, increases, and the resulting motional pressure p_t increases in an even greater proportion. However, the latter is almost completely balanced out by the increased internal pressure p_i. At high temperatures, the scale of these forces may be roughly estimated by the pseudoideal gas formula $p_t \sim nkT$. One can see that motional pressure in solids is greater than that in gases by at least a factor equal to the ratio of the densities of particles per unit volume. At atmospheric pressure, a gas has a density $n \sim 10^{19}$ cm^{-3}, while the density of atoms in solids is 10^{23} cm^{-3}. Therefore, at the same temperature, their pressure is greater by a factor 10^4. In order to balance this the internal pressure p_i must be 10^4 atm as well. The atmospheric pressure compressing a crystal from the outside is insignificant in the balance of these enormous opposing forces. Even doubled or trebled, it cannot affect considerably the shape and size of a solid which responds only to pressures comparable with p_i and p_t. This is also true for other condensed media (liquids, amorphous matter, polymers). This is why, trying to quantify the state of condensed matter, it is reasonable to give most consideration to these outwardly invisible but vary large and abruptly changing forces: motional and internal pressures.

With this in mind, the equation of state is merely the balance condition which requires that the external pressure p together with internal one counterbalance the motional pressure acting on the crystal surface from the inside

$$p + p_i = p_t. \quad (3.4.1)$$

The ideal gas is an exception in the sense that there is no need to distinguish between its external and motional pressures, since $p_i = 0$ by definition. However, in a real gas, the external pressure becomes less than the motional because intermolecular forces are taken into account. The difference in the two pressures is equal to the van der Waals internal pressure $p_i = A/V^2 = an^2$ which partially dampens the force of real molecules against the walls.

Whereas in a real gas $p_i \ll p \approx p_t$, in solids the reverse situation is observed: $p \ll p_i \approx p_t$ (Fig. 3.13). The intermolecular forces are so strong that even at high temperatures they can prevent the medium in contact with the crystal from being actively bombarded by the surface particles. Thus not all impacts reach the target, and the pressure directed outward is considerably less than the motional pressure. So the balance is primarily attained between p_i and p_t, although the quantity observed is still p.

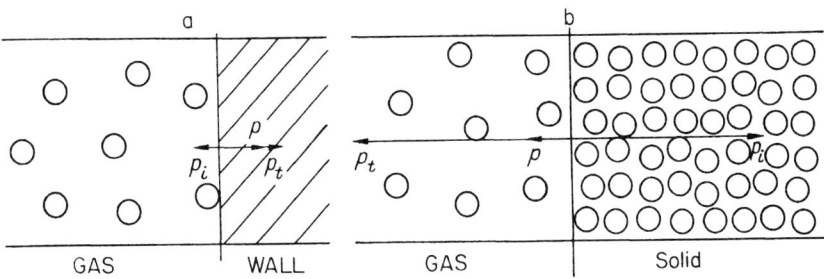

Figure 3.13 (a) Gas pressure applied to the wall and (b) the wall pressure applied to the gas (p_i and p_t are internal and motional components of the total pressure p which is the same in both cases).

This qualitative picture can be presented rigorously, provided that the components of equation (3.4.1) are well-defined. For solids this may be achieved by dividing the total lattice energy into the rest energy U_0 and the energy of motion $\mathcal{E}_D$

$$\mathcal{E} = U_0 + \mathcal{E}_D. \tag{3.4.2}$$

The first term is simply the potential energy of a crystal with its atoms fixed at the lattice sites. In the nearest-neighbor approximation,

$$U_0 = \frac{cN}{2} u(a), \tag{3.4.3}$$

where c is the number of the closest particles (in the first coordination sphere of radius a). This coincides with the first component of the potential energy (1.10.31). It is clear that $U_0(V)$ is not the total potential energy of atom in a cell, but just the constant fraction. The remaining potential energy is due to departures from regular packing caused by motion of the particles. Along with the kinetic energy, this makes up the Debye energy of motion $\mathcal{E}_D$. In the harmonic approximation, both energies contribute equally to $\mathcal{E}_D$.

The same breakdown is also appropriate for the crystal free energy

$$\mathcal{F} = \mathcal{E} - TS = U_0 + \mathcal{F}_D, \tag{3.4.4}$$

where

$$\mathcal{F}_D = \mathcal{E}_D - TS. \tag{3.4.5}$$

Using (3.4.4) in the general definition of pressure (1.10.8a), we have

$$p + \frac{dU_0}{dV} = -\left(\frac{\partial \mathcal{F}_D}{\partial V}\right)_T.$$

Identifying this result with (3.4.1) term by term, one obtains the following definitions:

$$p_i = \frac{dU_0}{dV}, \tag{3.4.6a}$$

$$p_t = -\left(\frac{\partial \mathcal{F}_D}{\partial V}\right)_T. \tag{3.4.6b}$$

Internal Pressure

Since $U_0(V)$ is made up of the energies of bonds which tie atoms (molecules) together into a solid, an internal pressure equal to dU_0/dV arises as the elastic response of the lattice to its extension or compression.

Imagine for simplicity the crystal to be a combination of parallel atomic chains. Then

$$U_0 = N\left[u_x(a_x) + u_y(a_y) + u_z(a_z)\right], \qquad (3.4.7)$$

where a_x, a_y, a_z are the periods of a simple cubic lattice along various axes. Neglecting the crystal anisotropy, $u_x = u_y = u_z = u$ and $a_x = a_y = a_z = a$, while $u(r)$ has the same meaning as in (3.2.1a). Using (3.4.7) in (3.4.6a), we have

$$p_i = N\left.\frac{du}{dr}\right|_a \cdot \left(\frac{da_x}{dV} + \frac{da_y}{dV} + \frac{da_z}{dV}\right) = 3N\left.\frac{du}{dr}\right|_a \cdot \frac{da}{dV}.$$

Taking into consideration that under isotropic deformation $V = Na_x \cdot a_y \cdot a_z = Na^3$, we find

$$p_i = \frac{1}{a^2} \cdot f(a), \qquad (3.4.8)$$

where $f(r)$ is defined in (3.2.2a).

The last result is easy to understand if one recalls that each crystal atom, even in its equilibrium position, experiences forces (3.2.3) if the lattice is compressed or extended. Even when occupying their regular lattice sites ($x_{n+1} = x_n + a = x_{n-1} + 2a$), the neighbors of the nth atom act on it with equal though opposite forces:

$$f(a) = f_+ = -f_- = e\Delta a. \qquad (3.4.9)$$

Inside a solid, these forces balance one another, as in (3.2.3a), and do not affect the crystal vibrations. Nevertheless, the lattice on the whole is under stress. This can be inferred from the state of the surface atoms. As they are acted on only from the inside, these atoms experience a force into a solid given by (3.4.9). Altogether there are $1/a^2$ atoms per unit area of the surface. Extension of the lattice period by Δa is responsible for inwardly directed pressure exerted normal to the solid surface and equal to

$$p_i = \frac{1}{a^2}f(a) = \frac{e}{a^2}\Delta a = E\frac{\Delta a}{a} = \frac{1}{\kappa}\frac{\Delta V}{V}. \qquad (3.4.10)$$

This is the internal pressure. However, this estimate is valid only in the harmonic approximation, when crystal deformation increases linearly with loading, and E is nothing but the Young modulus, while $\kappa = 3/E$ is the coefficient of isotropic compression.* The expansion (3.2.2b) shows that this is possible, as long as $\beta(a - a_0) \gg (g/2)(a - a_0)^2$, that is

$$\Delta a \ll 2\beta/g \ . \tag{3.4.11}$$

Then

$$\frac{3}{\kappa} = E = \frac{e}{a} = \frac{\beta_0}{a} - \frac{g_0 \Delta a}{2a} \approx \beta_0/a \ .$$

If condition (3.4.11) is not met it is necessary to use the more general definition $f(a)$ given in (3.2.2a).

It is obvious that $p_i(a)$ follows the dependence $f(a)$ depicted in Fig. 3.5. If p_i is complemented by external pressure, the isobaric force characteristic is merely shifted along the ordinate axis by p to form the family of curves shown in Fig. 3.14. Evidently, p can also take on negative values, since there is a possibility of extending a solid.

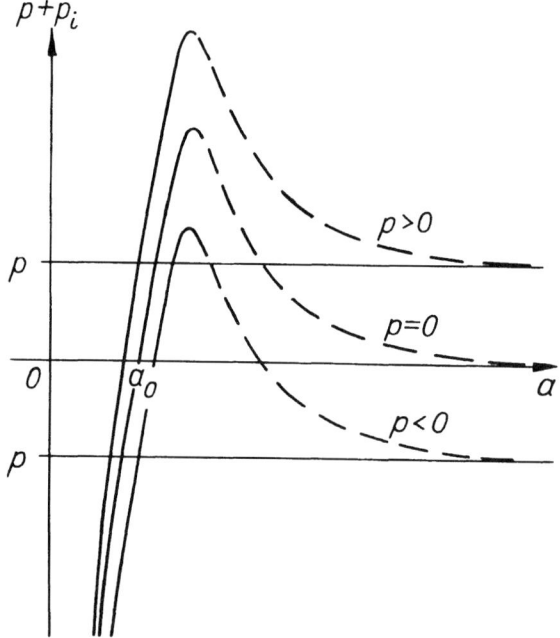

Figure 3.14 The total force $p + p_i$ acting on crystal at different external isotropic loads p ($p > 0$ is compressional, $p < 0$ is stretching) as a function of the lattice period a.

---

*Actually, $\kappa = 3(1 - 2\sigma)/E$ where σ is the Poisson coefficient defining the magnitude of the relative cross-compression of a solid under linear extension. In the linear chain model this effect is ignored.

The most essential difference between the internal pressure of solids and its van der Waals analog is that repulsive forces contribute to it along with attractive ones. If a crystal is considerably extended, attraction dominates, and the internal pressure is positive and inwardly directed, as in a real gas. However, in the case of a severe compression whereby the lattice period becomes less than a_0, the internal pressure is negative. It is directed outward, and its scale is specified by repulsive forces. The zero of internal pressure is of primary importance, for in this case the counteracting forces are at balance, and the lattice is not exposed to stress while its period is equal to a_0. The quasiharmonic approximation most commonly used in studies of solids is applicable only in the vicinity of this *tensionless state*.

Motional Pressure

In order to relate motional pressure to vibrational energy, it is necessary to use the well-known thermodynamic relation (5.5.11)

$$\mathcal{F} = \mathcal{E} + T \left(\frac{\partial \mathcal{F}}{\partial T} \right)_V,$$

which relates the free energy to the total energy. Substituting (3.4.2) and (3.4.4) into the above relation shows that the Debye fractions of $\mathcal{F}$ and $\mathcal{E}$ are connected by the same relation, identical to (5.5.12)

$$\mathcal{E}_D = \mathcal{F}_D - T \left(\frac{\partial \mathcal{F}_D}{\partial T} \right)_V = \left[\frac{\partial \left(\frac{\mathcal{F}_D}{T} \right)}{\partial \left(\frac{1}{T} \right)} \right]_V. \quad (3.4.12)$$

Note that, according to (3.3.15), the ratio $\mathcal{E}_D/\Theta$ is a universal function of the single variable Θ/T. Thus, dividing (3.4.12) by Θ and bearing in mind that $\Theta = \Theta(V)$, we can write

$$\frac{\mathcal{E}_D}{\Theta} = \frac{1}{\Theta} \left[\frac{\partial \left(\frac{\mathcal{F}_D}{T} \right)}{\partial \left(\frac{1}{T} \right)} \right]_V = \frac{d \left(\frac{\mathcal{F}_D}{T} \right)}{d \left(\frac{\Theta}{T} \right)} = \left(\frac{\partial \mathcal{F}_D}{\partial \Theta} \right)_T. \quad (3.4.13)$$

It follows from the above identities that $\mathcal{F}_D/T$ is also a the universal function of Θ/T as well as $\mathcal{E}_D/\Theta$. Thus the dependence on volume can appear in $\mathcal{F}_D$ only through the Debye temperature $\Theta(V)$ which, according to (3.3.12) and (3.2.9), is

$$\Theta = \frac{v_0 h}{k \cdot a} \left(\frac{3}{4\pi} \right)^{1/3} = \left(\frac{3}{4\pi} \right)^{1/3} \frac{h}{k} \left(\frac{\beta}{m} \right)^{1/2}. \quad (3.4.14)$$

It depends on the volume via $\beta = \beta(V)$. With this in view, the motional pressure defined in (3.4.6b) can be expressed as follows

$$p_t = -\left(\frac{\partial \mathcal{F}_D}{\partial \Theta}\right)_T \frac{d\Theta}{dV}.$$

Eliminating $(\partial \mathcal{F}/\partial \Theta)_T$ from the above equation with the use of the last expression in (3.4.13), we obtain the desired relation

$$p_t = -\frac{\mathcal{E}_D}{\Theta}\frac{d\Theta}{dV} = \gamma \cdot u, \qquad (3.4.15)$$

where $u = \mathcal{E}_D/V$. Formally, it is identical to (1.1.8) or (2.5.6). From this point of view, the Grüneisen constant

$$\gamma = \frac{d(\ln \Theta)}{d(\ln V)} \qquad (3.4.16)$$

is merely the numerical coefficient of the proportionality which relates the pressure to the energy density u.

However, on closer examination, we shall see that the situation is not so simple. Substituting the definition of the Debye temperature from (3.4.14) into (3.4.16) gives

$$\gamma = -\frac{V}{2\beta}\cdot\frac{d\beta}{dV}, \qquad (3.4.17)$$

but for isotropic deformation $dV/V = 3(da/a)$, and according to (3.2.2c)

$$\beta = \beta_0 - g_0(a - a_0). \qquad (3.4.18)$$

Thus, performing the differentiation in (3.4.17), we find

$$\gamma = -\frac{a}{6\beta}\frac{d\beta}{da} = \frac{ag_0}{6\beta} \approx \frac{a_0 g_0}{6\beta_0} = \text{const} > 0. \qquad (3.4.19)$$

This estimate of γ shows that, generally speaking, motional pressure arises due to the anharmonic nature of bonds. If there were a crystal composed of atoms obeying a purely harmonic potential ($g_0 = 0$), p_t would be equal to zero for any u since $\gamma = 0$.

If we return to the analogy between the phonon gas and a molecular one, the absence of pressure exerted on the wall seems paradoxical. It may be attributed to the fact that phonons do not have momentum in the ordinary sense. However, when the phonon gas is compared with the photon one, the paradox is naturally resolved. Photon pressure is applied to the surface of the container, but the surface of a solid should be considered absolutely transparent to an ideal phonon gas. Naturally, a surface permeable in both directions experiences no pressure, since elastic reflection is impossible. Thus in the harmonic approximation, a crystal can be treated as a transparent reservoir exposed to a phonon wind blowing in all directions and with equal strength.

In view of the preceding remarks, we can easily explain why motional pressure is absent in the harmonic approximation, but arises when anharmonic corrections to the potential are taken into account. Briefly, the solid surface is transparent to an ideal phonon gas, but impermeable, or only partially permeable to a real gas. The analysis of heat conduction shows that as soon as anharmonic corrections are allowed for, phonons become particles of finite cross-section. Anharmonicity is also responsible for motional pressure which is naturally interpreted as the pressure of real phonons on the surface that has become partly opaque to them. So a solid may be considered as a reservoir filled with phonons which move in all directions and are elastically reflected from all boundaries, thus creating motional pressure p_t (Fig. 3.15). One may say that the phonon gas exerts pressure on the crystal surface from the inside, while an ordinary gas does so from the outside.

Like the volume density of heat energy u, motional pressure depends on temperature and volume. To determine this dependence, we make use of (3.3.17), (3.3.19), (3.4.14), and find from (3.4.15)

$$p_t = \begin{cases} \gamma n \left[\frac{9}{8}\left(\frac{3}{4\pi}\right)^{1/3} h \left(\frac{\beta}{m}\right)^{1/2} + \frac{4\pi^5}{5h^3}\left(\frac{m}{\beta}\right)^{3/2}(kT)^4\right], & T \ll \Theta, \\ 3\gamma nkT, & T \gg \Theta. \end{cases} \quad (3.4.20)$$

It is clear that isochoric motional pressure increases, as shown in Fig. 3.10. However, if considered as an isothermal function of the crystal volume, it decreases in the same fashion as for an ideal gas: $p_t \sim n \sim 1/v$, provided that

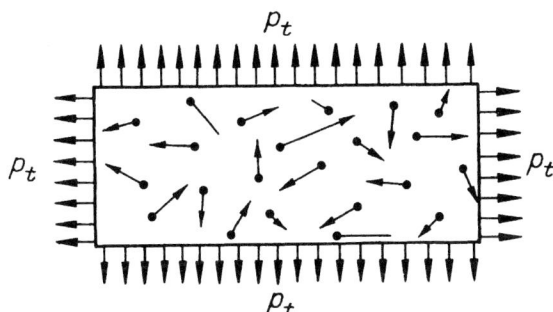

Figure 3.15 The motional pressure of phonons p_t applied to a crystal surface from inside.

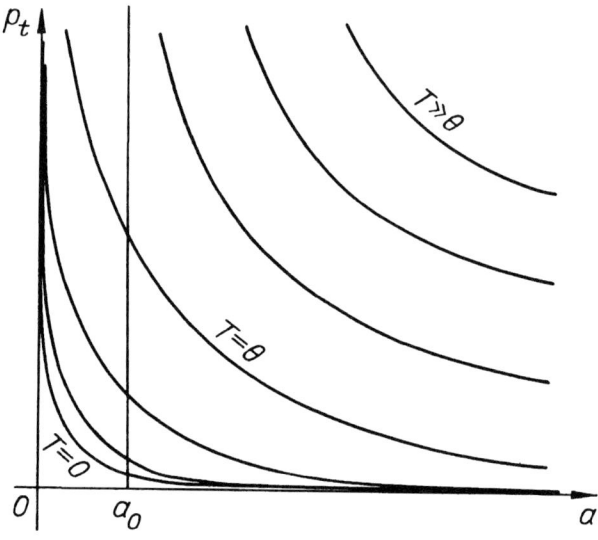

Figure 3.16 Isotherms of a phonon gas above and below the Debye temperature Θ.

the weak dependence $\beta(a)$ in (3.4.18) is neglected. That is why the isotherms $p_t(a)$ depicted in Fig. 3.16 form a family of hyperbolas, the lowest of which corresponds to zero temperature vibrations. The mere existence of zero energy which is not of statistical but dynamical origin indicates that the "pressure of motion" is only partially thermal below the Debye temperature. Only at $T \gg \Theta$, when all quantum corrections to the classical result of Dulong and Petit are negligible, p_t becomes genuinely thermal.

The Mie–Grüneisen Equation

Now we have at our disposal all the necessary tools to obtain the equation of state for solids. It remains only to substitute (3.4.6a) and (3.4.15) into (3.4.1) to obtain the desired result:

$$p + \frac{dU_0}{dV} = \gamma u. \qquad (3.4.21)$$

A classical variant of this equation was first derived by Mie for $T \gg \Theta$. Later Grüneisen extended this equation to the essentially quantum low-temperature region ($T \ll \Theta$). Let us recall its physical sense: the external pressure, together with the internal pressure (the response of the lattice to deformation), balances the motional pressure that tends to stretch the crystal from inside.

Under normal conditions the external pressure is negligible. Therefore, one can take $p = 0$ and obtain, in place of (3.4.1), $p_i = p_t$ and, in place of (3.4.21),

$$E \frac{\Delta a}{a} = \gamma u(a, T), \qquad (3.4.22)$$

where p_i, is estimated via Young's modulus as in (3.4.10). This form of the Mie–Grüneisen equation describes the isobaric heat expansion of gases:

$$\frac{\Delta L}{L} = \frac{\Delta a}{a} = \frac{\gamma}{E} u(T). \qquad (3.4.23)$$

The heat expansion is the result of a "compromise" between motional and internal pressures. As the temperature is raised, phonons of increasing number and energy cause an ever-growing surface pressure and corresponding stretching. The crystal lattice responds to this stretching as to any other attempt to increase its size. The lattice offers resistance depending on its elasticity. First, it creates an internal pressure of magnitude depending on the deformation which can, up to certain limits, oppose the motional pressure. Moreover, the elastic resistance first increases linearly, and a very slight stretching is sufficient to cope with the pressure of phonons.

These arguments can be illustrated by presenting the curves (Figs. 3.14 and 3.16) on a diagram. Such a diagram (Fig. 3.17) is rather unusual. The equilibrium states are the intersection points of phonon isotherms and the lines of total force, which is the sum of the external and internal pressures $p + p_i$. Inspecting the diagram one can make a number of conclusions of qualitative significance, which also follow directly from Eq. (3.4.21).

First, it is clear that a stable equilibrium is accomplished only if the phonon isotherms are cut by a branch of the total force with positive slope. Only then does any deviation from equilibrium give rise to a force that restores the system's previous state. If the same isotherm is cut by a negative sloping branch, then as soon as the point deviates to the right, motional pressure begins to predominate over the suppressing force, which leads to crystal decomposition. If the point deviates to the left, the total force dominates and tends to press the crystal. The process continues to develop along the isotherm until an intersection with the ascending branch is reached. There it remains in equilibrium.

In addition, for any given external pressure the total force obviously has a maximum, above which equilibrium is not possible because above there will be no intersection with an isotherm. This implies that in this range the crystal cannot withstand any loads and temperatures. For a given pressure there is a maximum temperature at which the isotherm touches the maximum of the total force. For isothermal crystal stretching the maximum load is that for which the total force is tangent to the isotherm of the process.

In both cases the limiting crystal lengthening should be about the distance which results in the maximum p_i. This lengthening is easy to estimate from the expansion (3.2.2b): setting $df/dr = 0$ at $r = a_m$ it can readily be seen that the maximum Δa is:

184 CRYSTALS

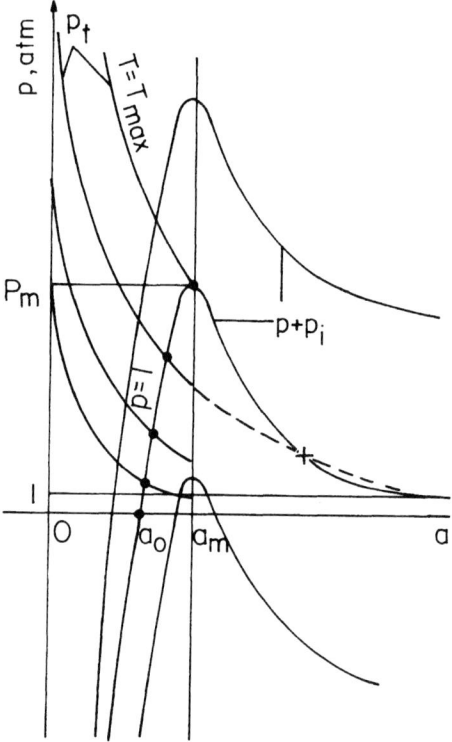

Figure 3.17 Stable (•) and unstable (+) balance of forces at cross-points of motional pressure isotherms $p_t(a)$ with the total withstanding force $p + p_i(a)$.

$$\Delta a_m = a_m - a_0 = \frac{\beta_0}{g_0} = \frac{a_0}{6\gamma} \approx \frac{a_0}{12}, \qquad (3.4.24)$$

where γ is defined as in (3.4.19). Its standard numerical value is approximately 2. Thus, one may stretch the crystal by at most an additional 8% relative to its initial size.* The required load grows with the desired stretching, but the lattice resistance increases nonlinearly, and when E becomes zero the rupture takes place. In reality the crystal begins to fail much earlier due to lattice imperfections. Still, an upper estimate of the maximum load can be obtained quite reliably using (3.4.24) in (3.4.10):

$$p_m < \frac{E}{a}\Delta a_m = \frac{E}{6\gamma}. \qquad (3.4.25)$$

---

*For the sake of clarity $\Delta a_m/a_m$ as well as $(a_0 - d)/a_0$ in the diagram of Fig. 3.17 are approximately five times as large as their real values.

This estimate is maximum also as it is valid for zero temperature, when $p_i \approx -p \gg p_t(0)$, and the lattice has to withstand only the force applied from outside. As the temperature is raised, phonons acting from inside assist in the process, and crystal destruction is achieved at a much lower cost. Taking into account that typically $E \approx 5 \cdot 10^5$ atm for metals, it is readily seen that $p_m \approx$ 40 000 atm. This explains why a solid bar undergoes considerable residual deformation and destruction upon stretching of the order of p_m, which is rather high.

From (3.4.25) it is possible to estimate roughly the temperature range within which solids can exist. Proceeding from the equality $p_m = p_t = \gamma 3nkT_{\max}$, which determines the highest available isotherm, we obtain

$$T_{\max} = \frac{E}{18\gamma^2 nk} = \frac{5 \cdot 10^{11}}{18 \cdot 4 \cdot 10^{23} \cdot 1.4 \cdot 10^{-16}} < 500 \text{ K} . \quad (3.4.26)$$

This is only an order of magnitude estimate and a particular melting point as well as p_m can be within a factor of 2–3 higher or lower.

Heat Expansion

Let us consider now in more detail what happens to a solid upon heating under normal pressures. Note, first, that crystals are already expanded at absolute zero temperature. Due to the quantum nature of lattice motion, it is still present even at $T = 0$. Such a motion provides crystal stretching from inside with pressure $p_t(0)$. The degree of stretching is easy to find from (3.4.23) using (3.3.11):

$$\frac{\Delta a}{a} = \frac{\gamma}{E} u_0 = \frac{9\gamma}{8E} nk\Theta . \quad (3.4.27)$$

For $E = 5 \times 10^5$ atm, $n = 10^{23}$ and $\Theta = 300$ K the relative extension is 1%, which is not so small when one considers that the maximum possible does not exceed 8%. From (3.4.10) it follows that for $\Delta a/a = 10^{-2}$ the lattice stands up to a pressure of about 10^3 atm. This fact justifies the neglect of atmosphere pressure, which is insignificant on this scale of force. Since stretching of the crystal is isotropic, the volume expansion is three times greater than the linear one:

$$\frac{v(0) - v_0}{v_0} = \frac{n_0 - n(0)}{n_0} = \frac{27k\Theta n_0 \gamma}{8E} = \frac{27\gamma h n_0^{2/3}}{8\sqrt{\beta m}} \left(\frac{3}{4\pi}\right)^{1/3}, \quad (3.4.28)$$

Here $v(0) = 1/n(0)$ is the specific volume at absolute zero, and $n_0 = 1/v_0$ is the density of the tensionless crystal, whose atoms are located in the minima of the pair potential $u(r)$. Provided that the crystal strength is related to β, from this formula it follows that the maximum zero-temperature stretching occurs in molecular crystals.

Crystal heating leads to a rapid increase in motional pressure. While the heat component in Eq. (3.4.20) proportional to T^4 remains low relative to a zero pressure, an increase in volume is negligible. As soon as the classical limit has been reached (at $T \gg \Theta$), the expansion becomes linear as a function of temperature:

$$\frac{\Delta a}{a} = \frac{3\gamma nkT}{E}. \qquad (3.4.29)$$

Of course, linearity holds within the limits of validity of the harmonic approximation. This implies that $E = e(a)/a = \text{const}$ ($e \approx \beta_0$). Thus, despite a significant change of p_t in the course of heating, the outward result is only a very slight crystal expansion. Over a short temperature interval $(T' - T \ll T)$ it can always be described by a linear law:

$$\frac{L' - L}{L} = \frac{d}{dT}\left(\frac{\Delta a}{a}\right) \cdot (T' - T) = \alpha \cdot (T' - T),$$

where

$$\alpha = \frac{\gamma}{E}\frac{du}{dT} = \frac{\gamma}{E} c_V \qquad (3.4.30)$$

is the linear thermal expansion coefficient. Evidently, the temperature dependence $\alpha(T)$ duplicates $C_V(T)$ as plotted in Fig. 3.11: first, α increases as T^3 and then tends to a constant, which is the classical limit. Near room temperature the limit has been attained and, therefore:

$$\alpha = \frac{\gamma}{E} 3nk \approx 10^{-5} \text{ K}^{-1}. \qquad (3.4.31)$$

Thus, the relative change of crystal size upon heating by 1 K is around a thousandth of one percent at the very most, although the competing forces are of tens of thousands of atmospheres.

To understand the sense of Eq. (3.4.31) at the microscopic level, we have to take $E = \beta_0/a$ and γ as defined in Eq. (3.4.19). Then we find that

$$\alpha = \frac{g_0 k a_0^2 n}{2\beta_0^2} = \frac{g_0 k}{2\beta_0^2 a_0}. \qquad (3.4.32)$$

This formula has a simple physical explanation which follows from the nature of the interatomic interaction. The point is that in the harmonic approximation, when the potential well is strictly symmetrical, heat expansion will not occur at all. As a matter of fact, for any oscillation energy an atom moving in a parabolic potential is, on the average, in the same position as an atom at rest. This observation provides an alternative interpretation of the fact that in a harmonic approximation phonons do not exert pressure on the surface of the crystal and, consequently, cannot cause its expansion. However, if the potential anharmonicity is taken into account, the atom, oscillating between the potential well edges, shifts to the right farther than to the left. Consequently, its average position $\bar{r} = a$ is different from a_0.

In fact, taking into account a very small anharmonic correction, we can calculate $\Delta a = \bar{r} - a = z$ with the use of the standard Boltzmann distribution in the potential (3.2.1a):

$$\Delta a = \frac{\int_{-\infty}^{+\infty} z \exp\left[-\frac{\beta_0 z^2}{2kT} + \frac{g_0 z^3}{3!kT}\right] dz}{\int_{-\infty}^{+\infty} \exp\left[-\frac{\beta_0 z^2}{2kT} + \frac{g_0 z^3}{3!kT}\right] dz} \approx$$

$$\approx \frac{\int_{-\infty}^{+\infty} z \left(1 + \frac{g_0 z^3}{6kT}\right) \exp\left(-\frac{\beta_0 z^2}{2kT}\right) dz}{\int_{-\infty}^{+\infty} \exp\left(-\frac{\beta_0 z^2}{2kT}\right) dz}.$$

If $g_0 = 0$, the integral in the numerator, as anticipated, vanishes. Only when the anharmonic term is taken into account is Δa different from zero:

$$\Delta a = \frac{g_0}{6kT} \frac{\int_{-\infty}^{+\infty} z^4 \exp\left(-\frac{\beta_0 z^2}{2kT}\right) dz}{\left(\frac{\pi \cdot 2kT}{\beta_0}\right)^{1/2}} = \frac{g_0 kT}{2\beta_0^2}. \qquad (3.4.33)$$

Also, $f(a) = \beta \Delta a \neq 0$, and $p_i = (1/a^2) f(a) = p_t$ is in agreement with (3.4.29), using the definitions of E and γ. Similarly, from (3.4.33) one derives the standard linear expansion law:

$$\frac{\Delta L}{L} = \frac{\Delta a}{a_0} = \frac{g_0 k}{2\beta_0^2 a_0} T,$$

with the coefficient defined in (3.4.32). However, we may now recognize that this latter result, obtained within the classical description of oscillations of independent atoms, is just an asymptotic high-temperature approximation to the general formula (3.4.30).

Knot of Isotherms

Now we can return to the question of isotherm intersection on the $F(v)$ diagram (Fig. 1.29) which looks like a knot situated in the region of the crystalline phase (Fig. 1.26). Since both pressure and temperature in the vicinity of the knot are quite high, in the balance of forces (3.4.21) neither of the terms is superfluous. On the other hand, the classical limit of this equation is applicable when u is determined in the Dulong–Petit approximation:

$$p + p_i(v) = 3\gamma \frac{kT}{v}. \qquad (3.4.34)$$

With a harmonic approximation for p_i this is Mie's equation, which can be expressed as:

$$F = \frac{pv}{kT} = 3\gamma - \frac{(v - v_0)}{\kappa kT}, \qquad (3.4.35)$$

Here v_0 is as before the specific volume of a tensionless crystal with lattice period a_0. Hence, the compressibility factor for all tensionless solids should be equal to 3γ, provided that Grüneisen's constant is independent of temperature. According to (3.4.19) the latter condition is indeed satisfied, which means that the intersection point of the isotherms (*knot point*) should be obtained at $v = v_0$. With such an interpretation of the isotherm knot one expects that its abscissa should correspond to the density of a tensionless state, $n_0 = 1/v_0$, and the ordinate to the tripled Grüneisen constant, that is, approximately 6. It is sufficient to look at Fig. 1.29 to be convinced of this fact. Moreover, the decrease in isotherm slope with increasing temperature seen at the knot point can also be explained quite satisfactorily. In fact, according to (3.4.35) this slope, equal to $1/\kappa kT$, should be inversely proportional to the temperature, as is the case.

Note, however, that we obtain agreement only in comparing theoretical predictions to numerical calculations made within the free volume theory that are plotted in Fig. 1.29. Although it is a generally accepted fact that this theory is good enough for crystals, a direct comparison with experiment is still preferable. With this in mind, the same diagram is reproduced from reliable data, available in full only for argon (Fig. 3.18). Comparing this diagram with the scheme displayed in Fig. 1.26 it is seen that the knot point of interest is located in a crystalline phase region. However, in reality not all isotherms converge to meet at one point. Isotherms with temperatures below the Debye temperature do not reach this point because they are not described by the Mie equation (3.4.34). On the other hand, isotherms of very high temperature are cut by the melting line before the crystal expands to its tensionless state. Nonetheless, the position of the knot point on the scale of Fig. 3.19 is indicated quite clearly by a set of isotherms that reach its vicinity.

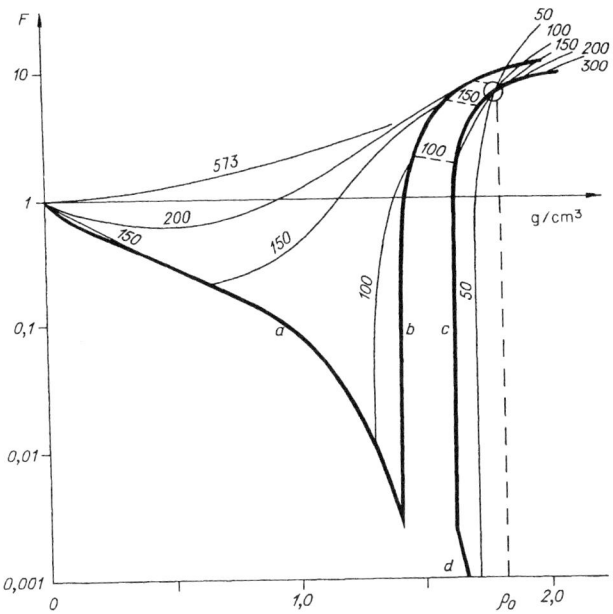

Figure 3.18 Isotherms of the compressibility factor for argon in semilogarithmic coordinates. The solid curve *a* is the vapour–liquid coexistence curve, while *b* and *c* indicate the borders of the "crystallization corridor" from the gas and the crystal sides respectively. The intersection point ("knot") in the circle corresponds to the density of the tensionless crystal (in g/cm³). The isotherms are labeled by their temperatures in K.

3.5 REAL CRYSTALS

All lattice vibrations so far discussed are only able to transfer energy, not matter. In an ideal crystal atoms oscillate about their equilibrium positions, and diffusion is improbable. Nevertheless, the substances of two solids brought in contact can penetrate into one another, and this is qualitatively observed in practice.

With this in mind, we need to proceed from the ideal model to the actual situation. When idealizing a gas, we considered it as consisting of freely moving point particles. Only later were their sizes and interactions affecting the free motion taken into account. With solids the situation is quite the opposite. First, a solid was treated as a perfect crystal; now, to approach reality, we must make allowance for the lattice defects—the only elements able to move in an ordered system.

Point Defects

The simplest example of a point defect is an impurity atom at a lattice site or between sites. Atoms and molecules of light gases, such as hydrogen, oxygen,

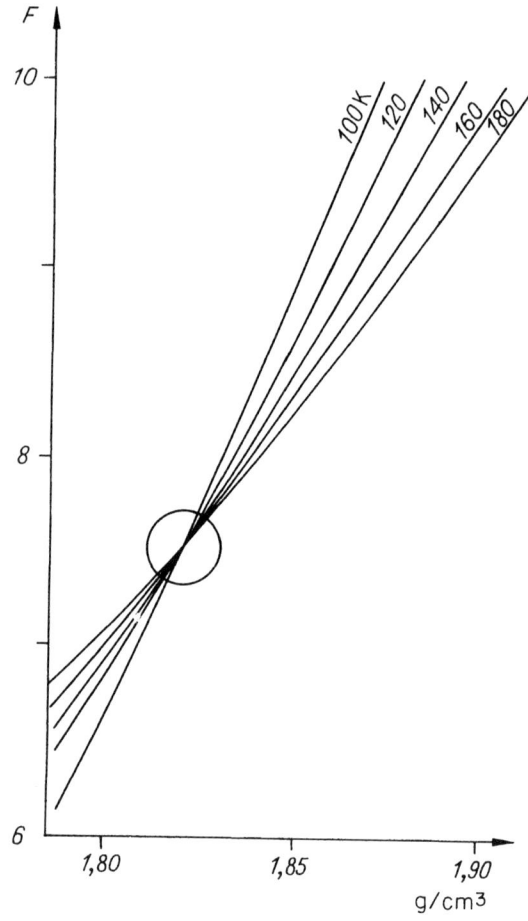

Figure 3.19 The vicinity of the argon knot (intersection point at tensionless density) on a larger scale.

nitrogen, and so on, can penetrate in considerable amounts into crystals, including metals. Not only alien, but also "native" atoms can find themselves in an interstitial site, that is, become dislocated. The lattice sites which remain vacant are also lattice defects. In some cases, their concentration accounts for 2% of the total number of places. Dislocated atoms and vacancies may arise simultaneously, when thermal fluctuations reach an energy sufficient to overcoming the barrier holding an atom in its site. This is how Frenkel defects are created. Having escaped from its potential well, an atom gives rise to a vacancy and becomes dislocated, thus acquiring some freedom of movement. In an accidental encounter with its own or another's vacancy, the atom can return to settled life, and both defects disappear.

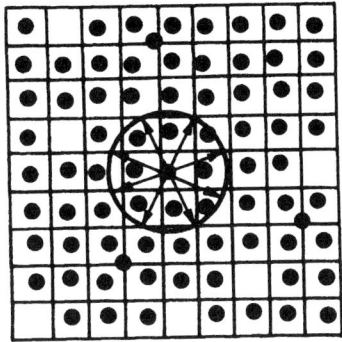

Figure 3.20 Interstitial atoms and vacancies. The arrows in the circle indicate the available sites for dislocation in the nearest neighborhood.

The greater the number of places suitable for occupation in the vicinity of the lattice site, the higher the chances of the atom stationed there to change its position. The total number of transitions per unit time is equal to $W_+ N' V_0$, where N' is the concentration of interstitial sites favorable for dislocation per unit volume, W_+ is the rate of transfer to one of them, and V_0 is the volume of a cell containing the vacancies closest to a given site (Fig. 3.20). Under equilibrium conditions, the defect formation rate $W_+ V_0 N' N$ (N is the concentration of normal atoms) is balanced by the recombination rate, equal to $W_- V_0 n_V n_\alpha$ (n_α is the concentration of dislocated atoms, n_V that of vacancies). It is assumed that defects are few in number, and most interstitial sites are vacant.

According to the principle of detailed balance

$$\frac{W_+}{W_-} = \exp\left(-\frac{\Delta \mathcal{F}}{kT}\right), \qquad (3.5.1)$$

where $\Delta \mathcal{F}$ is, roughly speaking, the difference in energy between the equilibrium and the dislocated positions. As $n_V = n_\alpha$, we find from the equilibrium condition $W_+ V_0 N' N = W_- V_0 n_V^2$ that

$$n_V = \sqrt{N' N} \exp\left(-\frac{\Delta \mathcal{F}}{2kT}\right). \qquad (3.5.2)$$

Thus at ordinary temperatures, some atoms of an initially perfect crystal must inevitably move to places between sites. Although the disruption of the crystal structure affects the ideal order of the atomic arrangement, the measure of this disorder is strictly related to the crystal temperature.

Penetration of atoms and vacancies into the crystal lattice may also proceed independently. Such defects owe their existence to the finite size of crystals. It is easier to move to an interstitial site from the solid surface, either inner or outer, that is, from any cracks, hollows, screw or edge dislocations (multiatomic disruptions of the lattice), than directly from a regular site. Any defect near the surface can move to the interior of a crystal and, finally, accumulate there in a concentration defined by the equilibrium statistics

$$n_V \approx N \exp\left(-\frac{\Delta \mathcal{F}_0}{kT}\right), \quad \Delta \mathcal{F}_0 < \Delta \mathcal{F}. \tag{3.5.3}$$

However, if hardened, that is, cooled so rapidly that the equilibrium is not shifted, a crystal will be overpopulated with defects which can not regain their original positions on the surface.

In ionic crystals it often happens that one or both of the ions are so large as to be practically unable to leave their equilibrium sites. In this case, it is very important that interstitial sites give not the ion but its vacancy the chance to move. In an ionic crystal a vacancy carries a charge equal but opposite in sign to the missing ion. This is seen from the fact that the same charge geometry may be obtained by placing an opposite charge on the ion instead of removing it from the site. The mosaic of alternating charges is electrically balanced, thus neutralization of any of them induces a Coulomb field centered at the neutral site (Fig. 3.21).

Vacancies in crystals which are not caused by dislocated atoms are known as Schottky defects. These may arise from diffusion of vacancies from the surface, or by the dissociation of coupled vacancies of metal and halogen ions initially located nearby. The existence of Schottky defects manifests itself as a significant increase in the volume of a crystal containing a lot of vacant sites. Coupled (associated) ion vacancies and individual (dissociated) ones are in a dynamic equilibrium similar to that attained among Frenkel defects. More complicated disruptions of the structure are also possible: triple vacancies,

Figure 3.21 The potential field around a vacancy and a neutral atom showing that they have the same sign.

vacancies filled with impurity atoms or ions, and so on. This complicates the picture, but does not affect the essence: the formation of any defects requires the expenditure of energy. That is why their concentration always increases with temperature.

In outline, the situation appears to be as follows: Frenkel defects prevail in metals and electron semiconductors. Away from the melting point they are few in number and their role is rather insignificant. In ionic crystals, defects of both types play an important part in the processes of matter and charge transfer. Schottky defects are bound to prevail in molecular crystals. This is because molecules considerably bigger than intermolecular distances cannot find suitable places for dislocation, and their chances to move throughout the crystal are very limited.

Diffusion

The motion of interstitial atoms and vacancies in solids has little in common with the free movement of gas particles. This motion is only occasionally interrupted by collisions with molecules or against the wall. The path of atoms wandering throughout a crystal is more intricate and complicated. Displacement even by one cell requires that the potential barrier ΔU be overcome. The barrier is created by atoms held at their own sites which are not inclined to facilitate the job of the dislocated atom. Each step in the diffusion process requires an amount of energy which considerably exceeds the available thermal resources of an atom ($\Delta U \gg kT$). That is why an atom remains in its original position for rather a long time, waiting until the energy accidentally captured from the surroundings proves to be sufficient to overcome the barrier. As a result, all displacements are of random rather than dynamic nature and may be defined by the rate of transition from one interstitial site to the neighboring one.

Let us consider two adjacent crystal layers of the thickness of one lattice period a (Fig. 3.22). If the densities of interstitial atoms in these layers differ, random motion of the type described above will lead to their gradual equalization. Macroscopically, this will appear as diffusion from the region rich in defects to that with fewer. Clearly, the flux of particles through a unit cross-section perpendicular to the x axis can be expressed

$$j = an_1 W_{12} - an_2 W_{21}, \qquad (3.5.4)$$

where W_{12} is the rate of transition from the first to the second layer, while W_{21} is the rate of returning back. The separation of the neighboring layers a is small compared with the macroscopic distance $[d \ln n/dx]^{-1}$ where the variation of n is noticeable. So, we can expand $n(x)$ as a series and determine the concentration of defects in these layers in the first approximation:

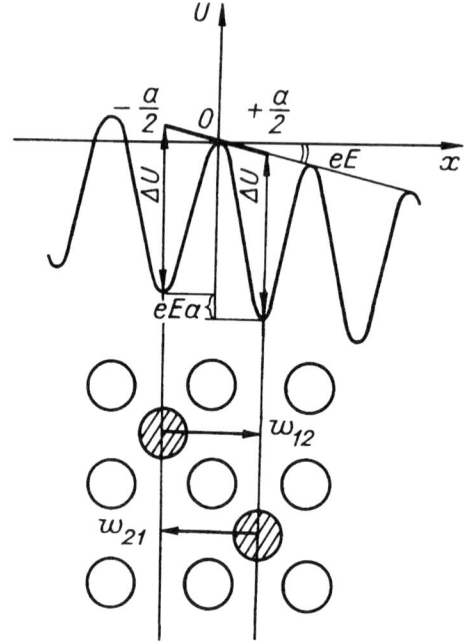

Figure 3.22 Overbarrier transitions between the nearest sites with the rates $w_{12} \neq w_{21}$; the atom energy in the field E is plotted above showing that the barrier is lower in the direction of the field.

$$n_1 = n - \frac{dn}{dx} \cdot \frac{a}{2}; \quad n_2 = n + \frac{dn}{dx} \cdot \frac{a}{2}. \quad (3.5.5)$$

Then, (3.5.4) gives

$$j = an(W_{12} - W_{21}) - \frac{(W_{12} + W_{21})}{2} a^2 \frac{dn}{dx}. \quad (3.5.6)$$

If there is no reason to prefer any one direction, the probability of transition to any of six neighboring positions (right, left, up, down, forward, back) is the same value W, so

$$W_{12} = W_{21} = W = \frac{1}{6\tau}, \quad (3.5.7)$$

where τ is the length of time spent by the atom in a given interstitial site. In this case, we obtain from (3.5.6)

$$j = -D \frac{dn}{dx}, \quad (3.5.8)$$

where

$$D = Wa^2 = a^2/6\tau, \qquad (3.5.9)$$

and

$$W = \nu_0 \exp\left(-\frac{\Delta U}{kT}\right). \qquad (3.5.10)$$

The structure of formula (3.5.10) is determined by obvious physical considerations. The rate of overbarrier transitions is the number of successful attempts per unit time. The total number of attempts is equal to the inverse frequency of atomic vibrations in the potential well, because this coincides with the number of attacks of one of its opposite sides. One would expect that any such attempt would certainly overcome the barrier if the energy of the oscillating atom is greater than ΔU. Thus the proportion of successful attempts is equal to the fraction of excited states with energy exceeding ΔU. From this it follows that the process of diffusion in a solid is activated. Accordingly, (3.5.9) and (3.5.10) yield

$$D = \nu_0 a^2 \exp\left(-\frac{\Delta U}{kT}\right). \qquad (3.5.11)$$

Both the exponential dependence of diffusion coefficient on temperature and its proportionality to $1/\sqrt{m}$ (following from the conventional definition of vibrational frequency in the potential well $\nu_0 = (1/2\pi)\sqrt{\beta/m}$) are experimentally confirmed. Thus, for very large particles diffusion proceeds very slowly, as it involves overcoming high barriers. And, vice versa, the smaller the mass and sizes of impurity atoms, the faster their random walk through the crystal lattice.

The motion of vacancies in a solid proceeds by the same mechanism as diffusion of interstitial atoms. It is in fact the particle adjacent to a vacancy that moves; moreover, at each time, a different particle moves. However, the vacancy appears to move in the direction opposite to the transition which actually occurred. In molecular crystals the motion of vacancies initiates both self-diffusion and diffusion of impurities residing at the lattice sites. Molecules may move to the neighboring site only when it contains a vacancy. Compared with (3.5.10), the probability of this process is lower by the factor p_V, where $p_V = n_V \cdot v$ is a fraction of vacant sites. Therefore,

$$D = D_V p_V, \qquad (3.5.12)$$

where D_V is the vacancy diffusion coefficient defined in (3.5.11). This formula has a simple physical meaning: impurity migration at the expense of vacancies

Ion Conduction

In ionic crystals electric current is often transferred by heavy particles, interstitial ions, or vacancies. When power is turned on, anions and cations start drifting in opposite directions. Once neutralized, they are deposited at the appropriate electrode in their pure phase. This process is known as solid state electrolysis: the decomposition of a chemical compound into components, as happens in liquid accumulators. The deposition of reaction products at the electrodes is unambiguous evidence in favor of the ion mechanism of charge transfer.

As has already been mentioned, ions and their vacancies move in a crystal by rare short jumps only to the neighboring cell or to the nearest site, rather than by long free paths. Thus thermal motion promotes conduction instead of retarding it. While in the case of free ions in a gas it was a handicap limiting the mean free path, ions fixed in a crystal lattice would be unable to move at all but for thermal activation of diffusion.

However, what is the role of the field? It turns out that the field merely facilitates jumps in one direction and makes them more rare in the opposite one. In the presence of the field the total potential $U(x) - eEx$ is decreasing in the direction of the field (see Fig. 3.22). Therefore the magnitude of the barrier to the right and to the left is not the same as before, but is equal to $\Delta U - eEa/2$ in one case, and to $\Delta U + eEa/2$ in another. For this reason, equality (3.5.7) assuming equal rates of transitions in any direction is violated. In the presence of the field, the transition rate (3.5.10) in the forward direction is

$$W_{12} = \nu_0 \exp\left(-\frac{\Delta U - eEa/2}{kT}\right) = W \exp\left(-\frac{eEa}{2kT}\right), \quad (3.5.13a)$$

while in the opposite direction it is

$$W_{21} = \nu_0 \exp\left(-\frac{\Delta U + eEa/2}{kT}\right) = W \exp\left(\frac{eEa}{2kT}\right). \quad (3.5.13b)$$

Although the difference in potential over a distance of the order of the lattice period is rather small ($eEa/2 \ll \Delta U$), nonetheless the transition rates are affected. Furthermore, it is possible that ΔU is not small compared with kT. In this case, though still small, the rates of forward and back transitions will differ by many orders.

The difference in barrier heights brought about by the field violates the balance between the opposing fluxes of particles which now prefer to take the easier way. As a result, we get from (3.5.6) and (3.5.13)

$$j = 2anW\,\text{sh}\left(\frac{eEa}{2kT}\right) - Wa^2\text{ch}\left(\frac{eEa}{2kT}\right)\frac{dn}{dx}.$$

For relatively weak fields ($eEa \ll 2kT$) the exponents may be expanded up the first nonvanishing term. Then the current is given by

$$i = ej = eunE - eD\frac{dn}{dx},$$

where D is the same as in Eqs. (3.5.9) or (3.5.11), and

$$u = \frac{eWa^2}{kT} = \frac{ea^2}{kT}\nu_0 \exp\left(-\frac{\Delta U}{kT}\right). \tag{3.5.14}$$

The general relationship between u and D,

$$u = \frac{eD}{kT}, \tag{3.5.15}$$

is called the Einstein relation.

4

LIQUIDS

4.1 PHENOMENOLOGY

As a phase, liquids occupy a specific position intermediate between the gas and solid states. As with all solids, liquids fall into the category of condensed media: they are not able to expand, and their molecules, held together by strong attractive forces, are in the immediate vicinity of each other. However, close packing of molecules does not prevent liquid, much like a gas, from taking the form of whatever vessel contains it, or any body that is plunged into it. Like a solid, a liquid is almost impervious to compression and to omnidirectional stretching. However, similarly to gases, liquids transfer pressure in all directions without the anisotropy typical for crystals.

This peaceful coexistence of diverse properties in liquids is possible only due to the fact that this state of matter is realized in a temperature range where the energy of thermal motion is high enough to disrupt the regular crystal structure, but is not sufficient to completely rupture the bonds that hold molecules within the finite volume. As is shown by X-ray structure studies, liquids are characterized by a so-called "short-range" order and the absence of a "long-range" ordering. Although the positions of the nearest neighbors of any molecule do not deflect significantly from a regular, pseudocrystal arrangement, these slight departures from ideal packing accumulate rapidly, and one can find an interstitial site instead of a lattice point even at the distance of several periods. In a similar manner, in a heap of wheat grains large voids are improbable: the distance between grain centers is approximately equal to the grain diameter. At the same time, there is no order in the arrangement of more or less distant grains. This chaotic packing is responsible for the ability of granular substances to rearrange themselves to make way for a hand dipped into them, and to assume the shape of a container.

Brick masonry, even without cement, has no property of looseness exactly due to its regular structure, which gives it the capacity for collective resistance. This is how crystals are structured: in a highly ordered fashion, with minimum entropy. However, this harmony is violated with rising temperature. Molecules become able to rotate and vibrate; moreover, above the Debye temperature they move practically independently, and further heating leads to complete rupture of the crystal structure. If the building walls are destroyed by an earthquake and become just a heap of stones and sand, the properties of the structure are radically changed, despite having the same volume and density.

The above analogies make it clear why a liquid, though as difficult to compress as a solid, is unable, or almost unable, to resist shearing. However, in order to account for its resistance to rupture, it is necessary to allow for attractive forces along with repulsive ones. Despite the presence of coupling forces, the liquid is readily partitioned into fractions and drops. The reason is that the layers slide easily past one another, and this results in the sequential rupture of bonds: one after another instead of all of them together. So, one can judge the liquid strength only under conditions of omnidirectional extension, when sliding is impossible. For example, water is only four times less difficult to rupture than lead. The coupling forces manifest themselves firstly in that they hold the liquid molecules into a finite volume almost without any outside help, and, second, they are responsible for specific features of the surface which resembles a film in tension.

Viscosity is the most prominent property of a liquid. Liquids are easily distinguishable by this property even by touch. We can change the viscosity by many orders by varying the temperature, the pressure, or the composition of the solution. As viscosity increases, some liquids gradually gain hardness, thus entering the so-called "amorphous state." Amorphous solids differ from crystals in that they have no long-range order. When viscosity is not very high, they are readily pugged (as clay, paste, etc.); however, on solidification, they are often indistinguishable from crystals by their mechanical properties. Many organic solutions, polymers, glasses and some other compounds are of this nature. In this case, there is no clear dividing line between the liquid and solid phases, as in structure and thermal motion amorphous substances are merely solidified liquids.

Thermal motion in liquids is of a very complicated, locally collective nature. The motion of an isolated particle cannot be considered free, since, being strongly bonded to the surrounding molecules, it is unable to leave its cell. As in solids, its motion is vibrational rather than translational in character. However, in liquids, the motion of a particle is correlated only with that of the nearest neighbors rather than with a whole ensemble of molecules.

Though close packing and bonds holding the molecules together transform their translational motion into something similar to vibrations, in most cases, quantization is not necessary. At ordinary temperatures liquid molecules behave classically and their distribution in velocities is still Maxwellian. In this sense, a classical liquid does not differ from gas, and the average kinetic energy of its atoms is $(3/2)kT$. The same also refers to the potential energy of vibrations. The total thermal energy is $\bar{\mathcal{E}} = 3kT \cdot N_0$, while $C_V = 3R$.

Rotational motion in liquids is retarded by steric obstacles which convert it into rotational vibrations (librations). The sole exceptions are light diatomic and spherical molecules, such as N_2, CO, CH_4, SF_6, and so on, which rotate, though chaotically, in their liquid phase. As for molecular hydrogen, it can remain in a state of quasifree rotation even in a crystal.

Simplifying the picture to some extent, we can assume that a liquid molecule undergoes a chaotic vibrations within a cell and participates in occasional structural rearrangements which transfer it in another cell. As an intermediate case, equally distant from the alternative models of gas and solid, liquid is less amenable to exact studies. The ideal gas model is of no use when strong interactions between particles produce a considerable effect on the ensemble behavior.

4.2 EQUATION OF STATE

At first sight, it is not a problem to distinguish between liquid and gas. However, proceeding to a quantitative description of these phases, we raise this problem once again. From the view of thermodynamic considerations, this is not an idle question. As seen in the diagram of Fig. 4.1 only the crystal state

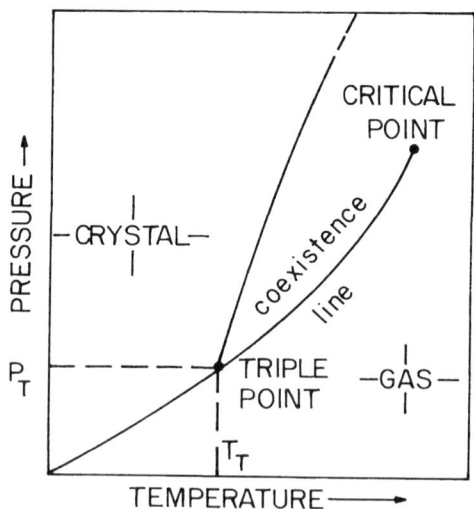

Figure 4.1 p–T diagram of state.

is everywhere separated from the other phases by the solid boundary extrapolated to infinite pressure by the dashed line. The equilibrium coexistence of ordered and disordered phases is possible only at this boundary. Crossing this curve leads either to sublimation or to melting. The curve increases monotonically over the whole accessible pressure range. On the other hand, the liquid–gas coexistence curve terminates at the critical point. In other words, the spatial separation of the liquid and its vapor which serves to discriminate between them is possible exclusively in a narrow pressure and temperature range, from the triple point to the critical point. Crossing the curve between these points results in a phase transition that proceeds discontinuously, by liquid evaporation. However, this transition can also be done smoothly, without crossing the coexistence curve, but bypassing it above the critical point. Along this route it is impossible to draw a clear border between liquid and gas.

Thus the difference between liquid and gas, which is evident under conditions of coexistence, becomes vague when we are dealing with a single phase. At all points of the diagram falling outside the coexistence curve, the question as to whether a disordered phase is a liquid or gas cannot be answered unambiguously. In the region above critical ($p > p_c$, $T > T_c$) this question has no sense at all.

Let us compare various approaches to the description of the liquid state. The van der Waals equation is appropriate for a qualitatively correct description of condensation, but considering the microscopic meaning of constants, it proves to be inapplicable for a quantitative description of liquids. This is obvious from the fact that even at $V < B = 4N_0v_d$, (still in the gaseous state) its right-hand side reverses sign, thus making the whole equation meaningless. The validity range of the virial equation of state (1.9.17) is much wider. At present, the relationship between all virial coefficients and the type of intermolecular interaction is rigorously established. However, even the virial series diverges at $V > V_c$. As is seen from the phase diagram in p–V coordinates (Fig. 4.2), the region of the liquid state is beyond the limits of this approximation as well.

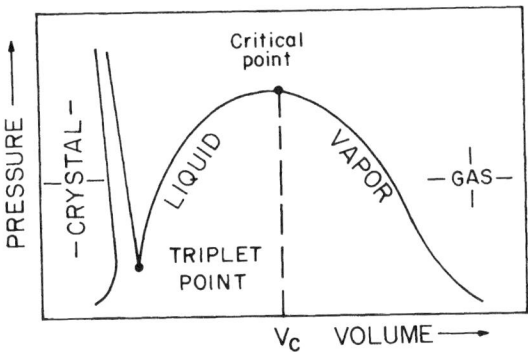

Figure 4.2 p–V diagram of state.

The theory developed by Kirkwood and Bogoliubov which makes allowance for correlated motion and treats the location of particles by correlation functions describes the dense state of matter more adequately. In principle, the theory is able to allow for correlated motion of three particles, and even larger ensembles; however, computational difficulties rapidly increase with the increasing number of molecules involved in the interaction. As the theory is too complicated to apply it to the case of an actual intermolecular potential, most analytical calculations were done for the hard sphere model, that is, when repulsion is taken into consideration in the simplest way. This introduced the necessary corrections to the right-hand side of the van der Waals equation. Depending on the approach, the left-hand side either remained unchanged (Katz, Uhlenbeck, etc.) or made allowance for attraction as a small correction (Zwanzig). In both cases, reliable results were obtained at $T > T_c$, when the attractive forces can be roughly estimated in the first approximation with respect to U/kT. This makes it possible to advance to the region of quite high densities $V < V_c$. However, in this region the substance is still a dense gas prevented from expanding solely by enormous external pressure. Liquid, in the ordinary sense, is able to avoid expansion by itself, but the low temperature region, where it exists, is beyond the limits of such an approach.

The only alternative is to approach the liquid phase region from the side of the crystal state rather than from the high pressure and high temperature gas side. The crystal state is similar to the liquid one in its density, but they are separated by the phase corridor (see Fig. 4.2). This corridor makes the fundamental difference between ordered and disordered systems even more evident. Nevertheless, Lennard-Jones and Devonshire attempted a uniform description of the crystal and liquid states in the context of the free volume theory outlined in Section 1.10. This theory exaggerates the similarity between solids and liquids, but allows the chance to enter to the region of the true liquid state.

It is particularly attractive that the limits of this theory may be extended so as to bring the equation of state into the form common for all phases:

$$p + p_i = \Gamma \frac{kT}{v}. \qquad (4.2.1)$$

Both the van der Waals equation (1.9.10) describing a real gas and the Mie equation (3.4.34) related to crystals may be represented exactly in the above form. Since the equation is applicable to these more extreme cases, one can hope that it will also be appropriate in the intermediate case, provided that p_i and p_t are defined in a proper way. Indeed, the Zwanzig equation valid for the high-temperature condensate can also be reduced to this form. The place of the free volume theory among other theories is determined by the fact that it suggests a physically meaningful estimate of $p_i(V)$ and $p_t(V, T)$ in the lowest part of the coexistence curve, in the neighborhood of the solid. As we move away from this region towards the critical point, the voids ("holes") increase considerably in number, and the difference between solid and liquid phases becomes more and more essential. When the structure becomes so loose that this assumes a fundamental significance, the free volume theory gives way to a more flexible, hole theory. However, we shall not consider the latter in detail, but just try to reveal the meaning of internal and external pressures in the dense liquid phase.

Internal Pressure

According to the free volume theory, the internal pressure (1.10.37a) is specified by the shape of the potential only in the center of the cell. That is why the internal pressure is highly sensitive to the details of the intermolecular interaction. Thus, in proceeding to a qualitative description of the liquid phase, it is necessary to make the problem more specific. Here we shall restrict ourselves to the consideration of simple liquids, such as inert gases, most diatomic molecules (hydrogen, oxygen, nitrogen, etc.) and certain simple hydrocarbons (methane, benzene, and so on).

As a subject of investigation, simple liquids are especially attractive, because the intermolecular interaction is determined most reliably by the Lennard-Jones (6–12) potential. Its general form

$$u = 4\epsilon \left[\left(\frac{\sigma}{r}\right)^{12} - \left(\frac{\sigma}{r}\right)^{6} \right] \tag{4.2.2}$$

resembles that depicted in Fig. 1.12 with the minimum at the point $a_0 = \sqrt{2} \cdot \sigma$. However, the intermolecular interaction potential alone is not sufficient for a complete determination of the cage field. We also need to know the location of particles neighboring a given molecule and the distance separating them. In a close-packed cubic lattice the nearest neighbors are at the distance a, the particles in the "second coordination sphere," at a distance $a\sqrt{2}$, while those in the third one at a distance $a\sqrt{3}$. As has already been mentioned, this face-centered packing is stable even within the nearest-neighbors approximation. This approximation is quite sufficient for rough estimates; however, nothing prevents us from making allowance for interactions with farther-removed shells.

The first shell involves 12 particles, the second 6, and the third 24. The total potential of the three shells at the cage center is additive over shells:

$$\Phi(0) = 12 u(a) + 6 u\left(a\sqrt{2}\right) + 24 u\left(a\sqrt{3}\right) + \dots . \tag{4.2.3}$$

If we were dealing with crystals, this series could be extended to infinity. However, in a disordered structure, errors in the arrangement of particles are accumulated so rapidly that very soon the particles are difficult to associate with one or another shell. Actually, only the nearest surroundings remain similar in structure to a crystal ("short-range order"), while in the second and the third shells, coordination of particles may turn out to be different from that expected in (4.2.3). Fortunately, the potential of the cage is almost unsusceptible to the structural details of the periphery shells, since the Lennard-Jones interaction decreases rapidly with distance. The contribution of particles residing in the second sphere is less than that of the nearest neighbors by at least a factor of $(\sqrt{2})^6 = 8$, and the influence of the third sphere molecules is negligible.

If we confine our considerations to the nearest neighbors, only the first component will be retained in formula (4.2.3), and it will become identical to that derived from (1.10.39) with $c = 12$. This approximation is a liquid version of the chain model of the interaction applied in Section 3.2 to crystals. A complete analogy between the elastic reactions of solid and liquid media in the nearest-neighbor approximation is easily traced if we pass from the general definition of p_i given in (1.10.37a) to its one-dimensional analog

$$p_i = \frac{1}{2}\frac{d\Phi}{dv} = \frac{c}{2}\frac{du}{dv} = \frac{1}{a^2}\frac{du}{da} = \frac{f(a)}{a^2}. \qquad (4.2.4)$$

To ease the comparison with the preceding chapter, it has been assumed here that the cell is structured as a simple cubic lattice: one neighbor to the right and one to the left along each of the coordinate axes, so that $c = 6$ and $v = a^3$. The obtained result coincides identically with (3.4.8), and the force $f(a)$ is defined in the same way as in (3.2.2a). This comparison demonstrates that, for any phase, either solid or liquid, the internal pressure is simply the force acting on the surface particles (per unit area) from the inside of the semispace filled with substance.

However, the typical cell for simple liquids is face-centered rather than simple cubic. Due to the closer packing of particles, the internal pressure is greater than (4.2.4) by a factor of $2\sqrt{2}$. Further, when the interactions with the second and the third shells in this structure are taken into account, the relative contribution of attraction and repulsion will change. The magnitude of the change can be inferred from the final result following from (4.2.3) and (4.2.2)

$$p_i^* = \frac{\sigma^3 p_i}{\epsilon} = 24\left[\frac{\lambda}{v^{*3}} - \frac{\mu}{v^{*5}}\right], \qquad (4.2.5)$$

where $v^* = v/\sigma^3 = a^3/\sigma^3\sqrt{2}$. The deviation of $\lambda = 1.14$ and $\mu = 1.01$ from unity is the only manifestation of the inclusion of the outer shells. This indicates that the nearest-neighbor approximation is quite suitable for simple liquids. One should only bear in mind that the relative error increases with expansion. The estimate of p_i by formulae (4.2.4) and (4.2.5) is reliable as long as the average separation of particles is much greater than their rms deviations from their equilibrium positions.

For gas, the idea of the "nearest neighbors" is senseless and is meaningless and is replaced by the "action sphere," whose population depends on the gas density and temperature (see Section 1.9). The internal pressure of a real gas is due to the combined effect of particles residing in the semisphere adjacent to the wall (see Fig. 1.15). With the same potential (4.2.2), the internal pressure in the van der Waals equation (1.9.10) is

$$p_i = \frac{\mathcal{A}^*}{v^{*2}} = \frac{\text{const}}{a^6}, \qquad (4.2.5a)$$

where $\mathcal{A}^* = A(T)/(\sigma^6 N_0^2)$. It is seen that for the gas phase $p_i(a)$ decreases with specific volume more smoothly than for condensed media, given in (4.2.5). The latter dependence includes the harmonic approximation (3.4.10), but remains valid far beyond its limits, until gas-like estimate (4.2.5a) proves to be better.

The free volume theory makes use of this reserve, allowing a to take values up to sizes typical for the liquid state. As the internal pressure is due only to the immediate surroundings it is indifferent to the packing of remote molecules and can be continuously extended to the transition region where the solid phase turns into the liquid one. As $u(r)$ is the same in the solid and the liquid phases, one can gain a qualitative idea of the variation of $p_i(a)$ from Fig. 3.5. In the case of solids we have only been interested in the quasiharmonic approximation of this curve. Now the other, descending branch becomes equally important. This wider application of relation (4.2.4) is possible only in the free volume theory. In the solid state, the restoring force increases with increasing intermolecular distance, as for a spring. For the liquid, its strength decreases under extension if $a > a_m$, that is, the interaction force decreases as molecules move away from each other.

The relative length of the ascending and descending branches of p_i can be judged by Fig. 4.3 which depicts the internal pressure calculated using formula (4.2.5). The quasiharmonic approximation which forms the basis of the solid state theory is just a power expansion of $p_i(v)$ to second order in $v - v_0$:

$$\frac{p_i^*}{24} = b\,(v^* - v_0^*) - \frac{c}{2}(v^* - v_0^*)^2. \qquad (4.2.6)$$

All parameters of this expansion are readily found from (4.2.5):

$$v_0^* = \sqrt{\frac{\mu}{\lambda}} = 0.94, \quad b = \frac{2\lambda}{v_0^{*4}} = 2.9, \quad c = \frac{9b}{v_0^*} = 28. \qquad (4.2.6a)$$

The zero of p_i coincides with the abscissa of the knot point in Fig. 1.29, and b specifies the slope of the intersecting isotherms in accord with (3.4.35). The harmonic approximation of p_i is valid while the quadratic term in Eq. (4.2.6) is relatively small, that is, when $|v^* - v_0^*| \ll 2b/c$. This region is rather narrow in the scale of the Fig. 4.3 and cannot be extended by retaining more terms in the power expansion. Indeed, the power series is useless in the region to the right of the maximum of p_i just as the virial expansion is unsuitable in the harmonic region. Dispersion asymptotics of internal pressure $p_i = 24\lambda/v^{*3}$ is an alterna-

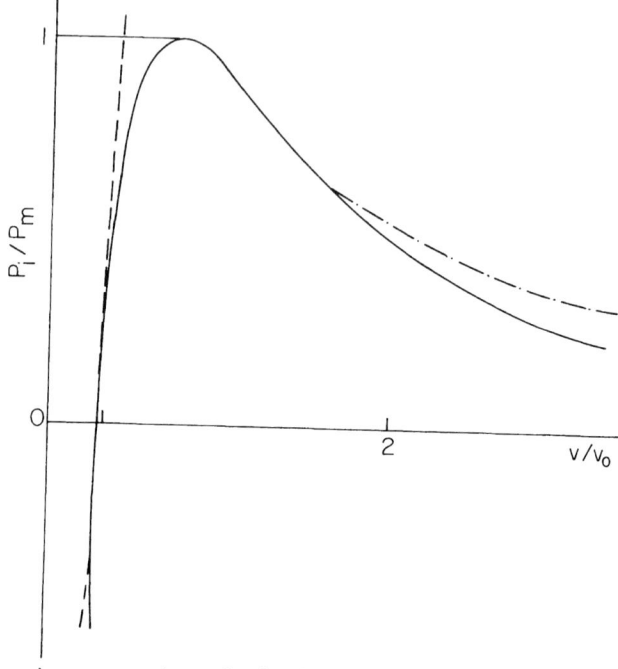

Figure 4.3 Internal pressure p_i (normalized to the maximum value) in the free volume theory. The dashed line shows the harmonic and the dash-dotted line the van der Waals approximation (v_0 is the specific volume of tensionless state).

tive to the harmonic approximation valid at large v^*. In between, the better approximation of internal pressure is given by its power expansion near the maximum:

$$p_i^* = p_m^* - 12\beta (v^* - v_m^*)^2. \qquad (4.2.7)$$

According to (4.2.5),

$$v_m^* = \sqrt{\frac{5\mu}{3\lambda}} = 1.217, \quad p_m^* = \frac{48\lambda}{5v_m^{*3}} = 6.05, \quad \beta = \frac{5p_m^*}{8v_m^{*2}} = 2.5. \quad (4.2.7a)$$

Around the maximum and to the right of it the phase state of substance is either liquid or gas.

Motional Pressure

As a solid expands to a liquid (the sequential stages of this transition are shown in Fig. 4.4) the cell potential changes its form. Imagine that particles creating a one-dimensional potential are pushed apart (Fig. 3.4). Such an occurrence

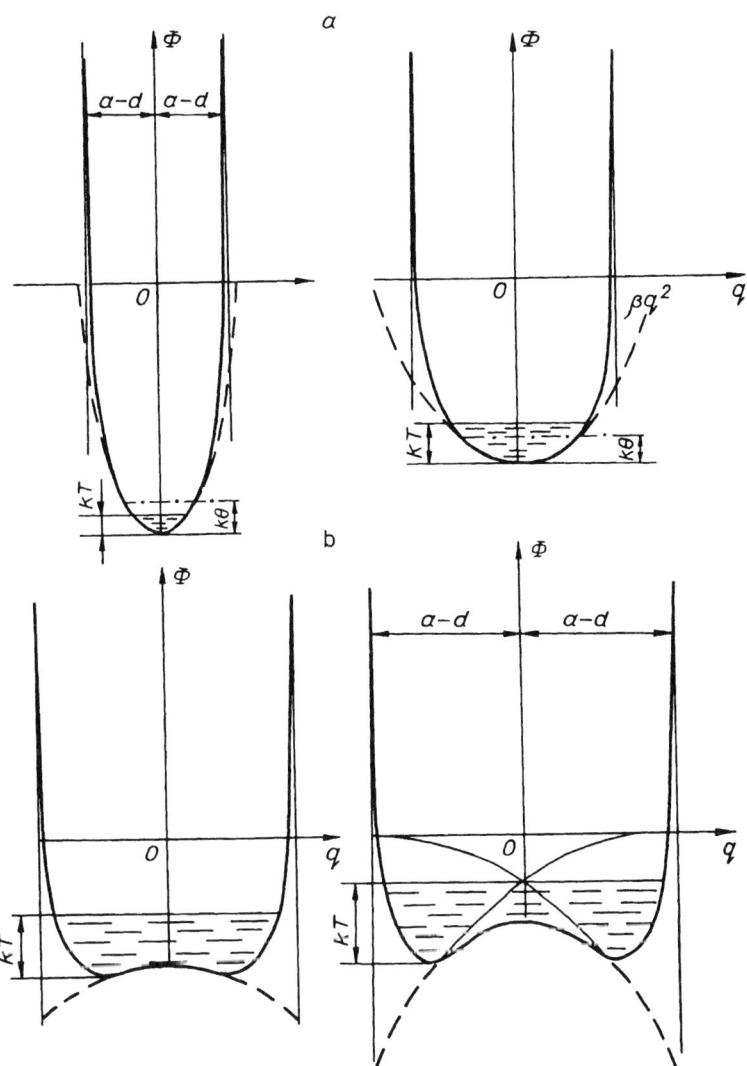

Figure 4.4 The cage potential transformation under thermal extension of crystal (a) and liquid (b).

inevitably results from thermal expansion of the medium (lengthening of a). As long as the lattice period is close to a_0, and the energy of heat motion is low, the particle oscillates in the lower part of the well where its shape resembles a parabola. For this part of the potential, the harmonic approximation is reasonable and very profitable in the theoretical description of solids. However, as thermal expansion progresses it loses validity as the anharmonicity of the process increases. The amplitude of vibrations with higher energy is limited by a sharp repulsive potential that more closely resembles a vertical potential wall rather than a parabola. Simultaneously, the bottom of the well in the cell center changes curvature, which depends on the sign of

$$\left.\frac{d^2\Phi}{dx_n^2}\right|_a = 2\frac{df}{da}.$$

When $f(a)$ is a maximum, the bottom becomes flat. With further expansion, in the region of the descending branch of the force, it becomes hump-shaped, forming a double well.

As the hump height is less than the mean energy of thermal motion, it is not a serious obstacle to the majority of particles, but their vibrations in the split well are completely anharmonic. Molecules in almost flat potential execute almost free motion. The hump-shaped well bottom has an insignificant effect on this motion and vibration in the cell is determined by reflections from the almost vertical walls of the potential. Idealizing this situation, we can replace the actual cell potential by a rectangular well model (Fig. 4.5). Within this model, each molecule of the liquid executes free motion within its cell, reflecting elastically from its boundaries.

Generally speaking, in the free volume theory there is no need for additional simplification of the potential. The motional pressure is defined by expression (1.10.37b) and the meaning of "free volume" clarified in (1.10.33). However, it is not so easy to perform the integration in (1.10.33), and fundamental problems arise in the account of passages to the neighboring cells left between the surrounding particles. Only after the actual cell potential is "smeared," as in (1.10.38), the theory becomes consistent: the cell walls become perfectly impermeable to the particle, and the integration of (1.10.33) is considerably simplified. However, even in this case, the results of calculations using the Lennard-Jones potential are too cumbersome to give here. Thus we return to the harmonic and rectangular approximations. This will simplify the estimations and make the comparison between solid and liquid phases much easier.

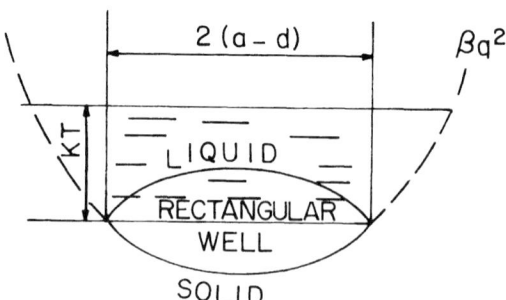

Figure 4.5 Harmonic and rectangular approximations of the cage potential.

First of all, we should check that the free volume theory reproduces the result for the motional pressure of solids derived in Section 3.4. In the harmonic approximation the potential is quadratic in the displacement from the cell center:

$$\Delta \Phi = \beta \left(q_x^2 + q_y^2 + q_z^2 \right),$$

where $\beta = \beta_0 - g_0 \Delta a$, and $\Delta a = a - a_0$ is the equilibrium lengthening of the lattice period corresponding to the given temperature. The free volume calculated in this approximation is as follows

$$v_f = \left[\int_{-\infty}^{+\infty} \exp\left(-\frac{\beta q_x^2}{kT}\right) dq_x \right]^3 = (\pi kT/\beta)^{3/2}. \qquad (4.2.8)$$

Extending the integration far beyond the cell limits is appropriate at temperatures such that the anharmonic contribution to the energy of heat motion is negligible: $kT \simeq \overline{\beta q_x^2} \gg \frac{1}{3} g \overline{q_x^3}$. Therefore, the estimate is valid in the temperature range $k\Theta < kT < 9\beta^3/g^2$.

At higher temperatures, the energy of thermal motion is sufficient for a molecule to reach the well walls created by the sharply increasing repulsion of the neighboring particles. In this region the rectangular approximation for the potential well will be more appropriate

$$\Delta \Phi = \begin{cases} 0, & |q| < a - d, \\ \infty, & |q| = a - d. \end{cases}$$

In this case, the free volume is simply

$$v_f = \frac{4\pi}{3} (a - d)^3. \qquad (4.2.9)$$

Now let us express the motional pressure determined in (1.10.37b) as

$$p_t = \Gamma \frac{kT}{v} = \Gamma \cdot nkT. \qquad (4.2.10)$$

Here the numerical coefficient Γ is determined by

$$\Gamma = \left(\frac{\partial \ln v_f}{\partial \ln v}\right)_T + \left(\frac{\partial \ln \bar{\sigma}}{\partial \ln v}\right)_T. \qquad (4.2.11)$$

It contains components of quite different nature. The first component is the thermal pressure resulting from the interaction between the particle and its nearest neighbors. The second component is determined by the state of the entire ensemble of particles. If the substance expansion has no effect on the packing of particles, the entropy remains unchanged, and its derivative goes to zero. This means that for the expansion of a crystal structure, when $\bar{\sigma} = 1$, the entropy contribution to Γ is absent. The same is true for a gas occupying a large volume, when $\bar{\sigma} = e$. However, in the intermediate region, $\bar{\sigma} = \bar{\sigma}(v)$ varies between 1 and e, and, generally speaking, the second term in (4.2.11) cannot be ignored. Unfortunately, reliable methods for calculating it have not yet been developed, and we have to restrict ourselves to the calculation of the first component only. The second term is equal to zero in both limiting situations, and is also rather insignificant in the intermediate case, but is nonetheless of fundamental importance for understanding the order–disorder phase transition (including melting) which will be discussed at the end of the section.

Substituting the free volume (4.2.8) calculated within the parabolic well model into (4.2.11), we have for $\bar{\sigma} =$ const

$$\Gamma = -\frac{1}{2} \cdot \frac{d(\ln \beta)}{d(\ln a)} = \frac{g_0 a}{2\beta} = 3\gamma, \qquad (4.2.12)$$

where γ is the Grüneisen constant defined in (3.4.19). Using this result in (4.2.10), we arrive at

$$p_t = 3\gamma \cdot nkT = \gamma u, \qquad (4.2.13)$$

where u is defined by the Dulong–Petit law. The complete identity of this with the result obtained from the estimate of the thermal pressure (3.4.15) is readily established. This supports the conclusion that above the Debye temperature the free volume theory describes the crystal state quite well. In the region of the harmonic approximation for internal pressure, the free volume theory gives us back to the Mie equation (3.4.35), obtained on quite different grounds.

Now let us consider the liquid phase. Within the rectangular well model, direct substitution of (4.2.9) into (4.2.11) yields (at $\bar{\sigma} =$ const)

$$\Gamma = \frac{d\ln(a-d)}{d\ln a} = \frac{a}{a-d} = \frac{a(a^2 + ad + d^2)}{a^3 - d^3} = \frac{1 + y^{1/3} + y^{2/3}}{1-y}. \qquad (4.2.14)$$

Assuming that $v = a^3 \sqrt{2}$, and $v_d = \pi d^3/6$, we can approximately estimate $d^3/a^3 \approx v_d/v = y$. In a dense medium,

$$\Gamma \approx \frac{3}{1-y} = \frac{3v}{v - v_d}. \qquad (4.2.15)$$

This expression is free of the problem inherent in the van der Waals estimate,

$$\Gamma = \frac{1}{1 - 4y} = \frac{v}{v - 4v_d}, \qquad (4.2.16)$$

which diverges too early, at $v = 4v_d$. However, although we may prefer to use (4.2.15) in dense media, one should remember that its advantage is lost in passing to real gas. The result given in (4.2.15) is an alternative to the van der Waals Eq. (4.2.16), rather than its generalization. Though both allow for repulsive forces only the former describes the isotherm $p_t(T,v)$ at $v \sim v_d$, while the latter is valid at $v \gg v_d$.

Is there any other way which is free of the limitations of the above approaches? If we consider the hard-sphere fluid, the advantages of the free volume theory are no longer significant, while its flaws are still of importance. In this case, the hyperchain theory of liquids based on the Percus–Yevick equation is more appropriate (see the review of Rowlinson in *Physics of Simple Liquids*). Within this approach, one obtains the following equivalent formulae:

$$\Gamma = \frac{1+y+y^2}{(1-y)^3} \quad \text{or} \quad \Gamma = \frac{1+2y+3y^2}{(1-y)^2}. \tag{4.2.17}$$

These dependencies, shown as the curves (b) in Fig. 4.6, differ little from one another. At high densities, both have a singularity at nearly the same point as the Γ-factor (4.2.14) found in the free volume theory (curve (c)), and are in good agreement with the van der Waals Γ-factor (curve (a)) when $y \ll 1$. On the other hand, the hard-sphere model, though described more adequately by formulae (4.2.17), is inapplicable to real liquids, and even less to crystals.

The solidification boundary is shown as the hatched region in Fig. 4.6. Approaching and crossing this boundary, the hyperbolic rise of the Γ-factor must cease, as the Grüneisen result (4.2.12) becomes valid. In the free volume theory, this transition, associated with the increasing contribution of the attractive forces, may be easily expressed semiquantitatively. Under moderate compressions, the cell potential changes from rectangular to parabolic. As a result, the hyperbolically increasing Γ-factor flattens out into Grüneisen's curve (Fig. 4.7), whose slight negative slope is due to the anharmonic $\beta = \beta_0 - g_0 \Delta a$ in (4.2.8). When the variation of the collective entropy is neglected ($\bar{\sigma}$ =const), the second component in (4.2.11) is equal to zero, and Γ is a continuous function giving no indication of where the liquid turns into solid. Actually, there is a discontinuity at the phase transition point to be discussed later.

Deformation of the potential affects considerably the nature of particle vibrations in the cell. As long as its motion is restricted to reflection from the parabola branches, the vibrational amplitude A increases with increasing thermal energy ($\beta \overline{A^2} = kT$), while the frequency remains constant and approximately equal to the Debye one. To put it another way, the amplitude of vibrations increases linearly with increasing mean square velocity of motion. This is the case as long as $A \ll a - d$. At higher temperatures (lower densities), particles moving with the average energy experience almost vertical walls. That is why the maximum amplitude of vibrations,

212 LIQUIDS

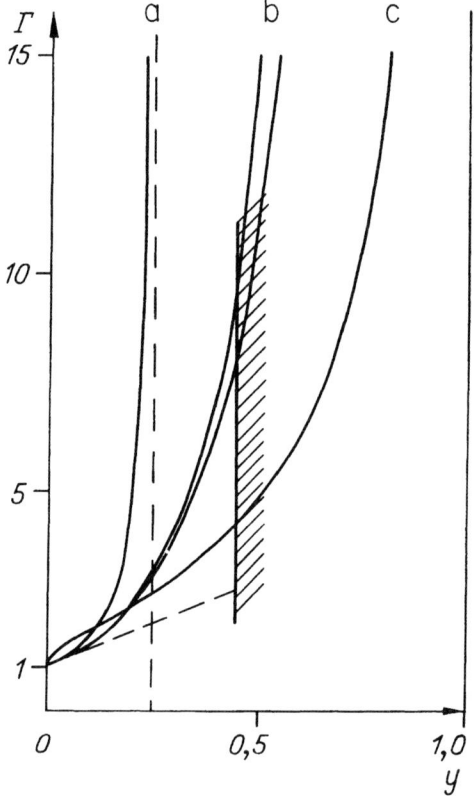

Figure 4.6 Density dependence of the Γ-factor as a function of $y = v_d/v$ in the (a) van der Waals theory, (b) rigid sphere model, and (c) free volume theory of a rectangular cage. The border of solidification is shaded.

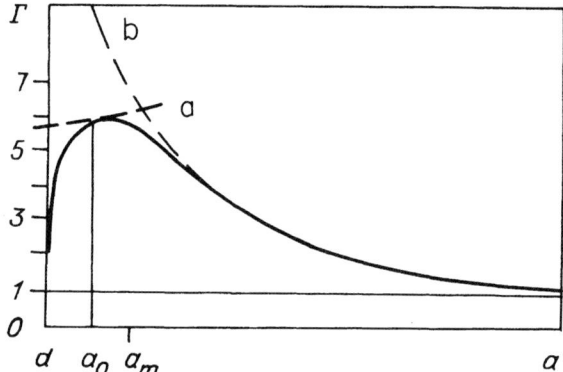

Figure 4.7 Schematic change of the Γ-factor under extension within the free volume theory. The dashed lines show the same in the (a) quasiharmonic and (b) rectangular approximations of the cage potential.

$$A = a - d, \quad (4.2.18)$$

remains constant, while the frequency increases linearly with increasing velocity of motion:

$$\nu = \frac{\bar{v}}{4(a-d)}. \quad (4.2.19)$$

Thus, the behavior of the particle in harmonic and rectangular wells is of radically different character. In the latter case, the particle moves within a limited volume the shape and size of which are randomly varying. So, despite the uniform nature of motion in the cell, the velocity changes magnitude and direction upon each reflection from the boundaries, just as for collisions of particles in gases. In fact, one cell crossing in any direction is a free path, and the cell is traversed each time with different velocity. Therefore, the time interval between sequential encounters with the cell walls is approximately equal to the relaxation time of the velocity,

$$\tau_c \approx \frac{1}{6\nu} = \frac{2(a-d)}{3\bar{v}}. \quad (4.2.20)$$

The notion of randomly vibrating particles is as typical for heat motion in the liquid, as the model of chaotic uniform motion is for gases, or harmonic vibrations are for the crystal lattice. Leaning upon this concept, one can reproduce result (4.2.14) from elementary considerations, just as the motional pressure of an ideal gas was calculated in Section 1.3, and that of a real gas in Section 1.9. Executing vibrations with the frequency $\nu = \overline{|v_x|}/2(a-d)$, a surface molecule collides against the wall limiting the liquid, thus transferring the momentum $2m\overline{|v_x|}$ to it. Assuming that there are $1/a^2$ cells per unit area, and each of them contains a particle encountering the wall ν times per unit time, the total number of collisions with the wall is

$$j = \nu \frac{1}{a^2} = \frac{\overline{|v_x|}}{2a^2(a-d)}. \quad (4.2.21)$$

Thermal pressure created by these impacts may be roughly estimated as

$$p_t = 2m\overline{|v_x|}j = \frac{m\overline{|v_x|}^2}{a^2(a-d)} \sim nkT \frac{a}{a-d}. \quad (4.2.22)$$

Acting from the inside, this pressure uniformly extends the liquid in all directions. However, due to the factor $\Gamma = a/(a-d)$, the above pressure proves to be greater than that of an ideal gas of the same density by approximately an order of magnitude. The identity of this factor with (4.2.14) points to the semiquantitative reliability of the lattice model. Taking the free volume fluctuations into consideration and using two varying parameters, Vörtler reached an adequate description of both solid and liquid branches of the hard sphere equation of state.

Balance of Forces

In order to make a qualitative analysis of the situation we require a graphic representation of the equation of state. Fig. 4.8 shows how the motional pressure expanding the substance from the inside and the opposing coalition of internal and external pressure forces (the force characteristic) reach a balance. As before, motional pressure is represented as a family of isotherms, while the force characteristic $p_i + p$ by isobars. Balance is attained at the intersection points. Mathematically, it is expressed as the equation of state (4.2.1).

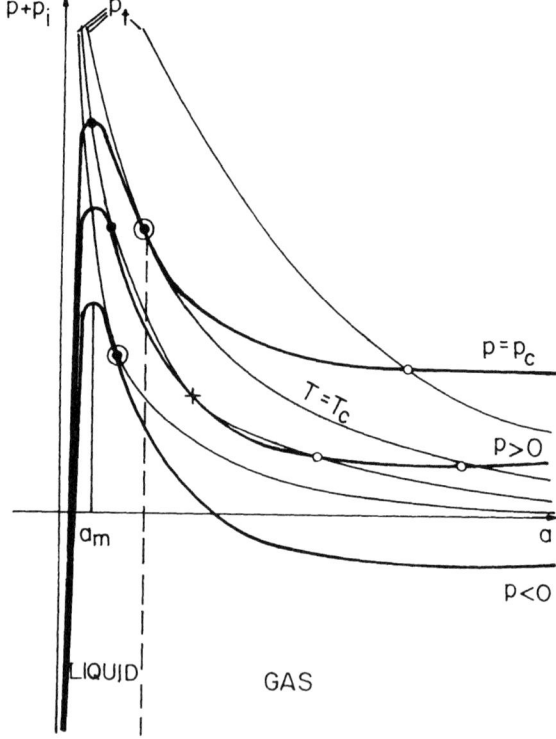

Figure 4.8 A graphical representation of the balance of forces in liquid (•), gas (○) and in the metastable region (+). The thick lines are isobars of force characteristic $p + p_i(a)$ and thin lines are motional pressure isotherms $p_t(a)$. Points of tangency are shown by ⊙.

Balance of forces is a necessary but insufficient condition for the attainment of thermodynamic equilibrium. In addition, it must be stable, and this is possible when

$$\left(\frac{\partial p}{\partial v}\right)_T < 0 \quad \text{or} \quad \frac{dp_i}{da} > \left(\frac{\partial p_t}{\partial a}\right)_T. \qquad (4.2.23)$$

This additional condition ensures that the system will regain its original position on removal of an infinitesimal external load dp. If this is not satisfied any balance violation results in instability.

Stable states of the crystal phase arise when the phonon isotherms intersect with the ascending branch of the force characteristic (see Fig. 3.16). On this branch $dp_i/da > 0$, and, as $dp_t/da < 0$ in any case, inequality (4.2.23) is satisfied. With liquids, the situation is not so simple. In this case a different equilibrium is possible which is established when the isotherm crosses the descending branch of the force characteristic. Liquid isotherms are steeper than crystal ones. Upon expansion, the thermal pressure decreases more rapidly than the internal pressure to the right of its maximum. That is why the isotherm crosses the isobar in a downward direction and gives rise to stable states marked by dots in Fig. 4.8. Though the derivatives dp_i/da and dp_t/da have the same sign, condition (4.2.23) is met, since $|dp_i/da| > |dp_t/da|$. However, this inequality is reversed with increasing a, and the same isotherm emerges from under the isobar after an unstable intersection (in the upward direction) marked by a cross. Eventually, these curves intersect again in a stable manner (circle) at a large distance where the isobar approaches asymptotically the horizontal line $p =$ const. This intersection takes place in the gas phase, where the internal pressure is a small correction to external one.

Not all curves can intersect three times. It is seen from Fig. 4.8 that each high temperature isotherm has a single, stable intersection, as it must be above the critical point where the liquid and gas states are indistinguishable. Fig. 4.9 shows the isothermal derivatives of the internal and motional pressure. According to (4.2.23), the stability of the state depends on their relative values. Only low temperature isotherms cross the negative loop of dp_i/da, and, the lower the temperature, the greater the section of the isotherm which is confined within the loop. The intersection points a_p and a'_p where

$$\frac{dp_i}{da} = \left(\frac{\partial p_t}{\partial a}\right)_T \qquad (4.2.24)$$

correspond to the extrema of the van der Waals loop. The locus of these points form a spinodal. The unstable section of the isotherm confined between spinodal branches narrows with temperature and finally contracts to the a tangent point where the first and the second derivatives of internal and motional pressure coincide:

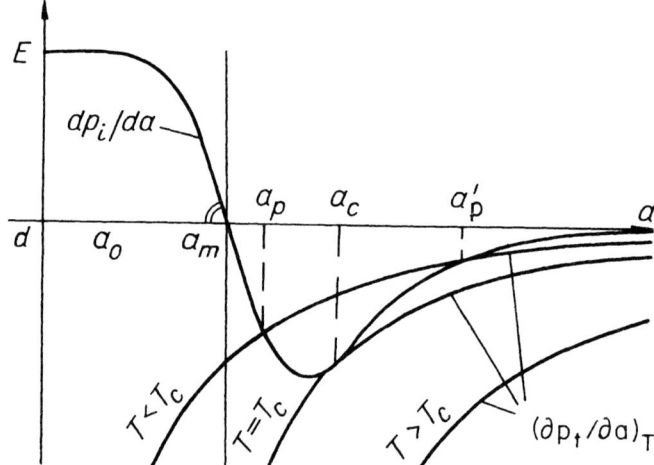

Figure 4.9 First derivatives of internal pressure $p_i(a)$ and motional pressure $p_t(a)$ at different temperatures. The unstable region between binodal points a_p and a_p' narrows with increase of temperature and disappears above critical temperature T_c.

$$\frac{d^2 p_i}{da^2} = \left(\frac{d^2 p_t}{\partial a^2}\right)_T. \qquad (4.2.25)$$

This is the critical state. The density of substance in this state is much less than near the maximum p_i. Therefore the asymptotic estimate

$$p_i = \frac{24\lambda}{v^{*3}} = \frac{\text{const}}{a^9}, \qquad (4.2.26)$$

which includes only attraction, or even the long-range van der Waals asymptotics (4.2.5a) may be valid in the near-critical region. These are qualitatively similar approximations of the internal pressure decreasing hyperbolically with volume.

However, this does not mean that internal pressure in a liquid always decreases under expansion. The triple point may be located in the region of the internal pressure maximum, or even in the ascending branch. Thus it is best plan to avoid any specific approximations for $p_i(v)$ when deriving the equation of state from (4.2.1) and (4.2.15):

$$p + p_i(v) = \frac{3kT}{v - v_d}. \qquad (4.2.27)$$

This is important, as the internal pressure is much more susceptible than the motional pressure to the details of the intermolecular interaction. For example, the rectangular simplification of the potential well for the estimation of the right-hand side of (4.2.27) makes p_i vanish if used in (4.2.4). Fortunately, there is no need for this. Formulae (4.2.5) and (4.2.5a) define the internal pressure much more reliably than the motional.

Boiling

Now let us consider the conditions required for equilibrium coexistence of the liquid and its vapor. Because the phase boundary is permeable, there exist oppositely directed fluxes of evaporating and condensing particles which must be in balance. However, there is a difference between the two fluxes. A molecule encounters no difficulties in passing from the gas to the liquid, while in the opposite case it has to overcome a potential barrier created by attractive forces. To do this, it must have an energy $\psi_0 = |\Phi(\infty)|$ which is considerably greater than the average heat resource (Fig. 4.10). Among the a^{-2} particles per unit surface of the liquid, there are only $a^{-2} \exp(-\psi_0/kT)$ able to cope with the problem. All of them hit the barrier ν times per second, executing ordinary thermal motion within the well. In view of the estimate of ν given in (4.2.19), we can readily see that the flux of particles leaving the liquid is as follows

$$j_+ = \frac{\nu}{a^2} e^{-\psi_0/kT} = n\bar{v} \frac{a}{4(a-d)} e^{-\psi_0/kT} . \qquad (4.2.28)$$

As for the opposing flux, according to (1.4.12), it is equal to

$$j_- = \frac{1}{4} n_V \cdot \bar{v} , \qquad (4.2.29)$$

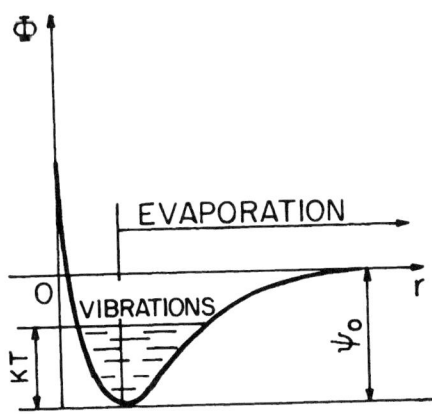

Figure 4.10 The cage potential Φ of a surface molecule in the direction perpendicular to a phase boundary. At energies larger than $\psi_0 = -\Phi(\infty) \gg kT$ the molecule is able to escape and thus to evaporate.

where n_V is the vapor density. Equating both expressions, we find the equilibrium condition

$$n_V = \frac{n a}{a - d} e^{-\psi_0/kT}. \qquad (4.2.30)$$

Its physical meaning is quite clear. Two phases located on different sides of the step barrier are in equilibrium only because of the higher density of the phase from which it is more difficult to escape.

At low temperatures, when the vapor above the liquid can be treated as ideal, the mere multiplication of (4.2.31) by kT gives the following

$$p_V = p_t \exp\left(-\frac{\lambda}{RT}\right), \qquad (4.2.31)$$

where p_V is the vapor pressure, p_t is the motional pressure of the liquid defined in (4.2.22), and $\lambda = n_0 \psi_0$ is the molar latent heat of evaporation. Equality (4.2.31) is just the equation of the liquid–vapor coexistence curve. One can see that the change in motional pressure in passing through the phase boundary is determined by the energy required for this passage. For example, under normal conditions, the vapor pressure of mercury is about $16 \cdot 10^{-5}$ atm, and motional pressure of liquid mercury is close to $4 \cdot 10^4$ atm. Thus, according to (4.2.31), we have for the heat of evaporation

$$\lambda = RT \ln \frac{p_t}{p_V} = 2 \cdot 273 \cdot 2.3 \lg \frac{4 \cdot 10^4 \cdot 760}{1.6 \cdot 10^{-4}} = 13.8 \text{ kcal/mole},$$

which is close to the observed result. Over a limited temperature range, λ and p_t may be considered constant as compared to p_V. Then, on differentiating (4.2.31), we obtain the relation

$$\lambda = RT^2 \frac{d(\ln p_V)}{dT}, \qquad (4.2.32)$$

known in thermodynamics as the "Clausius–Clapeyron equation" (5.3.5). The recovery of this equation points to correctness of the above semiquantitative estimates.

Melting

A much more complicated situation arises with the order–disorder transition which occurs when a crystal is transformed into liquid or dense gas. Within the free volume, theory the liquid and the crystal phases are indistinguishable while the collective entropy variation is ignored. It is precisely this term in Eq. (4.2.11) that is responsible for the crystallization of a real substance or even a hard-sphere fluid. As there is no attraction in the hard-sphere model only pressure is sufficient to order the system, just as billiard balls pressed into shape form a lattice. Since the crystallization is still possible when p_i disappears one may expect $\Gamma(n)$ to undergo itself a nonmonotonous (S-like) or even discontinuous alteration with increasing density. While uncertainty in the collective entropy precludes us for getting something like the van der Waals loop in $\Gamma(n, T)$, computer experiments reliably indicate that there are two separate stable branches of this isotherm, for the ordered and the disordered sphere system. The compressibility factor of such a system,

$$F = \frac{pv}{kT} = \Gamma,$$

which is just the Γ-factor, has been calculated a several times since it became possible in the late 1950s.

There are two ways of performing this calculation for an ensemble of several hundred hard spheres, which imitates a macrosystem. One of the methods is to integrate in time the mechanical equations of motion of the entire ensemble at a given external volume and total energy of the system. This method, known as "molecular dynamics" reproduces the actual trajectory of the system from which all observed quantities are obtained by time averaging. Another technique is based on the concept of the canonical ensemble and is known as the Monte Carlo method. A computer chooses arbitrarily a sequence of microstates, each of which is a particular configuration of particles and their velocities. These are weighted with the Boltzmann factor for each configuration, and the statistical sum is calculated. Both recipes allow one to determine all parameters of the equilibrium state and to follow their variations under change of external conditions. It turns out that for both methods the isotherm of the hard-sphere system is discontinuous. One of its branches belongs to the crystal phase, while the another, to the disordered one (Fig. 4.11). Moreover, the molecular-dynamic method enables one to see in the phase transition region the erratic abrupt transitions of the system between ordered and disordered states.

Taking these results into account to correct the behavior of the motional pressure of a real substance, we can easily account for crystallization (Fig. 4.12). This transformation differs from crystallization of hard spheres only in that the pressure condensing the liquid into an ordered state is internal rather than external. At low temperatures, only the branch of the isotherm corresponding to the ordered state intersects the isobar $p + p_i$, and at high temperatures, only the disordered branch, while at the melting temperature T_m, there are stable intersections with both of them. The spacing between the two intersection points along the abscissa axis corresponds to a slight increase in the volume accompanying the phase transition. In Fig. 3.18 this spacing appears as the corridor that separates the crystal phase from any others.

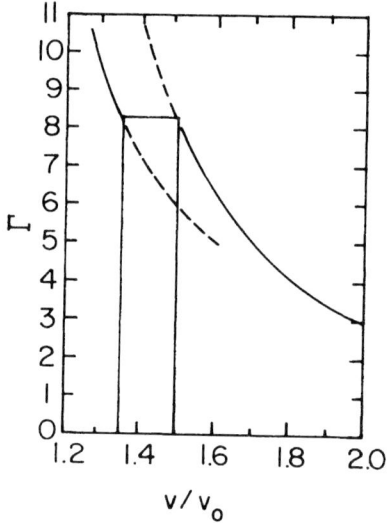

Figure 4.11 The solid (left) and fluid (right) branches of the hard sphere equation of state. The horizontal line is a guess at the tie line connecting the two phases. The dashed extensions of the solid and fluid branches represent metastable states generated on the computer. ($v_0 = d^3/\sqrt{2}$ is the close-packed specific volume.) (From B. J. Alder and W. G. Hoover, *Physics of Simple Liquids*, edited by H. N. V. Temperley, J. S. Rowlinson and G. S. Rushbrooke, North-Holland Publishing Company Amsterdam 1968.)

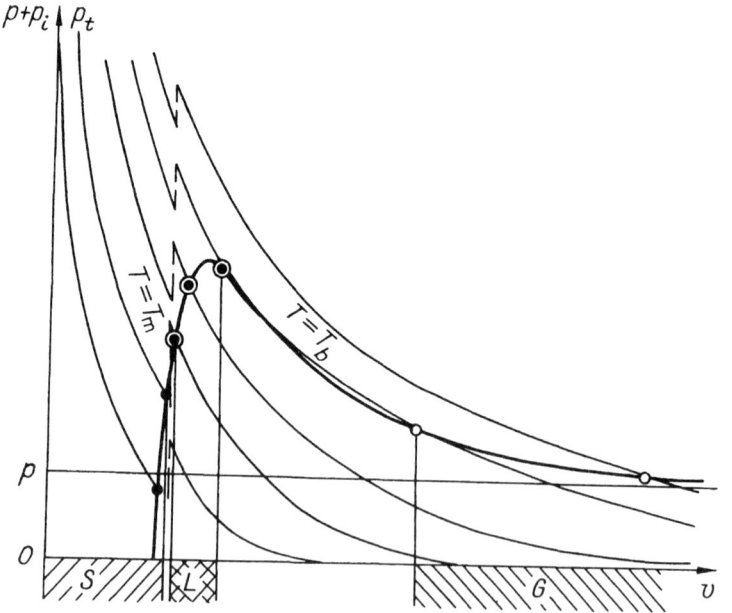

Figure 4.12 The intersections between isotherms $p_t(v)$ and total force at pressure p in solid (●), liquid (⊙), and gas (○). Corresponding volume regions (S, L, G) are marked by different hatching.

Thus the isotherm discontinuity allows one to account for all phenomena related to melting without qualitative contradictions. For consistency with the free volume theory, one must treat seriously the last component from formula (4.2.11). This term represents the state of the ensemble on the whole. Unfortunately, all that we know for sure is that $\bar{\sigma}$ varies from 1 in ideal crystal state to e in expanded liquid or gas. Originally, Lennard-Jones and Devonshire, the originators of the free volume theory, assumed this variation to proceed abruptly and only upon melting. With this assumption they even estimated the melting heat. However, this point of view proved to be invalid. Partial disordering of crystals always precedes melting. In this state the liquid is much more ordered than rarefied matter. Therefore, the total entropy is bound to increase gradually in a finite density interval. It would be more reasonable to assume that both the entropy and its derivative have a discontinuity at the melting point (Fig. 4.13). Just the break in the derivative is needed to abruptly change the motional pressure, while the entropy increases gradually with increasing disordering of the system. If this is the case, then it is clear why the free volume theory developed for $\bar{\sigma} = \text{const}$ is adequate for crystals and acceptable for liquids, but inapplicable in the vicinity of the melting point. It is also obvious that the internal pressure is unaffected by this problem. As a local measure of tension, it is indifferent to a cooperative change of the ensemble on the whole.

Orthometric Line

Now let us return to the problem of the orthometric curve (briefly outlined in Section 1.9). The orthometric curve is the locus of points corresponding to the pseudoideal states of matter with $F = 1$ (the abscissa axis on Fig. 3.18). The free volume theory is able to describe only a small section of this curve in solid

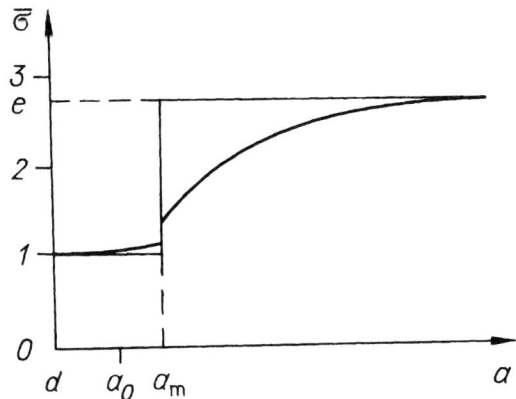

Figure 4.13 The change of collective entropy $\bar{\sigma}$ under expansion with a break at the melting point a_m (thick line), and its stepwise approximation according to Lennard-Jones and Devonshire (thin line).

state, while the virial expansion is restricted to the region adjacent to the ideal gas state: the coordinate origin in Fig. 3.18. Here the hard-sphere model is inapplicable because within its limits $F \equiv \Gamma > 1$ for all $p > 0$, and orthometric states are inaccessible. Surprisingly, the van der Waals theory, despite its demerits, allows one to correctly determine the form of the orthometric curve from the ideal gas to the highly condensed state. As there are no miracles in natural sciences, we expect and should check that this success is accidental.

To do this, let us reexpress (4.2.1) to represent the equation of state as

$$\frac{1}{F} = \frac{\Phi}{p} + \frac{1}{\Gamma}\left[1 - \frac{\Phi}{p}\right], \qquad (4.2.33)$$

where

$$\Phi = \frac{p_i}{\Gamma - 1} \qquad (4.2.34)$$

is the ratio of the van der Waals corrections for real gases. According to (4.2.33), the orthometric curve specified by $F = 1$ is given by the equation

$$p = nkT = \Phi. \qquad (4.2.35)$$

In the van der Waals theory the fundamental characteristic of a substance $\Phi(n, T)$ is temperature-independent and quadratic in density, that is, has a form of "parabolic arch":

$$\Phi = \frac{an^2}{\frac{1}{1-bn} - 1} = \frac{a}{b}(1 - bn)n, \qquad (4.2.36)$$

where $a = A/N_0^2$ and $b = B/N_0 = 4v_d$. One zero of this function is due to the more rapid decrease of $p_i(n) = an^2$ compared with $(\Gamma - 1)$ as $n \to 0$, while the other from the divergence of $\Gamma(n)$ at the point $n = 1/b = 1/4v_d$ (Fig. 4.14). However, we have already seen (Fig. 4.6) that the Γ-factor has no singularity at the point $n = 1/4v_d$. The divergence of the Γ-factor in the van der Waals theory demonstrates the inadequacy of this model at high densities. Nevertheless, this divergence is responsible for the fact that the van der Waals orthometric curve,

$$p = \frac{a}{b}(1 - bn)n = nkT, \qquad (4.2.37)$$

becomes linear in the coordinates (T, n), thus recovering (1.9.21):

$$T = T_B(1 - 4v_d n), \qquad T_B = \frac{a}{bk}. \qquad (4.2.38)$$

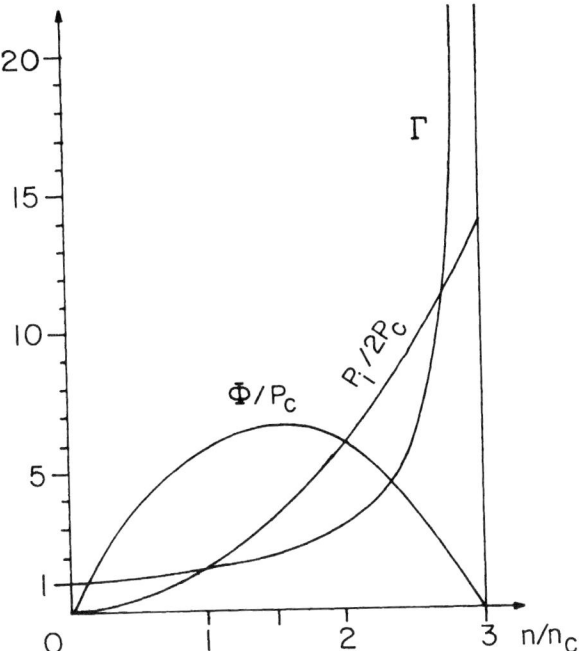

Figure 4.14 The orthometric arch Φ, the inner pressure p_i, and Γ-factor as functions of reduced density $n/n_c = 3\,b\,n$ in the van der Waals theory.

Though the linearity of $T(n)$ is an experimental fact, the slope of the orthometric curve obtained experimentally does not agree with that derived from (4.2.38). As is seen from Fig. 4.15, the slope specified by the orthometric density at $T = 0$ is less than half than that established experimentally (for nitrogen). Meanwhile, the density at $T = 0$ is just the coordinate of the singularity of the van der Waals Γ-factor which is physically meaningless. Such an explanation of the orthometric curve cannot be considered satisfactory. At the same time, modification of the van der Waals equation to remove the non-physical divergence at $n = 1/4v_d$ makes matters worse. For example, if we use one of formulae (4.2.17) without a divergence at this point, but retain the internal pressure in the original van der Waals form, then the orthometric line becomes the curve (Fig. 4.15). Furthermore, the second zero of the function $\Phi(n)$ in this case is also associated with the divergence of the Γ-factor but at $n = 1/v_d$. This point is located four times higher along the density scale than its previous position. As a result, the orthometric density at $T = 0$ is greater than the true value by the same factor it was previously less than it.

The way out is prompted by the free volume theory. According to this theory, it is the internal pressure rather than the Γ-factor that causes the second zero of $\Phi(n)$, as it vanishes long before the Γ-factor goes to infinity. As a result,

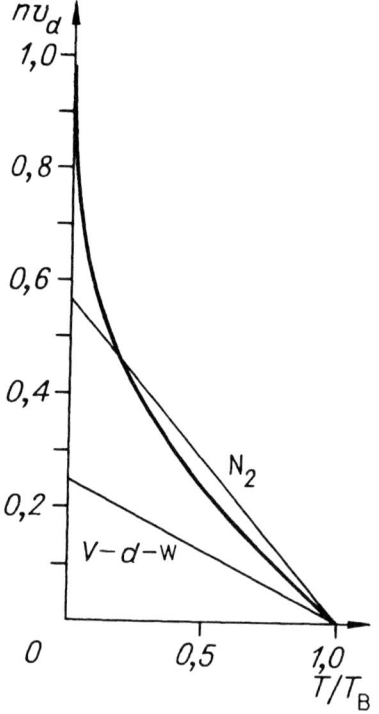

Figure 4.15 The real orthometric line for nitrogen in comparison with that of van der Waals (marked V-d-W) and within the model of attracting hard spheres (thick curve).

the orthometric curve expressed in coordinates (T, n) can be reliably determined not only at $n = 0$ but also at $n = n_0$ because $p_i(n_0) = \Phi(n_0) = 0$. Indeed,

$$kT = \frac{\Phi(n)}{n} = \begin{cases} kT_B, & n = 0, \\ 0 & n = n_0. \end{cases} \qquad (4.2.39)$$

The orthometric straight line joining these points is then of the form

$$T = T_B \left(1 - \frac{n}{n_0}\right). \qquad (4.2.40)$$

The density of an ideally packed face-centered structure, where the internal pressure and Φ become zero, is

$$n_0 = \frac{\sqrt{2}}{a_0^3} = \frac{1}{\sigma^3}. \qquad (4.2.41)$$

This is approximately twice as large as $1/4v_d = 3/2\pi\sigma^3$ and half of $1/v_d = 6/\pi\sigma^3$, just as shown in Fig. 4.15.

The equation of the orthometric curve in the solid phase follows from (3.4.35) at $F = 1$:

$$T = \frac{v - v_0}{(3\gamma - 1)\kappa k} \approx \frac{1}{(3\gamma - 1)\kappa k n_0} \cdot \left[1 - \frac{n}{n_0}\right] \qquad (4.2.42)$$

The linear dependence is valid at $T \gg \Theta$, so (4.2.42) is only the high temperature asymptotic of the orthometric curve for crystal. In the case of argon, for which the Debye temperature is rather high (93 K), this asymptotic is tangent to the orthometric curve at the melting point. It intersects with the orthometric straight line of the liquid at $T = 0$ (Fig. 4.16). This is no surprise: it follows immediately from a comparison of\,(4.2.40) and (4.2.42) that the orthometric curves of liquid and crystal meet at $T = 0$, $n = n_0$. In fact, the intersection of solid and liquid orthometric lines testifies that the position of the zero of internal pressure is independent of the aggregate state. Fig. 4.17 allows one to judge what is the accuracy of this intersection. Since the Debye temperature of xenon is well below the melting point, the linear high temperature asymptotic of the crystal orthometric curve (4.2.42) is well-marked.

Finally, the coincidence of the zeros of the Φ-function and the internal pressure becomes evident if we resolve the equation of state (4.2.1) into components. Before we discuss this procedure let us compare the result of such a decomposition (Fig. 4.18) with that of the van der Waals equation (Fig. 4.14).

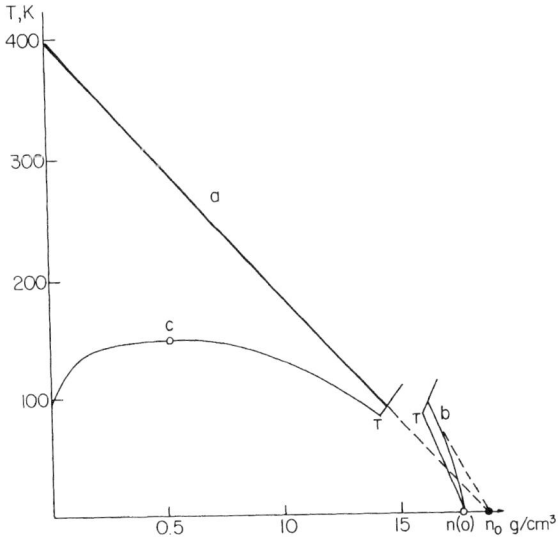

Figure 4.16 Orthometric curves of (a) liquid and (b) solid argon separated by the crystallization corridor (c is the critical point, T the triple point). The dashed lines are the extrapolation of the linear parts of the curves to low temperatures.

226 LIQUIDS

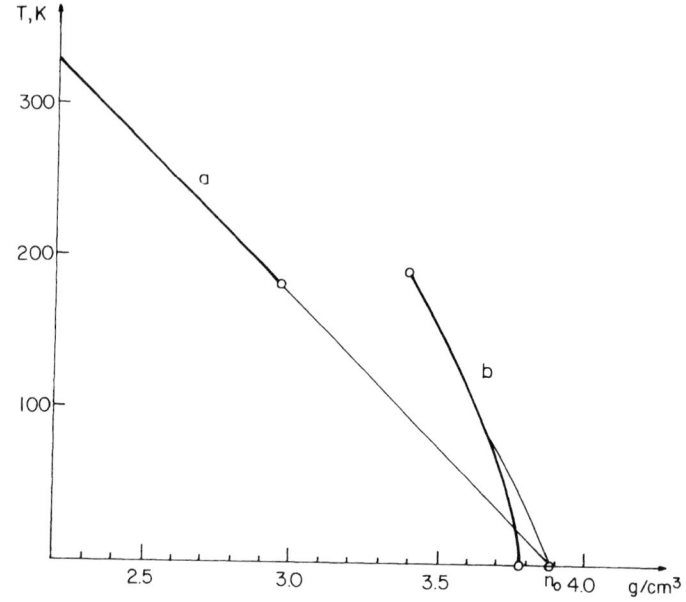

Figure 4.17 Intersection of the orthometric straight lines of (a) liquid and (b) solid xenon (the thin lines are their extrapolations to $T = 0$).

Evidently, the zero of the internal pressure is responsible for the fact that the orthometric arch $\Phi(n)$ passing through the gas and the liquid phases terminates at the point corresponding to an ideally packed crystal. The peculiarities of the behavior of the Γ-factor have nothing to do with this.

As the orthometric arch has two zeros, it must have a maximum located between them. The maximum orthometric pressure $p_0 = \Phi(n_x)$ is the coordinate of the "knot" in the Hougen–Watson chart on Fig. 1.18. In the vicinity of the extremum the following expansion is valid

$$\Phi = p_0 - \Delta (n - n_x)^2 . \qquad (4.2.43)$$

However, in simple liquids this expansion holds over the whole interval between the two zeros, since, according to (4.2.39) and (4.2.40),

$$\Phi = kT_B \left(1 - \frac{n}{n_0}\right) n . \qquad (4.2.44)$$

By redefining the parameters according to

$$\Delta = \frac{kT_B}{n_0} , \quad n_x = \frac{n_0}{2} , \quad p_0 = \Delta n_x^2 = \frac{kT_B n_0}{4} , \qquad (4.2.45)$$

we can identify (4.2.43) with (4.2.44).

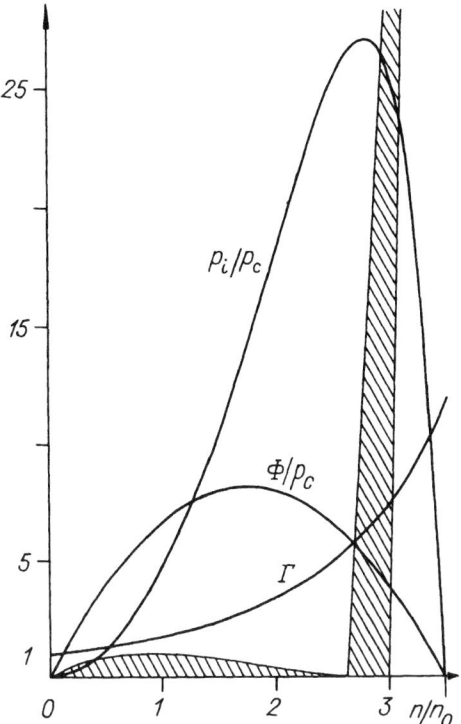

Figure 4.18 The Γ-factor of argon together with the orthometric arch $\Phi(n)$ and internal pressure $p_i(n)$ normalized by p_c. The regions below the liquid–gas coexisting curve and in the crystallization corridor are shaded.

Decomposition of the Equation into Components

None of the above models of matter (real gas, hard spheres, ideal crystal) is sufficiently universal to serve as a basis for a general description of the liquid state. Nevertheless, it is desirable to universally resolve the equation of state into two components p_i and p_t in order to compare all aggregation states on a common ground. What are our expectations based on the preceding considerations? They are collected in Table 4.1.

According to van der Waals the internal pressure of real gas (4.2.5a) is quadratic in particle density n while the Γ-factor is a rational fraction (4.2.16) of reduced density $y = nv_d$. In the hard-sphere model there is no internal pressure but the more complex functions $\Gamma(y)$ in (4.2.17) are valid at higher densities. At the highest densities, in a crystal phase, the internal pressure in harmonic approximation (3.4.10) is linear in density while $\Gamma = 3\gamma$ is approximately constant according to (3.4.19). In between, at moderate densities, where the liquid phase is expected to be, the internal pressure is given by Eq. (4.2.5) of the lattice (free volume) theory (see also (4.2.55)) which includes the harmonic

Table 4.1

p_i	Theories	Γ
an^2	van der Waals	$(1-4y)^{-1}$
—	Hard-sphere model	$\dfrac{1+y+y^2}{(1-y)^3}$
$\alpha n^3\left[1-\left(\dfrac{n}{n_0}\right)^2\right]$	Lattice model	
$\dfrac{1}{\kappa}\left(1-\dfrac{n}{n_0}\right)$	Mie–Grüneisen	3γ

approximation as a high density limit. The lattice estimate of the Γ-factor leads also to a rational fraction (4.2.14) but this estimate is too rough to be considered. Although the free volume theory is heuristically rather useful, it cannot be extended to a liquid density range when the contribution from collective entropy to Γ was neglected. The extrapolation of the hard-sphere result to this region looks more reasonable and the break between this result and the crystal estimate 3γ is natural. However, one may combine the lattice p_i with the hard sphere Γ only for semiquantitative calculations.

The lack of universality indicates that there is no unique method to resolve the equation of state into components. Neither internal nor motional pressure is observable, and neither may be rigorously defined within thermodynamics. However, one may proceed phenomenologically. Knowing that the compressibility factor becomes unity on the orthometric line $T = T_0(n)$, we bring the equation of state into the following form:

$$F = 1 + Xn\left[1 - \frac{T_0}{T}\right], \qquad (4.2.46)$$

where $X(n, T)$ is an unknown function. The advantage of such a representation lies is that $X(n, T)$ will have a much simpler form than $F(n, T)$, and will be easier to approximate. At the same time, the identification of (4.2.46) with a phenomenological expression

$$F = \Gamma - \frac{p_i}{nkT} \qquad (4.2.47)$$

suggests the simplest way of resolving equation of state into components

$$p_i = X(n, T)\, n^2\, kT_0(n) \qquad (4.2.48)$$

and

$$\Gamma = 1 + X(n, T)n .\qquad(4.2.49)$$

If $X(n, T)$ remains finite as $n \to 0$, and $T_0(0) = T_B$, then in dilute gases

$$p_i = bn^2 kT_B = an^2 \quad \text{and} \quad \Gamma = 1 + bn\qquad(4.2.50)$$

depend on density as given by the van der Waals equation with the parameters

$$b(T) = X(0, T), \qquad a(T) = b(T) \cdot kT_B .$$

The last relation is the direct generalization of\,(1.9.20). According to (4.2.46) and (1.9.17), the second virial coefficient is

$$B_0 = b(T) \cdot N_0 \left(1 - \frac{T_B}{T}\right),\qquad(4.2.51)$$

and is related to the parameter $A = aN_0^2$ by

$$A = \frac{RT_B B_0(T) T}{(T - T_B)} .\qquad(4.2.52)$$

This parameter is positive over the whole temperature range, because both the denominator and $B_0(T)$ reverse sign at $T = T_B$.

Now let us check that these phenomenological definitions of p_i and Γ are also in good agreement with the theory of condensed matter. To this aim, let us substitute the explicit expression $T_0(n)$ from (4.2.40) into (4.2.48). It is seen that p_i becomes zero when $n = n_0$:

$$p_i = X \cdot n^2 kT_B \left(1 - \frac{n}{n_0}\right)\qquad(4.2.53)$$

in agreement with its actual behavior in condensed matter. Since $\Gamma = 1 + Xn$ the orthometric arch

$$\Phi = \frac{p_i}{\Gamma - 1} = nkT_0 = kT_B n \left(1 - \frac{n}{n_0}\right)\qquad(4.2.54)$$

does not depend on X and coincides with that defined in (4.2.44).

For $n \to n_0$, expression (4.2.5) can be expected to adequately represent the dependence of p_i on density,

$$p_i = \text{const} \cdot n^3 \left[1 - \frac{\mu}{\lambda} n^2\right] = \alpha n^3 \left[1 - \frac{n^2}{n_0^2}\right]. \qquad (4.2.55)$$

Using this information in (4.2.53), we can predict the behavior of $X(n)$ in dense media as well as in gases:

$$X = \begin{cases} b(T) & \text{at } n = 0, \\ \dfrac{\alpha n}{kT_B}\left(1 + \dfrac{n}{n_0}\right) & \text{at } n \approx n_0, \end{cases} \qquad (4.2.56)$$

where $\alpha = \text{const}$.

Now we are ready to completely recover the form of the internal pressure and the Γ-factor at any density. Information on both of these can be obtained from direct comparison between the phenomenological equation (4.2.46) and experimental findings. At present there exist many empirical multiparametric equations determining the state of simple liquids to within experimental accuracy. Using these equations, one can rigorously determine the form of $X(n, T)$ at any density and temperature and then both p_i and Γ defined in (4.2.48) and (4.2.49) in terms of $X(n, T)$. Such a decomposition of the equation into components was presented in Fig. 4.18.

The actual data thus obtained show how the alternative approximations, gas-like and solid-like, work at low and high densities, respectively. As seen from Fig. 4.19, the van der Waals approximation gives an estimate of internal pressure correct to an order of magnitude, but describes neither the bending nor the maximum that occurs in the region of the liquid phase. On the contrary, the lattice model is valid near the maximum and indicates correctly the zero position, but it underestimates the internal pressure in real gases. The boundary between the two approaches lies near $n = n_0/2$, that is, approximately midway between the critical and triple points.

As for the Γ-factor, it increases monotonically up to the crystallization boundary (Fig. 4.20). If we compare it with the hard-spheres approximation, by choosing v_d such that their low-density asymptotics coincide, it will turn out that the actual Γ-factor increases less steeply with density. For the crystal phase (dashed lines), the distinctions increase and also become qualitative: within the hard-sphere model the Γ-factor continues to increase with increasing density, while in fact it is approximately constant. This can be attributed to the fact that the hard-sphere model may not be extended to the region where repulsion and attraction are of equal importance.

4.2 EQUATION OF STATE   231

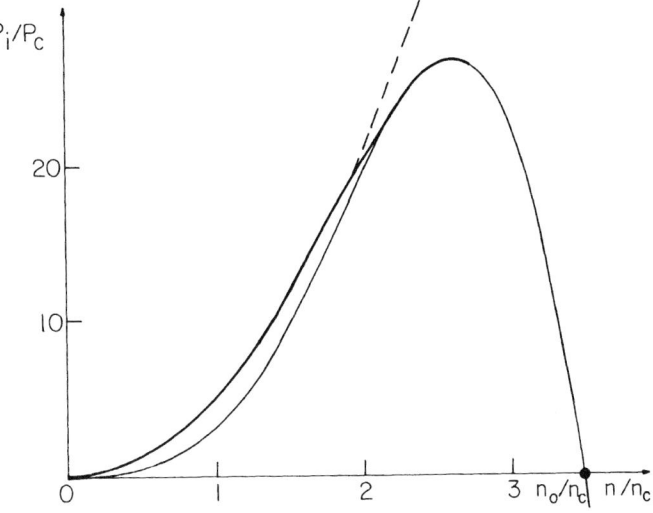

Figure 4.19 Argon internal pressure at $T = 2T_c$ as a function of density (bold curve), its lattice approximation (thin curve) and the van der Waals asymptotic (dashed line).

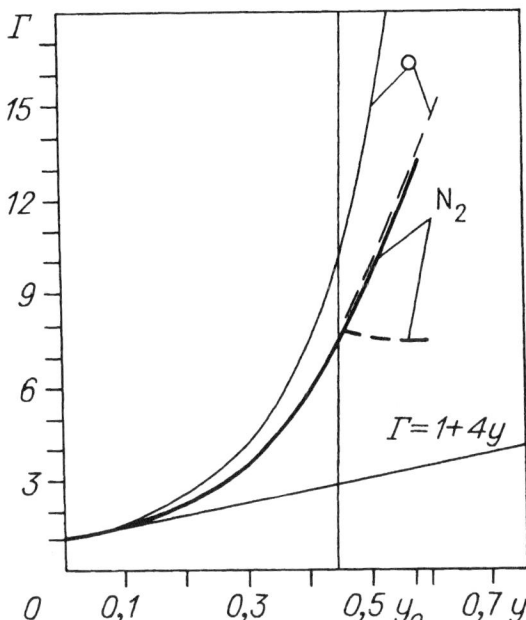

Figure 4.20 The two-branch Γ-factor for N_2 (thick lines) and hard sphere fluid (○) as function of reduced density $y = nv_d$. Solid lines correspond to disordered and dashed lines to ordered states of matter. The straight line $\Gamma = 1 + 4y$ is a low-density asymptotic similar to that of the van der Waals gas.

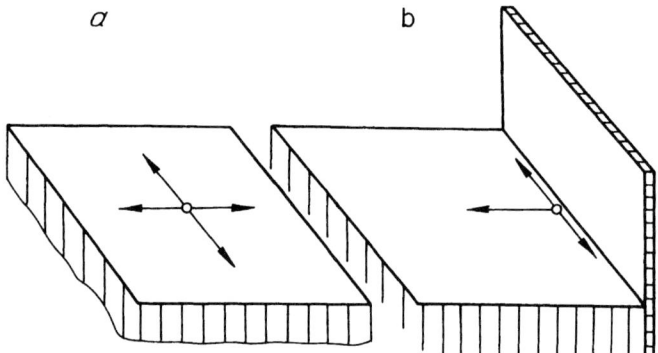

Figure 4.21 Forces acting on the molecule (a) in the surface layer and (b) at the edge of it coming into contact with a wall.

4.3 SURFACE PHENOMENA

A salient feature of liquids is their ability to twist their surface quite freely. When not contained in vessel and not affected by gravity, the liquid rolls itself into a ball, assuming the shape with minimum surface area. This phenomenon is well-known and is due to the fact that the liquid surface resembles an elastic film uniformly extended in all directions which tends to reduce its area to a minimum.

If the liquid surface is flat, there are no external manifestations of its tension. It can be observed, however, using a plane partially submerged in the liquid (Fig. 4.21). The surface molecules in contact with the plane will experience the action of the liquid from one side only. Obviously, the resultant of the force of tension is perpendicular to the line of contact, and its magnitude is directly proportional to the length of this line

$$F = \gamma l. \qquad (4.3.1)$$

The force acting per unit length is the "surface tension" of the liquid γ.

If the plane in question is a mobile bar fixed in the frame filled with a layer of liquid, it intersects two surfaces (Fig. 4.22), the edges of which pull it in the same direction with equal force. The mediators of the force are particles adjacent to the bar and bound to it by intermolecular attraction. Owing to these coupling forces, the bar is "stuck" to the liquid, and, if not held in a fixed position, begins to move under the action of both liquid surfaces. This motion stops if a load of weight equal to the sum of the two forces is suspended on the bar. If the bar still moves, though extremely slowly and isothermally, the work executed on lifting the weight to the height h is

$$A = -\Delta \mathcal{F} = 2Fh = \gamma \cdot 2lh = -\gamma \Delta S. \qquad (4.3.2)$$

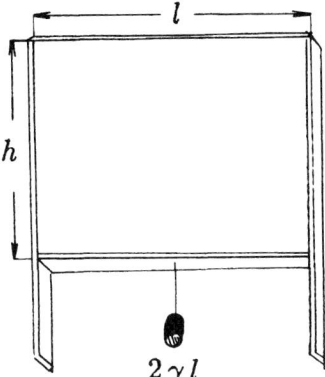

Figure 4.22 A frame enclosing a liquid film, with one side a mobile plank. The suspended load compensates the surface tension of the two surfaces of the film.

This occurs at the expense of the free energy of liquid whose decrease $-\Delta\mathcal{F}$ is proportional to the reduction of the surface area $-\Delta S = 2lh$. Clearly

$$\gamma = \left(\frac{\Delta\mathcal{F}}{\Delta S}\right)_{T,V,N}$$

has the meaning of the excess of free energy of the surface layer per unit area.

The origin of this excess is easily traced. Molecules which find themselves in the surface layer are half separated from the liquid, since there are no neighbors above. They execute vibrations transverse to the surface, in a one-sided well which is less deep than that inside the bulk (Fig. 4.23). The average potential energy of surface particles is larger than for those deep inside the liquid. If the surface particles pass into the bulk, the energy difference is converted into work and heat. Thus the existence of surface tension can readily be accounted for in terms of energy.

The origin of the forces responsible for surface tension is more difficult to understand. Evidently they are due to intermolecular interactions binding together the particles of the surface layer. A quantitative measure of this interaction is the internal pressure acting along the surface. As seen from Fig. 4.23 this pressure is less than that acting in a bulk. However, in the surface as well as in the bulk, the internal pressure competes with the motional pressure. If the latter dominates, the surface will exert pressure on the wall, rather than pull it. This is what happens inside the bulk where $p = p_t - p_i > 0$ and just the opposite situation occurs at the interface with the pressure acting along the surface. Due to a space anisotropy created by the surface, this pressure component, known as the *tangential pressure* of liquid, p^T, differs from that in perpendicular direction, which is the *normal pressure*, p^N. Away from the phase boundary (with either a gas or a liquid) both pressure components are equal to p.

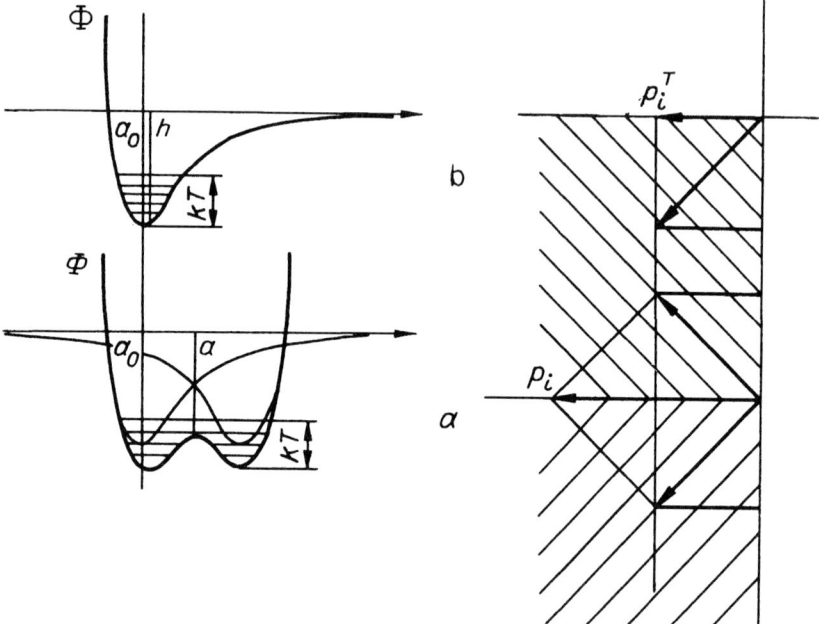

Figure 4.23 The cage potential Φ (left) and the tangential internal pressure p_i^T (right) (a) in the bulk, where both quarter-spaces contribute in p_i, and (b) in the surface, where only one of them is left and only one neighboring atom contributes in Φ.

However, in approaching the surface the tangential pressure decreases or even change sign, creating tension (5.6.26) directed towards the liquid:

$$\gamma = \int_{-\infty}^{+\infty} \left[p - p^T(z) \right] dz . \quad (4.3.3)$$

Hence, we come to conclusion that the motional pressure along the surface is even less than internal pressure though the latter is smaller than in the bulk.

As the surface tension is specified at the coexistence curve, it is a function of temperature only, decreasing with increasing T and going to zero at the critical point. At the same time, the interface is extremely sharp near the triple point, but it gradually spreads and becomes a wide region near the critical point. For this reason, it is necessary to invoke alternative models for the description of surface structure and tension at low and high temperatures. At low temperatures, the surface may be treated as monomolecular, and rather simple estimates for its tension may be used. At high temperatures, the transition region is so wide that it can be considered as a continuous medium, with its density $n(z)$ gradually decreasing in passing from liquid to gas. In this case, the calculation of the surface tension and structure must be self-consistent. This is achieved with the use of quasithermodynamics to be discussed in Section 5.6.

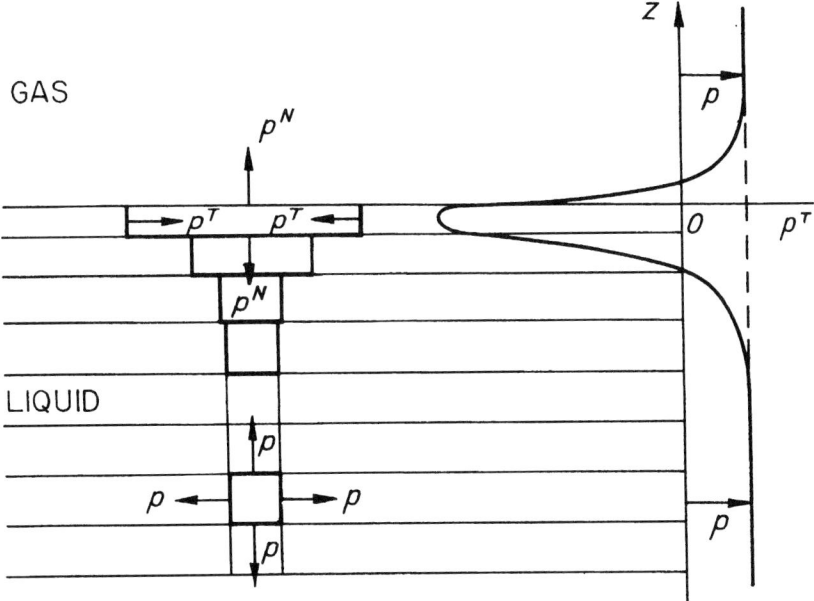

Figure 4.24 Deformation of the cage as the surface is approached (left); and the corresponding variation in tangential pressure $p^T(z)$ (right). The horizontal lines represent molecular layers.

Monomolecular Surface

In the case of a sharp liquid–vapor boundary, only the few layers adjacent to it are affected (Fig. 4.24). Simplifying the situation to some extent, we assume that only the upper layer of the thickness h is subject to deformation and makes a decisive contribution to γ. The calculated surface tension will agree with the experimental value only when the tangential pressure is negative and well exceeds p in magnitude. In this case, according to (4.3.3)

$$\gamma = [p - p^T] h \approx -p^T h \approx -p_i^T h . \qquad (4.3.4)$$

The last expression should be considered as the upper estimate for $p^T = p_i^T - p_t^T$, because the motional pressure p_t^T, although smaller than internal, is not necessarily negligible. The estimate (4.3.4) is actually valid, provided that $|p_i^T| \gg p_t^T, p$.

Deep inside the liquid the pressure is defined by Eq. (5.6.20). The components of the pressure tensor in the surface layer may be defined by analogy:

$$p^N = -\frac{1}{s}\left(\frac{\partial f}{\partial h}\right)_{T,s}, \quad p^T = -\frac{1}{h}\left(\frac{\partial f}{\partial s}\right)_{T,h}, \qquad (4.3.5)$$

where f is the free energy per molecule in the surface layer, h is the height, and s is the area of the surface cell. Internal pressure components can be defined by analogy with their isotropic value (4.2.4)

$$p_i^N = \frac{1}{2s}\left(\frac{\partial \Phi}{\partial h}\right)_s, \qquad p_i^T = \frac{1}{2h}\left(\frac{\partial \Phi}{\partial s}\right)_h. \qquad (4.3.6)$$

Assuming that the number of the nearest neighbors in the surface layer is m, and $s = \xi l^2$ where l is the distance between them, we have

$$p_i^T = \frac{m}{4\xi hl} \cdot \frac{du}{dl}. \qquad (4.3.7)$$

As has already been mentioned, at average distances between molecules typical for the liquid phase, attractive forces dominate over repulsion. This is especially true for the surface where $l > a$. Thus, neglecting repulsion and assuming attraction to be of dispersion origin, we use $u = -4\epsilon(\sigma/l)^6$ in (4.3.7) and obtain

$$p_i^T = \frac{6m\epsilon\sigma^6}{\xi l^8 h}. \qquad (4.3.8)$$

Substituting this result into (4.3.4), we finally have

$$\gamma = \frac{6m\epsilon\sigma^6}{\xi l^8} = \frac{24\epsilon\sigma^6\sqrt{3}}{l^8}. \qquad (4.3.9)$$

The above expression is obtained on the assumption that the (111) plane of a closely packed cubic lattice with the lowest particle density ($m = 6$, $\xi = \sqrt{3}/2$) is exposed at the surface.

Eq. (4.3.9) gives the order-of-magnitude estimate of γ. However, l is still unknown, that is, the question as to how the surface is extended as compared to the volume remains open. To answer this question rigorously, it is necessary to consider the conditions for equilibrium coexistence of the surface layer and the liquid volume. As follows from Eq. (5.6.18), this calls for the equality of normal pressure and chemical potential in the bulk and at the surface. In principle, the two parameters (h and s) can be chosen such that both conditions are met. Actually, the problem is more complicated. One cannot proceed from the monomolecular nature of the surface to determine its structure. This can be obtained by considering equilibrium in several contacting surface layers. To avoid this complication, we merely assume that the liquid remains "incompressible" up to the phase boundary, that is, the volume of a cell is not changed under deformation:

$$v = h \cdot s = \frac{\sqrt{3}}{2} \sigma l^2 .$$

The last expression makes use of the negligible thermal lengthening of h (estimated by formula (3.4.33)) so that $h \approx \sigma$. Using this in (4.3.9), we derive

$$\gamma = \frac{27\sqrt{3}}{2} \cdot \frac{\epsilon \sigma^{10}}{v^4} . \tag{4.3.10}$$

The surface tension thus represented is a universal function of the liquid specific volume. This conclusion is consistent with the "conservation law" of the quantity

$$\mathcal{P} = \frac{N_0 \gamma^{1/4}}{n_L - n_V}, \tag{4.3.11}$$

phenomenologically established by A. I. Batschinski and called *parachor* by Sugden. As now we deal with low temperatures, the vapor density n_V in Eq. (4.3.11) may be ignored in comparison with $n_L = 1/v$. Then, substituting (4.3.10) in (4.3.11), we obtain the following estimate of the parachor

$$\mathcal{P} = L\epsilon^{1/4}\sigma^{5/2} , \tag{4.3.12}$$

where $L = [(27\sqrt{3})/2]^{1/4} \cdot N_0 = 9 \cdot 10^{23}$. The existence of such a relationship between the parachor and the potential parameters was established by Lennard-Jones and Corner. As we see in the Table 4.2, taken from the book of Hirschfelder, Curtiss, and Bird, the constant L is practically the same for many simple liquids and differs little from the calculated value which is the upper estimate.

Smooth Interface

Now let us turn to the high temperature range, when the liquid and vapor densities are comparable and influenced by the phase boundary even at quite

Table 4.2

Liquid	$\mathcal{P}$	$L \cdot 10^{-23}$
Ne	25	7.6
Ar	54	7.0
N_2	60.4	6.8
CO	61.6	6.8
CH_4	73.2	7.1

large distances from it. To determine the character of this influence, let us apply quasithermodynamics (Chapter 5, Section 5.6) to find the unique density distribution $n(z)$ in the transition layer which provides a constant normal pressure (5.6.35), and, therefore, a constant chemical potential (5.6.34). To this aim, we rewrite the pressure conservation law (5.6.35) in the form of the differential equation

$$\frac{b}{2} \dot{n}^2 = m, \qquad (4.3.13)$$

the right-hand side of which is defined by

$$m = p - \mu n + f_0(n) \cdot n, \qquad (4.3.14)$$

and depends only on density (not on the coordinate). Integrating (4.3.13), on the assumption that b is independent of n, we find

$$z = \sqrt{\frac{b}{2}} \int_{\bar{n}}^{n} \frac{d\rho}{\sqrt{m(\rho)}}. \qquad (4.3.15)$$

It is reasonable to place the origin of coordinates of the z axis at the point where the medium density is equal to the arithmetic mean of the vapor and liquid densities: $\bar{n} = (n_L + n_V)/2$.

Formula (4.3.15) gives the density profile in the form of the dependence $z(n)$. To determine it, we need only know the function $m(n)$. Taking into consideration that, according to (5.6.19),

$$p_0 = (\mu_0 - f_0)n, \qquad (4.3.16)$$

and $\mu = \mu_0(n_L)$, this function may be redefined in the following way:

$$m = p - p_0 - [\mu - \mu_0]n = p - p_0 + n \int_{n_L}^{n} \frac{d\mu_0}{dn'} dn'. \qquad (4.3.17)$$

Differentiating (4.3.16) using $\mu_0 = n(df_0/dn) + f_0$, according to (5.6.21), we easily obtain that

$$\frac{d\mu_0}{dn} = \frac{1}{n} \frac{dp_0}{dn}.$$

Substituting this relation into (4.3.17) and integrating by parts, we eventually arrive at the result

$$m = n \int_{n_L}^{n} \frac{(p_0 - p)}{n'^2} dn'. \qquad (4.3.18)$$

The dependence $m(n)$ is completely defined by the above expression at any vapor pressure p corresponding to a given temperature T. The knowledge of the isotherm $p_0(n)$ corresponding to the same temperature is sufficient for the calculation of this dependence. In the simplest case, one can use the van der Waals isotherm; however, it is better to employ the currently existing multiparametric equations of state which describe the isotherms of particular gases within experimental error. This makes it possible to calculate $m(n)$ with the aid of (4.3.18), and, then, to use this information in (4.3.15) to determine the profile $z(n)$ characterizing the structure of the transition region. The latter depends on the constant b, the numerical value of which is unknown.

In order to remove this uncertainty, one need only take into consideration that the surface tension defined by formula (5.6.37) of the van der Waals theory is equal to

$$\gamma = b \int_{n_L}^{n_V} \dot{n}^2 \, dn . \qquad (4.3.19)$$

Since n must satisfy the equilibrium condition (4.3.13), this formula is easily reexpressed as

$$\gamma = \sqrt{2b} \int_{n_L}^{n_V} \sqrt{m(\rho)} \, d\rho . \qquad (4.3.20)$$

This indicates that both the surface tension and structure are unambiguously defined by the function $m(n)$. Taking information from the appropriate equations or directly from experiment, we can calculate γ through formula (4.3.20) and, comparing it with the observed value, determine the only unknown parameter of the theory b. Thus the density distribution in the transition layer $n(z)$ is unambiguously defined.

This program was carried out by Carey and coworkers with the use of model equation of state and by Burshtein and Shokhirev with the use of the analytical equations of state for nitrogen. At high temperatures the analytical equations describe adequately both the liquid and gas branches of the isotherm, and, more importantly, the transition region between them (see Fig. 1.17). The results are given in Fig. 4.25. The higher the temperature, the closer the densities of both phases to critical value, and the wider the transition region between them. This is also supported by the change in the pressure tensor near the phase boundary. This is characterized by the value $\Delta p = p - p^T \equiv p^N - p^T$. According to (5.6.35), (5.6.39) and (4.3.13), it takes the value

$$\Delta p = b\dot{n}^2 = 2m . \qquad (4.3.21)$$

As seen from Fig. 4.26, this quantity does not change sign but reaches a maximum near $z = 0$ and becomes zero at a distance several times greater than the molecular diameter of nitrogen (3.7 Å).

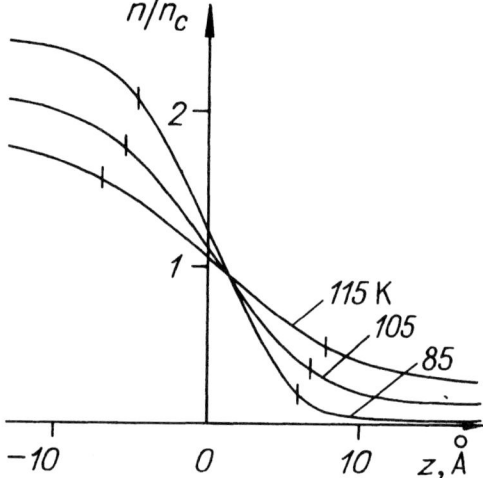

Figure 4.25 Density profiles normalized to the critical density n_c at different temperatures.

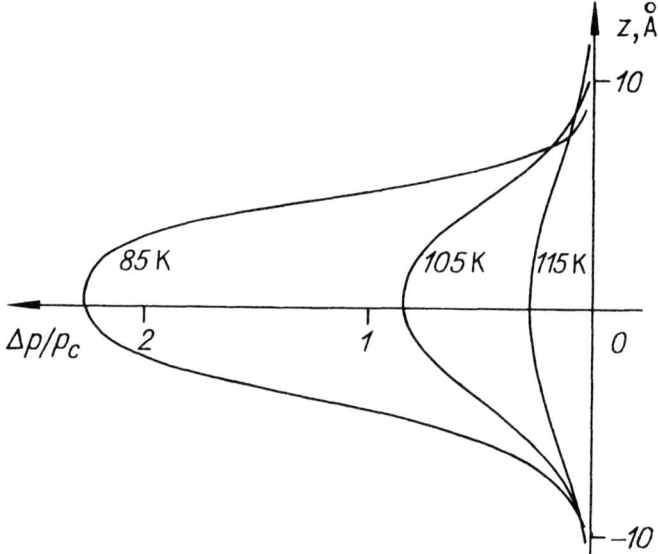

Figure 4.26 Excess of tangential pressure Δp defined in Eq. (4.3.21) within the interfacial region.

It should be noted that the quantity $\Delta p(z)$ is not unique. As shown in Section 5.6, the local free energy f and tangential pressure p^T (as well as Δp) are not unambiguously defined. Integrating (5.6.29) by parts, we can bring the free energy of the two-phase system into the following equivalent form

$$\mathcal{F} = \int \left[f_0 - \frac{1}{2} b \ddot{n} - \frac{1}{2} \cdot \frac{db}{dn} \dot{n}^2 \right] n \, dz . \qquad (4.3.22)$$

Identifying the integrand with fn, we find that the free energy is

$$f = f_0 - \frac{1}{2} b \ddot{n} - \frac{1}{2} \frac{db}{dn} \dot{n}^2 , \qquad (4.3.23)$$

and the tangential component of the pressure determined from (5.6.27) in view of (5.6.34), (4.3.23), and (4.3.16), takes the form

$$p^T = p_0(n) - \tfrac{1}{2} b(n) n \cdot \ddot{n} . \qquad (4.3.24)$$

As for the normal pressure (5.6.35), it remains the same, equal to p. However, using (5.6.34), (4.3.93), we can reduce it to the form

$$p^N = p_0(n) - b(n) \cdot n \ddot{n} + \frac{b(n)}{2} \dot{n}^2 - \frac{1}{2} \frac{db}{dn} n \dot{n}^2 = p . \qquad (4.3.25)$$

At $b = $ const the tensor anisotropy changes as follows:

$$\Delta p = \frac{b}{2} [\dot{n}^2 - n\ddot{n}] = \frac{1}{2} [m + p - p_0] . \qquad (4.3.26)$$

Unlike (4.3.21), this quantity can reverse sign (Fig. 4.27). Its maximum and minimum correspond to the extrema of the van der Waals loop, and the section between them would be unstable states, if the medium were isotropic (Fig. 5.25c).*

Since it is assumed that the substance density varies continuously from liquid to gas, such states inevitably occur, and the corresponding substance layer is located at the phase boundary. On the density scale its limits are marked by dashes on the curves (Fig. 4.25). Now let us suppose that the state of the substance in this layer remains unstable, and, hence, unrealizable. Then, an abrupt change in density is to be found near the phase boundary which is as sharp as the case of a monomolecular layer. However, in both directions away from the boundary the density varies gradually, asymptotically approaching its limiting values n_L and n_V (Fig. 4.28). This is how Tolmen, a pioneer of surface phenomena studies, visualized the transition layer and the behavior of the pressure tensor within it.

---

*As the isotherm section joining the maximum and minimum is obtained by extrapolation of experimental data to the unstable region, the extrapolation becomes less reliable for lower temperatures. In particular, this deficiency is responsible for the nonphysical bending at 85 K (see Fig. 4.27).

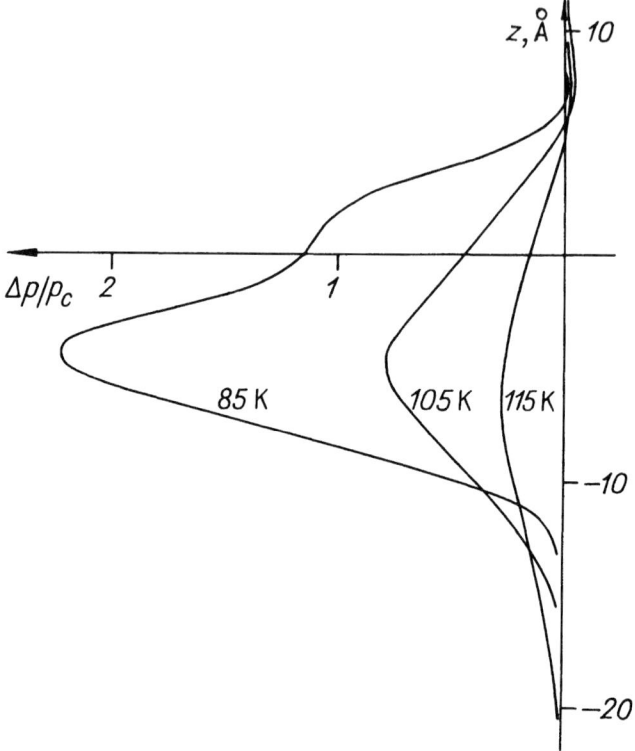

Figure 4.27 Excess of tangential pressure Δp redefined in Eq. (4.3.26) reproducing the van der Waals loop in coordinate space.

4.4 TRANSFER PHENOMENA

In order to describe transfer processes in gases, it is sufficient to know the free path time, that is, the velocity conservation time. To do the same for liquids, model theories often involve two times. One of them, the velocity correlation time (4.2.20), is a successor of the free path time, and the other is the length of time a particle spends in certain surroundings and in a certain place before changing its environment. The closest analog of the latter is the time that a dislocated atom wandering over crystal spends in any particular interspace. However, the difference is in that there is no order in the liquid, and therefore there are no defects. Thus nomadic life is typical for the entire ensemble, rather than for its isolated members.

Taking into consideration that in a dense medium $na^3 \approx 1$ we can represent formula (4.2.20) as

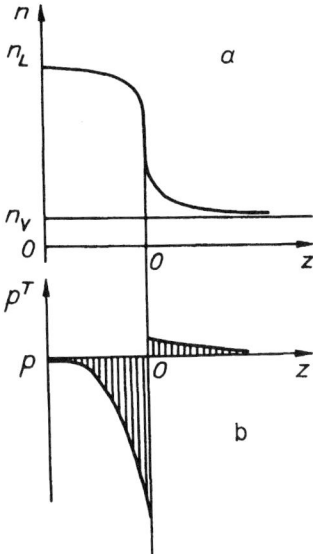

Figure 4.28 The discontinuity in (a) the density and (b) the tangential pressure p^T when the unstable region of the interface is excluded.

$$\frac{1}{\tau_c} \approx \frac{na^3 \bar{v}}{a-d} \simeq n\sigma\tilde{v}, \qquad (4.4.1)$$

where $\sigma \approx a^2$ is a cross-section. The only essential difference between the above estimate of collision rate and that given in Eq. (1.15.8), is the effective speed

$$\tilde{v} = \bar{v}\frac{a}{a-d}, \qquad (4.4.2)$$

which is considerably greater than the average heat velocity. When the finite volume of particles is properly taken into account, this alteration is inevitable. Just as the collision cross-section defines the effective transverse size of a particle, so the coefficient $a/(a-d)$ in Eq. (4.4.2) is related to its longitudinal size. This reduces each free path by d. In dilute gases, the correction $\lambda/(\lambda - d)$ is close to unity and may be ignored, but in dense media, where $\lambda \sim a$, $\tilde{v}$ is an order of magnitude larger than $\bar{v}$, it must be necessarily taken into consideration.

When the kinetic energy of particles is much greater than the potential energy, their free path is limited only by collisions. That is why at high temperatures, in the supercritical region, transfer phenomena are governed by τ_c only. This inference has been verified in hard-sphere systems by computer simulation experiments. The particle motion was traced in time by molecu-

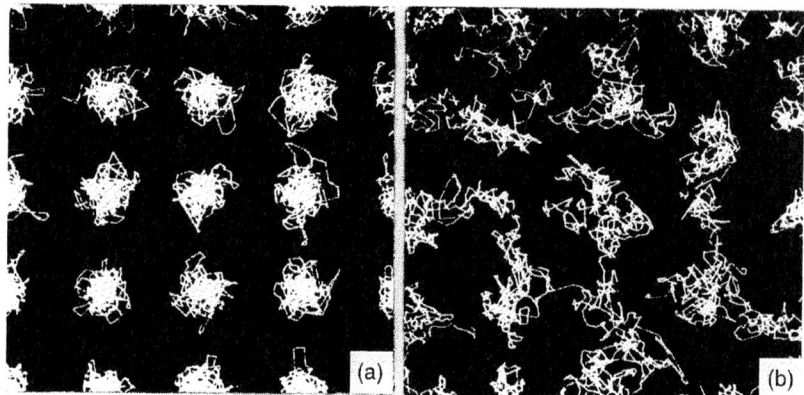

Figure 4.29 The molecular trajectories in a 32-particle hard-sphere system in a cubic box projected onto the xy plane in (a) solid phase, (b) liquid phase. (From T. W. Wainwright and B. J. Alder, *Del Nuovo Cimento*, 1958, Supplemento al vol. IX, serie X, No. 1, p. 116.)

lar-dynamic methods both in the ordered state corresponding to the crystal branch of the isotherm, and in the disordered one corresponding to the state of a gas of approximately the same density (Fig. 4.29). In the latter case rare displacements and permutations of particles are possible, despite the high density of the surroundings. However, it is established that these permutations proceed by sequential short steps obeying almost the ordinary statistics of free paths (1.15.13). Non-Poissonian deviations from the exponent occur at $t \approx 4\tau_0 \div 8\tau_0$ and do not exceed 20% at the most.

At low temperatures, when the free motion of molecules is limited by coupling forces, the liquid behaves differently. On the one hand, the nearest neighbors of a particle are able to prevent it from changing its position, even if there is a good chance to settle somewhere nearby. On the other hand, such an opportunity does not arise frequently.

The first difficulty is quite similar to that encountered by a dislocated atom or vacancy wandering throughout a crystal. In order to leave the cell a particle has to overcome the barrier ΔU at the expense of thermal energy. Assuming that overcoming the barrier is the only problem, and after this the particle will face no difficulties in settling down, it is natural to expect that the frequency of transitions can be expressed by a formula similar to (3.5.10). Under this assumption, Frenkel defined the time of a molecule's "settled life" as

$$\tau = \tau_c \exp\left(-\frac{\Delta U}{kT}\right). \tag{4.4.3}$$

The exponent appearing in this relation shows that there is a considerable increase in coordinate conservation time τ compared with the velocity conservation time τ_c. This time ratio makes it possible to treat a low-temperature liquid as a quasicrystal structure with its short-range order remaining unchanged for a fairly long time. Although a particle crosses a cell, on the average, in a time $\tau_c \sim 10^{-13} s$, it does not leave it more often than $10^{10} - 10^{11}$ times per second, executing approximately 100 vibrations in the same neighborhood. This is also true for amorphous solids which do not differ fundamentally from liquids, except for their longer rearrangement time τ (seconds, hours and more) due to the large value of $\Delta U/kT$. As a result, some plastic masses compete in their strength characteristics with crystal solids.

From this point of view, the separation of amorphous media into liquids and solids is merely conventional, and is defined in the kinetic sense only. Substances with $\tau \sim 10s$ are considered to be solids, since the plasticity of a substance may be tested by hand over this time scale. If one uses a fast-acting experimental apparatus to study plasticity, the criterion will be shifted towards shorter times, while in the case of a prolonged static load it will be shifted towards longer times. That is why, unlike the melting point, the glassing temperature is a kinetic characteristic, not an equilibrium one. Its position on the temperature scale is determined by the point where τ reaches a prescribed value. Any liquid, including such a low-viscosity one as water, has all properties of an amorphous solid under high-frequency tests. Under normal conditions, these are not manifested solely for the reason that the load variation proceeds slowly compared with $\tau \sim 10^{-11} s$. In usual situations when a man jumps into water, or a ship screw rotates in it, the time of the action of the force is always greater than τ; thus the liquid has enough time to move in an appropriate direction. In the case of a bullet, it enters the liquid so rapidly that the latter fails to respond to its penetration. The particles have no time to escape from their cells and change positions so as to make way for it. As a result, at such speeds the liquid on the whole behaves as a solid, breaking and cracking on sudden impact.

Heat Conduction

The heat motion in a monoatomic liquid, such as mercury or argon, consists only of random vibrations of atoms. The energy inherent in each atom is transferred freely from one to another at each collision. So only the heat transfer in the liquid has nothing to do with the rearrangement of structure, but is limited only by the collisional frequency (4.4.1). The mechanism is the same as in gases, except that the parameters in formula (1.16.3) should be correspondingly redefined

$$\kappa = kn\bar{v}\lambda \to kn\tilde{v}a . \qquad (4.4.4)$$

The high density of the substance is responsible for the reduction of the free path length to a and for the higher rate of energy transfer in accordance with formula (4.4.2). The increase in the rate of energy and momentum (sound) transfer in a dense media can be attributed to the sequential "relay" mechanism of this process. The velocity of a torch is also greater than that of the runners, since, in passing from hand to hand, it instantly covers a distance equal to their total length d. Using (4.4.2) in (4.4.4), we get

$$\kappa = kn^{2/3}\bar{v}\frac{a}{a-d} = kn^{2/3}\sqrt{\frac{3kT}{m}}\left(\frac{v}{v_f}\right)^{1/3},$$

where v and v_f are the specific and free volumes, respectively. Surprisingly, this rough estimate is in a rather good agreement with experimental data on the heat conduction of monoatomic liquids.

Viscosity

All other transfer processes related to the change of the particles' positions, such as diffusion, electric conduction, and viscosity are limited by the structural rearrangement time τ. According to Eyring, the excess Δu of the velocity of the liquid layer with respect to a neighboring layer closer to the solid surface can be attributed to the drifting of molecules under the action of the corresponding component of the shearing tension F

$$\Delta u = q \cdot (Fa^2), \qquad (4.4.5)$$

where q is the molecular mobility, and a^2 is the area per molecule in a moving layer. Dividing the difference in velocity by the distance between neighboring layers, we obtain the phenomenological law of viscous flow

$$\frac{\Delta u}{a} = qaF = \frac{F}{\eta}. \qquad (4.4.6)$$

In order to clarify the meaning of the mobility q or the viscosity coefficient η, let us estimate microscopically the velocity increment Δu which results from the presence and migration of vacancies in a flowing layer of the liquid.

Similarly to the migration of impurities in crystals, vacancies (holes) in the liquid structure move with and against the stream with different probabilities. It is easier for a vacancy to move against the stream, since the atom filling it is shifted in the direction of the impressed shearing force, while in the opposite situation the reverse takes place. Thus the height of the barrier which must be overcome either decreases or increases by the magnitude $\frac{1}{2}Fa^3 = F/2n$ respectively, and the probabilities of transitions against and with the stream take a form similar to (3.5.13)

$$K_+ e^{-F/2kTn} = K_- e^{-F/2kTn} = K = K_0 e^{-\Delta U/kT}. \qquad (4.4.7)$$

So the average velocity of drifting vacancies which are shifted by a as a result of each transition is equal to

$$c = a(K_+ - K_-) = aK \cdot 2 \cdot \sinh\left(\frac{F}{2nkT}\right). \quad (4.4.8)$$

However, this is certainly not equivalent to the relative velocity of the layer current Δu. Unlike vacancies, the motion of which is only limited by energetic barriers, molecules are not free in their movement and execute a jump only if there is a vacancy nearby. Obviously, the frequency of encounters with vacancies is determined by the concentration n_v of vacancies in the medium, and is equal to $c\, n_v\, a^2$. Assuming that on each encounter an atom is shifted streamwise by the distance a, we can estimate

$$\Delta u = a^3 c n_v = aK \cdot 2 \sinh\left(\frac{F}{2nkT}\right) p_v, \quad (4.4.9)$$

where $p_v = n_v a^3$ is the probability of finding a hole in the liquid structure. This dimensionless coefficient which distinguishes (4.4.9) from (4.4.8) is often ignored, although Eyring took it into consideration and believed that it could be estimated by the formula

$$p_v = \frac{v - v_0}{v_h} \geq \frac{v - v_0}{v_0}. \quad (4.4.10)$$

Here v_0 is the specific volume of a tensionless solid, and v_h is the volume of the vacancy that may be less than v_0 but sufficient for rearrangement.

The nonlinear relationship (4.4.9) between the stream velocity and the shearing force indicates that, generally speaking, the stream becomes Newtonian only at relatively small values of the external force. Under such a restriction, one may expand the sinh function to the first term only, and formula (4.4.9) is brought into one-to-one correspondence with (4.4.6), with

$$\eta = \frac{nkT}{W}. \quad (4.4.11)$$

Here

$$W = K p_v = K_0 e^{-\Delta U/kT} p_v \quad (4.4.12)$$

is the frequency of the displacement of a particle to its neighboring cell, and $6W = 1/\tau$ is the rate of structural rearrangements which result in a change of the particle temporary position. A particle attempts to leave it in any one direction K_0 times per second, and the total frequency of attempts is

$1/\tau_c = 6K_0$. That is why the coordinate and velocity conservation times are still related to one another as in (4.4.3)

$$\tau = \tau_c \frac{e^{-\Delta U/kT}}{p_v}, \qquad (4.4.13)$$

but with an important correction taking account the necessity to make the structure "loose" enough for a particle to be able to move.

As is seen from (4.4.6), both the fluidity of the medium and the mobility of particles in it are limited by the probability of structural rearrangement

$$q = \frac{1}{\eta a} = \frac{Wa^2}{kT}. \qquad (4.4.14)$$

Since the mobility of particles is connected with the diffusion coefficient by the Einstein relation, the latter can also be expressed in terms of W

$$D = qkT = Wa^2 = D_v p_v. \qquad (4.4.15)$$

As before, the reduction in the coefficient of self-diffusion with respect to the vacancy diffusion coefficient $D_v = a^2 K$ is determined by the fraction of vacancies. The equilibrium properties of the medium as defined by the equation of state affect essentially its kinetic characteristics via p_v. As all kinetic coefficients are related to each other, we restrict ourselves to the analysis of viscosity only.

The Free Volume Theory

Strictly speaking, the above reasoning refers to an ordered lattice, but the liquid does not fall into this category. In a liquid the free volume is not divided into an integer number of equally sized vacancies, but is dispersed almost continuously, that is, distributed among all particles in different amounts. The particles with little free volume remain more or less stationary, while others having more free space at their disposal migrate using the available backlashes. Developing this idea consistently, we arrive at the conclusion that p_v has the meaning of the fraction of particles possessing the free volume v^+ smaller than a molecular volume but sufficient to escape from the potential cage. To determine how large this fraction is, Cohen and Turnbull derived the distribution of particles over their free volume on the basis of entropy considerations only. We will now consider this approach.

Assuming the particles to be interacting hard spheres differing only in the volume of space they occupy, one may divide the volume into cages i and assign the spheres into groups of n_i particles occupying the cages of the same volume Δ_i. Then one may calculate the number of possible ways of making such a partitioning, under the conditions that the total volume and the total number of particles N remain unchanged. Since permutation of cages of equal size does not lead to a new state of the system, the total number of distinguishable states is

$$W = \frac{N!}{\prod_i N_i!}, \qquad (4.4.16)$$

and

$$\sum_i N_i = N, \qquad \sum_i \tilde{v}_i N_i = v_F \cdot N, \qquad (4.4.17)$$

where $\tilde{v}_i = \Delta_i - v_d$ is the free volume of the ith hole, the nominal volume of which is Δ_i, and

$$v_F = v - v_d \qquad (4.4.18)$$

is the average free volume per molecule. It is obvious that the total free volume of the system $v_F N = V - N v_d$ remains unchanged, as do V and N. It should be noted from the outset that the "free volume" used here is, in essence, the empty volume of a cell and does not coincide in its definition with the Lennard-Jones and Devonshire free volume v_f available for a particle occupying the cell. In the case of hard spherical molecules, there is a simple correlation between these notions:

$$v_f = \text{const}\left[v^{1/3} - v_d^{1/3}\right]^3 \approx \text{const}\,[v - v_d]^3/v_d^2 = \text{const}\,\frac{v_F^3}{v_d^2}, \qquad (4.4.19)$$

where the value of the constant depends on the packing of particles (the lattice type).

As the system's entropy is proportional to the quantity

$$\ln W = N \ln \frac{N}{e} - \sum_i N_i \ln \frac{N_i}{e}, \qquad (4.4.20)$$

our goal is to find its maximum under the additional conditions (4.4.17). Applying the Lagrange method, we define the quantity

$$\varphi = \ln W - \alpha \sum_i \tilde{v}_i N_i - \beta \sum_i N_i$$

and determine its extremum by varying the number of particles in the groups with the same free volume

$$\frac{d\varphi}{dN_i} = -\ln\frac{N_i}{e} - 1 - \alpha\tilde{v}_i - \beta = 0.$$

This immediately yields the desired distribution

$$N_i = \frac{\exp(-\alpha\tilde{v}_i)}{\exp(\beta)} = \frac{N\exp(-\alpha\tilde{v}_i)}{\sum_i \exp(-\alpha\tilde{v}_i)}, \qquad (4.4.21)$$

where β is determined from the normalization condition $\sum_i N_i = N$. Now taking into consideration that in fact the free volume assumes a sequentially continuous series of values, it is necessary to transform the distribution (4.4.21) correspondingly

$$dN = N\frac{\exp(-\alpha\tilde{v})d\tilde{v}}{\int \exp(-\alpha\tilde{v})d\tilde{v}} = N\cdot e^{-\alpha\tilde{v}}\alpha d\tilde{v}.$$

In this expression the integration extends from 0 to ∞, though, in fact, the magnitude of the free volume is bounded both above and below. Therefore, it is assumed that in the liquid phase the average free volume is equally distant from both limits, minimum and maximum. With the second condition (4.4.17) represented as $\int \tilde{v}dN = v_F N$, we can easily see that $\alpha = 1/v_F$, and, consequently, the distribution of free volumes,

$$dW = \exp\left(-\frac{\tilde{v}}{v_F}\right)\frac{d\tilde{v}}{v_F}, \qquad (4.4.22)$$

is identical in structure to the distribution of free path times and lengths in a gas. Using this distribution, we can easily determine the fraction of particles with the free volume $\tilde{v} > v^+$:

$$p_V = \int_{v^+}^{\infty} dW = \exp\left(-\frac{v^+}{v_F}\right). \qquad (4.4.23)$$

This is a quantity which appears in expression (4.4.12) and competes with the Arrhenius factor. The latter is usually less efficient and the temperature-dependence of fluidity almost does not manifest itself when the volume of liquid is fixed. On the other hand, when the pressure is fixed this dependence is well pronounced because the specific volume increases as a result of thermal expansion and essentially affects p_v. In glassing media, both the range and scale of viscosity variation are wider than in crystallizing liquids. Their behavior is even better described by the "free volume theory" of fluidity which allows for packing fluctuations, particularly important at low temperatures and high pressures.

At high temperatures when $v_F > v^+$, $p_v \approx 1$ and the flow is simply an activated process, exactly as Frenkel believed. However, at lower temperatures, the

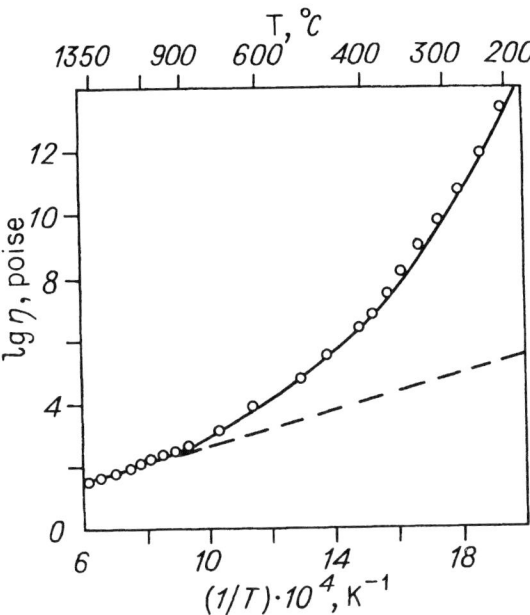

Figure 4.30 Superexponential increase of viscosity η in molten B_2O_3 with cooling (from T. A. Litovitz and C. M. Davis in *Physical Acoustic*, Part A, Vol. 2, p. 281 (1965).) The high-temperature Arrhenius behavior is shown by the dashed line.

lack of spare space causes the dependence $\ln \eta(1/T)$ to curve (Fig. 4.30). As a result, the effective activation energy increases with cooling, reaching values which are so large (up to 4 eV) that it is meaningless to associate them with actual barriers. However, within the free volume theory of viscosity this fact is not at all surprising, due to the sharp decrease in the volume during glassing. A phenomenological parameter v' in Eq. (4.4.23) is just an excess of free volume which is necessary to open a passage to neighboring cells. Entering a narrowing tunnel between its neighbors (see Fig. 1.28) an inertially moving particle continues deep inside with a good chance to reach the exit.

4.5 BROWNIAN MOTION

Nonstationary Diffusion

Now let us consider the time development of one-dimensional diffusion in a thermally uniform medium. In order to initiate it, one can pour some tea on to the surface of water in a deep glass. It is evident that, owing to diffusion, the colored strip will spread and fade monotonically in time, until the concentration of tea becomes the same throughout the vessel. This process is easily described quantitatively by taking into account that $Sj(x)$ dissolved particles enter any horizontal layer of thickness dx per second, while $Sj(x+dx)$ particles leave it, S being the area of the layer. Dividing the difference of these fluxes by the volume Sdx, we obtain the rate of change of concentration:

$$\frac{dn}{dt} = \frac{j(x) - j(x+dx)}{dx} = -\frac{dj}{dx}. \tag{4.5.1}$$

This is the "continuity equation" expressing the conservation of particle number. Using the formal definition of the diffusion flux (1.16.7), we get

$$\frac{\partial n}{\partial t} = D \frac{\partial^2 n}{\partial x^2}. \tag{4.5.2}$$

Though this is a partial differential equation, n is a function of one universal variable $z = x/\sqrt{t}$. By direct substitution of $n(z)$ into (4.5.2) statement, one may obtain the ordinary differential equation in terms of z:

$$D\frac{d^2n}{dz^2} + \frac{z}{2}\frac{dn}{dz} = 0. \tag{4.5.3}$$

The general solution of the equation is

$$n(x,t) = C_1 \int_{-\infty}^{x/\sqrt{4Dt}} \exp(-t^2)dt + C_2, \tag{4.5.4}$$

which involves the constants C_1 and C_2.

If initially the two liquids were separated by a plane, then the spreading of the boundary between them is described as follows

$$n = \frac{n_0}{\sqrt{\pi}} \int_{-\infty}^{x/\sqrt{4Dt}} \exp(-t^2)dt.$$

If the colored liquid originally lies in a thin layer $0 < x < x_0$ inside the clear fluid, and spread from there in both directions, then

$$n = \frac{n_0}{\sqrt{\pi}} \int_{(x-x_0)/\sqrt{4Dt}}^{x/\sqrt{4Dt}} \exp(-t^2)dt.$$

Decreasing the width of the layer x_0 and simultaneously increasing the initial concentration n_0 in such a way that $n_0 x_0 = N = \text{const}$, we obtain in the limit of an infinitely narrow colored layer

$$n = \frac{N}{\sqrt{4\pi Dt}} \exp\left(-x^2/4Dt\right). \tag{4.5.5}$$

The distribution of the colored particles at any instant of time after their start from the origin of coordinates is given by the following formula:

$$dW(x,t) = \frac{ndx}{N} = \frac{\exp(-x^2/4Dt)}{\sqrt{4\pi Dt}} dx . \tag{4.5.6}$$

The Gaussian density of this distribution expands monotonically and the average distance between the particles and their starting point increases linearly in time

$$\overline{x^2} = \int_{-\infty}^{+\infty} x^2 dW(x,t) = 2Dt . \tag{4.5.7}$$

Thus, the width of the region where the colored particles are approximately uniformly distributed increases in time as $\sqrt{t}$. This is a lengthy process: at room temperatures the strip takes many hours to increase up to 1 cm in width. It is even slower if Brownian particles rather than molecules are involved in the process. However, the rate is the only difference. Brownian motion is a visualization of thermal motion which obeys all appropriate statistical laws.

The Langevin Equation

Although now we accept the above statement as a truism, at some time it had to be experimentally verified. Those who were inclined to consider the diffusion equation phenomenologically saw no commonality between substance dissolution and Brownian motion. On the other hand, subscribers to the atomic concept believed that the arbitrary movements of Brownian particles and their intricate paths could be attributed to the action of some random force varying continuously and randomly in magnitude and direction. This force is the resultant of a large number of impacts acting on a microscopic particle from its surroundings. A large body submerged in the liquid practically does not respond to these fluctuations, which are smoothed by its inertia. Brownian particles are greater than atoms in size, but are still small enough to demonstrate the real existence of thermal motion through their own random behavior. This conclusion was a convincing argument in favor of the atomic concept, when laws predicted by the theory were revealed in the random motion of Brownian particles.

The surrounding medium produces a bifold effect on the motion of a Brownian particle. On the one hand, it presents a natural resistance to this movement, increasing linearly with the velocity of motion. The resisting force is $F = -\dot{x}/q$, where q is the mobility of a particle. On the other hand, the molecular surroundings stimulate the Brownian motion by providing a randomly oscillating force $f(t)$ acting on the particle. The fluctuations in force impart an acceleration to a particle in random directions, thus causing it to execute random motion even though $\overline{f(t)} = 0$. According to the mechanical laws, it responds to these fluctuations according to the Langevin equation

$$M\ddot{x} = -\frac{\dot{x}}{q} + f(t). \qquad (4.5.8)$$

The random nature of the force $f(t)$ causes the response $x(t)$ to be also a random function of time.

There is a definite relationship between the regular and random actions produced by the molecular surroundings. The regular influence manifests itself in retardation of the particle velocity, which is described by the averaged equation

$$\dot{v} = -\frac{1}{\tau_0} v,$$

where $v = \dot{x}$, and $\tau_0 = qM$ is the relaxation time of the particle velocity. According to current ideas, the relaxation rate is expressed in terms of the variance of the random force $\overline{f^2}$ by the formula

$$\frac{1}{\tau_0} = \overline{f^2} \tau_c / kTM, \qquad (4.5.9)$$

where τ_c is the force correlation time which is the successor of the collision duration. It is usually assumed that

$$\tau_c \ll \tau_0, \quad \text{and} \quad t \gg \tau_c, \qquad (4.5.10)$$

but these restrictions may be avoided if one turns to a generalized Langevin equation,

$$M\ddot{x} = -\int_0^t R(t-t')\dot{x}(t')dt' + f(t).$$

Its kernel is the so-called memory function $R(t-t') = \overline{f(t)f(t')}/kT$, which has decay time τ_c as with the force correlation function $\overline{f(t)f(t')}$. If conditions (4.5.10) are met, one can bring the more slowly decaying function $\dot{x}(t)$ outside the integral, and allow the upper limit to tend to infinity, thus obtaining the Langevin equation with $q = \tau_0/M = kT/\overline{f^2}\tau_c$.

Perrin's Experiments

Observing the motion of Brownian particles away from the starting line, Perrin has established that their motion proceeds uniformly in both directions, so that $\bar{x} = 0$. Thus, their displacement can be quantitatively described only through the variance of the spatial distribution $\overline{x^2(t)}$. In order to describe the long-time behavior of this quantity, let us use the Langevin equation (4.5.8). After multiplication by x, it can be identically reduced to the form

$$\frac{1}{2}\frac{d^2x^2}{dt^2} - \left(\frac{dx}{dt}\right)^2 = -\frac{1}{2qM}\frac{dx^2}{dt} + xf(t)/M . \qquad (4.5.11)$$

For any given realization of the random force $f(t)$, this equation makes it possible to determine accurately the path $x(t)$ of a particle subject to its action. Different particles experiencing a different action find themselves at different points of space at time t. They follow different paths, but the average distance from the starting line increases. In order to calculate the mean square deviation $\overline{x^2(t)}$, it is necessary to average equation (4.5.11) over all possible realizations of $f(t)$.

Due to the random and rapid change of $f(t)$, the correlation between f and the resulting displacement is relevant only for times of the order τ_0. For longer times, $f(t)$ and $x(t)$ may be considered as independent random quantities. In this case, the mean of the product is equal to the product of the means: $\overline{x(t)f(t)} = \overline{x(t)} \cdot \overline{f(t)} = 0$. In view of this fact, we have, after averaging (4.5.11),

$$\frac{1}{2}\frac{dz}{dt} - \frac{kT}{M} = -\frac{z}{2qM} \quad \text{at} \quad t \gg \tau_0 = qM . \qquad (4.5.12)$$

Here we have introduced the notation $z = \overline{dx^2}/dt$. The above expression takes into consideration that $M\overline{\dot{x}^2} = kT$, according to the equipartition law. This is due to the fact that, although Brownian motion involves larger particles than molecules, it is nevertheless ordinary thermal motion with the appropriate energy.

Although the velocity of Brownian particles increases in proportion to $\sqrt{kT/M}$ pointing to its molecular origin, the experimental measurement of the rapidly varying instantaneous velocity is practically not feasible. It is much easier to follow the position of particles, measuring their coordinates at sequential instants of time and averaging according to $\overline{x^2} = \frac{1}{N}\sum_{i=1}^{N} x_i^2$. The general solution of equation (4.5.12) is of the form

$$z = 2kTq + \text{const} \cdot \exp(-t/qM) . \qquad (4.5.13)$$

Neglecting the exponentially vanishing term, we obtain a very simple asymptotic result for $t \gg \tau_0 = qM$:

$$\overline{x^2} = 2kTq \cdot t . \qquad (4.5.14)$$

In view of the Einstein relation $\mathcal{D} = qkT$, it is readily seen that this result is identical to (4.5.7).

It is remarkable that the deterministic law of motion which controls the behavior of a particle in a medium eventually gives the solution of explicitly irreversible nature. This happens whenever a dynamical system is driven by a truly random force. The system's behavior acquires a random character, although, being mechanical, it always obeys Newton's equation of motion.

The fundamental law of the linear increase in time of the mean square fluctuation $\overline{x^2}$ was established by Einstein, who developed the theory of Brownian motion. If the Brownian particles are spherical with radius R, their mobility, according to Stokes, is equal to

$$q = (6\pi R \eta)^{-1}. \qquad (4.5.15)$$

In this case Eq. (4.5.14) yields

$$\overline{x^2} = \frac{kT}{3\pi R \eta} \cdot t.$$

This dependence can be experimentally verified in detail, provided that the liquid viscosity η and the particle radius R are known. Classical experiments were carried out by Perrin. He measured $\overline{x^2}$ through the position of Brownian particles as observed under the microscope. The process of their random wanderings provided convincing evidence in favor of the theory. This was a powerful confirmation, and the atomic concept became universally acknowledged.

5

THERMODYNAMICS

5.1 FIRST LAW

Basic Concepts

Thermal phenomena have attracted the attention of man probably since ancient times when he first began to use fire. Nevertheless, the science studying such phenomena—thermodynamics—appeared only when mere physiological perception of heat was supplemented by its physical measurement. The first step toward this was the invention of the thermometer, which gained acceptance in the seventeenth century due to its good design which had been first proposed by Torricelli and then improved at the Accademia del Cimento in Florence. After that temperature and heat became synonyms for almost a century. The situation radically changed only following experiments conducted by Georg Wilhelm Richmann (Rihman) in Petersburg. He established in 1750 that when equal amounts of water with different temperatures T'_i and T''_i are mixed, the resulting temperature T_f of the mixture is the arithmetic mean $(T'_i + T''_i)/2$, while in mixing liquids with different volumes V' and V'' the resulting temperature should be determined from the equation:

$$V'(T'_i - T_f) = V''(T_f - T''_i), \qquad (5.1.1)$$

which leads us to the Rihman formula: $T_f = (V'T'_i + V''T''_i)/(V' + V'')$.

When mixing different liquids the concept of *heat capacity* must be introduced as was already realized at Accademia del Cimento. The heat capacity of a unit volume of any substance is its specific heat c. Since each component in the mixture has its own value for c the equation (5.1.1) should be generalized:

$$c'V'(T'_i - T_f) = c''V''(T_f - T''_i). \qquad (5.1.2)$$

In place of the specific heat, a molar value of heat capacity $C = cV_0$ is often used. In this case the amount of substance is measured in moles $i = V/V_0$, while the equation (5.1.2) is rewritten in the form:

$$-Q' = c'i'(T'_i - T_f) = c''i''(T_f - T''_i) = Q''. \qquad (5.1.3)$$

Here Q' denotes the heat taken away the cooled body, which by universal convention is agreed to be negative, while Q'' denotes the heat acquired by the heated body and, therefore, is positive. With this definition of heat sign, all of the equations given above are just variations of the same law of heat conservation in its transfer from a hot body to a cold one. The law may be expressed as follows:

$$Q = Q' + Q'' = 0. \qquad (5.1.4)$$

Caloric

Nevertheless, yet another form of equation (5.1.2) is possible when something different is conserved:

$$q_i = c'V'T'_i + c''V''T''_i = (c'V' + c''V'')T_f = q_f$$

According to Black q is a quantitative measure of *caloric*, assumed to be an invisible but all-pervading elastic fluid. In the course of heat exchange caloric is conserved, that is, flows from one body into another but never disappears. The caloric theory existed for almost a century. It was maintained despite the fact that since ancient times up to the middle of the eighteenth century heat was dominantly thought of as an inherent state of a body or a measure of the movement of the small particles constituting it. The concept of caloric was supported by Boyle, Euler, and Lomonosov. In the transition to a quantitative description of heat phenomena, a self-consistent model explaining predictions of the theory became necessary. This model showed itself to be clearly illustrative as well as versatile. The caloric theory allowed the overcoming of difficulties associated with the interpretation of phase transitions in which heat is not conserved. Melting was thought as the coupling of caloric with ice transforming it into water. In 1759 Joseph Black established that in order to transform ice into water at the melting temperature it is necessary to apply to each mole of it "latent heat" λ. We now call it the heat of melting

$$Q = \lambda i. \tag{5.1.5}$$

Anyway, Black himself believed that chemical coupling of caloric with ice or water takes place in reactions:

$$Q + \text{ice} = \text{water} \tag{5.1.6a}$$

$$Q + \text{water} = \text{vapor} \tag{5.1.6b}$$

He was the first to measure the latent heat of melting and evaporation.

These results substantially facilitated solving the tasks of calorimetry. The quantity of heat yielded to water by a body immersed in it was measured by the increase in the temperature of the water. An increase of temperature of 1 gram of water by 1°C (from 19.5°C to 20.5°C) meant the acquiring by it of 1 calorie or "small" calorie (cal)—a unit of heat introduced by Johan Carl Wilcke in 1772. J. K. Wilcke confirmed Rihman's formula and started systematic measurements of specific heat of various bodies using the "mixing method." For these purposes he used a technique based on the melting of ice. This method was perfected by Lavoisier and Laplace, who actually invented the first "calorimeter." The practical success of calorimetry facilitated the establishment of a fluid theory of heat. It did not contradict the basic law of heat transfer formulated by Fourier: a stationary flux of heat q in a temperature-heterogeneous medium is described by the equation:

$$q = \frac{Q}{St} = -\kappa \frac{dT}{dx}, \tag{5.1.7}$$

where S denotes the cross-sectional area of x axis, t the heat travel time, and κ the thermal conductivity. Working with Eq. (5.1.7) and Lambert's laws for radiative heat transfer (Section 2.1), Fourier investigated spatial and temperature variations in the atmosphere at various altitudes and latitudes. The mathematical formalism which he developed allowed the analysis of heat transfer phenomena as rigorous as for mechanical equivalents.

Mechanical Equivalent of Heat

The caloric theory was defeated as soon as thermodynamics research moved beyond the frame of studying heat transfer phenomena. Crucial arguments against it were put forward by Benjamin Thompson (Count Rumford) in 1798, who proved the possibility of the unlimited production of caloric by means of friction. He boiled water by drilling an immersed cannon barrel. From this experiment he came to the conclusion that: "the entity, which may be supplied in unlimited quantity by an isolated body or system of bodies, can not be a material substance." Fifty years later J. Joule measured the mechanical equivalent of heat, J, by heating water in a calorimeter with the

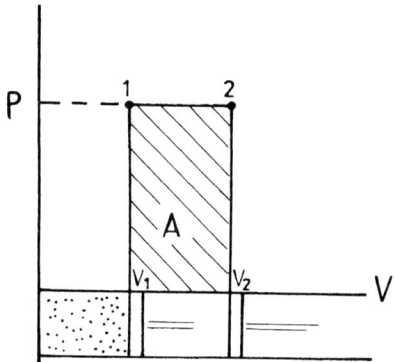

Figure 5.1 The isobaric expansion of a gas from state 1 to state 2 performs work A equal to the hatched area in the pV plane.

help of a rotating wheel with blades. Thus the quantitative relation between heat and work was established:

$$A = JQ. \qquad (5.1.8)$$

According to modern data $J = 427 \, \text{kgm/kcal}$ or $4.186 \times 10^{10} \, \text{erg/kcal}$. A short time before Joule's work this coefficient was approximately determined by J. R. Mayer. He was not only keenly aware that friction transforms work into heat, but had been able to justify the reverse: that work produced by a gas expanding at constant pressure (Fig. 5.1) is done due to the acquired heat. That is why J. R. Mayer was convinced that molar specific heat at constant pressure C_p would be higher than that at constant volume C_V. At fixed volume, the work done by the gas

$$dA = pdV \qquad (5.1.9)$$

is zero. Alternatively, if p is constant, then for one mole of gas

$$JdQ = J(C_p dT - C_V dT) = pdV = d(pV). \qquad (5.1.10)$$

According to the equation of Clapeyron describing the equilibrium state of an ideal gas,

$$pV = RT, \qquad (5.1.11)$$

where $R = 0.848 \, \text{kgm/mole} \cdot {}^\circ\text{C}$. Thus from Eq. (5.1.10) the Mayer equation was deduced

$$C_p - C_V = R/J. \qquad (5.1.12)$$

Mayer determined J from this equation using experimental data on C_p and C_V.

Internal Energy

In fact, J. R. Mayer had closely approached the formulation of the law of energy conservation, but only in the special case of isobaric processes in application to the particular system of the ideal gas. To formulate this law in its general form the introduction of an entirely new quantitative characteristics of the substance state, its energy, was necessary. This important step was taken by Hermann von Helmholtz, who in 1874 independently came to the conclusion of the existence of some conserved value—energy—which cannot appear or disappear but is only transferred from one body to another in such a way that its total amount remains unchanged. Energy is distinguished from matter in that it is weightless and from caloric by its inability to react with a substance and being only inherent to it. Elaborating Helmholtz's ideas R. Clausius came to the conclusion that Eqs. (5.1.6) should be rewritten in the form:

$$Q + \mathcal{E}_1 = \mathcal{E}_2, \quad (5.1.13)$$

introducing thereby a new concept, the internal energy of substance $\mathcal{E}$, which takes different values before and after a phase transition. A similar relation is valid for the isochoric process of heating of a substance. It is valid in the case when the change of internal state of a substance (its temperature, internal energy, etc.) is infinitely small:

$$\delta Q = d\mathcal{E}. \quad (5.1.14)$$

If the internal energy of a body acquiring heat does not change, it might transform completely into an equivalent work to prove that nothing had disappeared without a trace:

$$\delta Q = \delta A. \quad (5.1.15)$$

This equation is a differential form of Eq. (5.1.8) in which it is assumed that heat and work are measured using the same units (either kgm or kcal) and therefore $J = 1$. From here on we will follow this convention about units for measuring energy.

It is absolutely clear that in the general case heat is spent either in changing the internal energy of a body or to perform work. Therefore the direct generalization of (5.1.14) and (5.1.15) is the following equation:

$$\delta Q = d\mathcal{E} + \delta A. \quad (5.1.16)$$

This is the first law of thermodynamics. It is important to note that, formulated in such a way (by Hermann von Helmholtz), it assumes execution of any work (not necessarily mechanical) and by any body (not necessarily by gas) and, finally, in any process (not necessarily an isobaric one as in the Mayer case). Hence, the point is about conservation of energy in principle and in any form

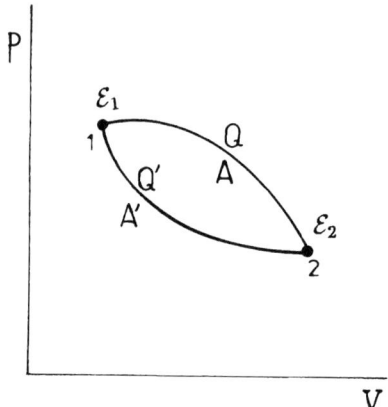

Figure 5.2 Two different quasistatic transitions between states 1 and 2 (with internal energies U_1 and U_2), accompanied by transfer of different amounts of heat Q and Q' and work A and A'.

including thermal. This is the total balance established by Eq. (5.1.16). It should be noted that such a formulation of the energy conservation law became possible due to the introduction of internal energy. The latter is characterized, as is known, by a complete set of independent thermodynamic parameters such as p and V (or p and T, etc.) and is denoted by a point on the appropriate diagram of state. Internal energy is a *function of state* which is unambiguously specified at any point on the diagram, whereas both heat and work are *functions of process*, depending on a particular path from one state to another (Fig. 5.2).

Reversible and Irreversible Processes

If the process is "quasistatic," that is, accomplished so slowly that one equilibrium state of a substance smoothly transforms into another, then this continuous sequence of states forms a trajectory on the diagram. The first law of thermodynamics in its differential form (5.1.16) is applicable to any infinitesimal segment of this trajectory. Alternatively, if one wishes only to compare the initial and final states, the transition between them may be described by the energy conservation law in its integral form:

$$Q = \mathcal{E}_2 - \mathcal{E}_1 + A. \qquad (5.1.17)$$

Quasistatic processes are reversible. This means it is possible to return quasistatically from the final state of a process to the initial one through the same sequence of intermediate states. If in the direct process the system acquired heat $dQ > 0$, which was partially or completely converted into work $dA > 0$, then in the reverse process it acquires the same work $dA < 0$ to return the same amount of heat $dQ < 0$. In other words, the direct and reverse quasistatic processes differ only by the direction of movement along the trajectory and in the signs of the components of equation (5.1.16).

Nevertheless, quasistatic, reversible processes are just the idealization of real ones. If a system is subjected to changes (expanding, heating, etc.) so fast that it will not be able to follow them in time, then space and temperature irregularities will appear. Such inhomogeneous states are nonequilibrium and there is no place for them on the diagram of state (Fig. 5.2). Nevertheless, precisely through these states a nonequilibrium transition is accomplished. The initial state 1 and the final one 2 are the only equilibrium points on this trajectory. Since the internal energy is defined only for equilibrium states, description of nonequilibrium processes is possible only in an integral form (5.1.17).

The transition between given states 1 and 2 can be accomplished via a variety of equilibrium routes and through an even greater number of nonequilibrium ones. In order to stress the dependence of heat and work from on trajectory of transition, their infinitesimal values in Eq. (5.1.16) are marked sometimes by a symbol different from the one denoting the increment of internal energy. The latter is an exact (total) differential of a function of several variables. Its integration over a contour always gives 0, because at return to the initial point the internal energy is restored to its initial value irrespective of the trajectory taken:

$$d\mathcal{E} = \mathcal{E}_1 - \mathcal{E}_1 = 0 \qquad (5.1.18)$$

In this sense the internal energy is similar to the gravitational one, which has the same magnitude at any point on the surface of the Earth, independent of the trajectory along which the point was reached.

The situation is completely different with heat and work, which are determined not at a point, but over a segment of a trajectory between the two points. Their infinitesimal values are "inexact" differentials and the integration over a contour does not result into 0. Therefore, in application to cyclic processes the first law of thermodynamics gives according to (5.1.16) and (5.1.18) the following:

$$Q = \oint \delta Q = \oint \delta A = A. \qquad (5.1.19)$$

Although further on we will use the conventional symbols of differentials to denote any infinitesimal values, one must keep in mind the prime difference between work and heat on the one hand and internal energy on the other. In particular, for a system accomplishing work through thermal expansion, the energy conservation law (5.1.16) will be rewritten in the form:

$$dQ = d\mathcal{E} + pdV. \qquad (5.1.20)$$

With the proviso that internal energy is conserved in thermal expansion ($U = $ const), all heat applied to a body transforms completely into work

$$Q = \int p dV = A. \qquad (5.1.21)$$

However, this transformation, as opposed to the one taking place in a cyclic process (5.1.19), is accompanied by changes in the state of a system: its volume and pressure change as well as the temperature, which may either decrease or increase depending on the sign of the derivative:

$$\left(\frac{\partial T}{\partial V}\right)_{\mathcal{E}} \lessgtr 0. \qquad (5.1.22)$$

Without the introduction of internal energy it is impossible to even formulate this alternative, that is, to pose correctly the question: "what happens in the process of adiabatic expansion of a gas into vacuum—heating or cooling?" A process is called *adiabatic* if the system is thermally insulated from the rest of the universe ($Q = 0$). If the system is a gas expanding in a nonequilibrium way from a filled part of a volume into an empty one after opening a plug, then work is not executed ($A = 0$). According to (5.1.17) this means that

$$\mathcal{E}_1 = \mathcal{E}_2 \qquad (5.1.23)$$

although $V_1 > V_2$. One may see that for a small expansion ($dV > 0$) the quantity (5.1.22) is positive if gas is being heated ($dT > 0$) and negative in the opposite case. Hence, to answer the abovementioned question one has just to calculate the sign of $(\partial T/\partial V)_{\mathcal{E}}$.

Ideal Gas

Only in the case of an ideal gas does the temperature after adiabatic expansion in vacuum remains the same. This conclusion, deduced from the throttling experiment of Gay–Lussac, attests that for an ideal gas the relation (5.1.23) could be restated as follows:

$$\mathcal{E}(T, V_1) = \mathcal{E}(T, V_2) \text{ or } \mathcal{E} = \mathcal{E}(T). \qquad (5.1.24)$$

This means that in an ideal gas the constancy of $\mathcal{E}$ is maintained by the constancy of T, that is, the internal energy is conserved along isotherms (Fig. 5.3).

It should be noted that the ideal gas is the simplest and therefore the most convenient system to illustrate general theory. The concept of the ideal gas gained acceptance through the study of noble gases at high temperatures and low pressures, studies which led to the establishment of the laws of

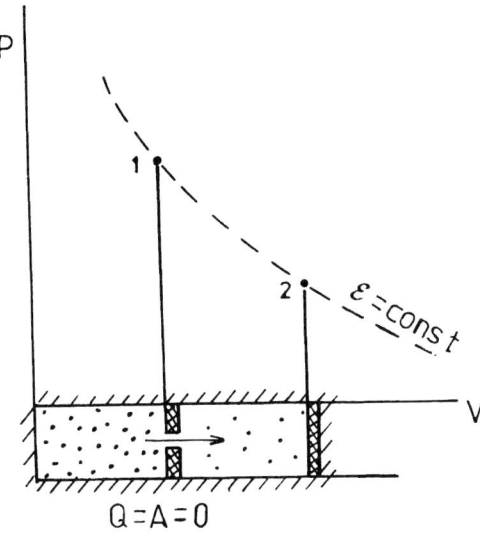

Figure 5.3 Adiabatic expansion of gas into vacuum (from state 1 to state 2) on the pV plot. The shading around the container indicates adiabatic isolation from the surroundings.

Boyle–Mariotte and Gay–Lussac. These were included in the most general form in Clapeyron's equation (or the Mendeleev–Clapeyron equation) for i moles

$$pV = iRT, \qquad (5.1.25)$$

that should be considered as empirically established. Another important feature of the classic ideal gas is constancy of its specific heat

$$c = \text{const}. \qquad (5.1.26)$$

Referring to the first law of thermodynamics (5.1.20) it can be easily seen that

$$c_V = \frac{1}{V}\left(\frac{\partial Q}{\partial T}\right)_V = \frac{1}{V}\left(\frac{\partial \mathcal{E}}{\partial T}\right)_V \qquad (5.1.27)$$

is a function of volume and temperature, as well as $\mathcal{E}(T, V)$. The exact differential of the latter entity is

$$dU = \left(\frac{\partial \mathcal{E}}{\partial T}\right)_V dT + \left(\frac{\partial \mathcal{E}}{\partial V}\right)_T dV = c_V V\, dT + \left(\frac{\partial \mathcal{E}}{\partial V}\right)_T dV.$$

Since $c_V = $ const and according to (5.1.24) $(\partial \mathcal{E}/\partial V)_T = 0$ it can be deduced that the internal energy of an ideal gas is

$$\mathcal{E} = c_V V T = i C_V T, \qquad (5.1.28)$$

assuming the constant of integration being zero.

The internal energy and volume are extensive properties meaning that they increase proportionally to the amount of substance, expressed in moles of gas. In order to characterize completely the state of a mole of a substance it is necessary to know two equations of state: a thermal one, establishing the dependence $p(V, T)$ and a caloric one, determining $\mathcal{E}(V, T)$. In this sense a mole of ideal gas is completely described by the equations:

$$p = RT/V \qquad (5.1.29)$$

and

$$\mathcal{E} = C_V T. \qquad (5.1.30)$$

Further we will see that the caloric equation for the ideal gas is a corollary of the thermal one. Using both for elaboration of the first law of thermodynamics (5.1.20) we get

$$dQ = C_V dT + RT dV/V. \qquad (5.1.31)$$

Now let us consider the adiabatic and isothermal expansion of an ideal gas.

In isothermal expansion $(dT = 0)$ Eq. (5.1.31) becomes the differential analogue of Eq. (5.1.21). Integrating with finite limits results in the following relation for equilibrium expansion:

$$Q = RT \ln \frac{V_2}{V_1}. \qquad (5.1.32)$$

When $V_2 > V_1$ the system acquires heat transforming it into work. Alternatively, when the gas is compressed $(V_2 < V_1)$ the reverse situation takes place. There is an essential difference between equilibrium and nonequilibrium expansion, especially expansion into a vacuum. In the latter case in the course of expansion $Q = A = 0$, but in order to return a system to its initial state by means of isothermal compression it is necessary to do work and extract the equivalent amount of heat (5.1.32). Irreversibility of expansion into a vacuum is manifested in the changes in the surrounding bodies which perform the work and extract heat on the way back. When expansion is done in an equilibrium manner then in reversal process the system itself and all surrounding bodies return to their initial states.

Using Mayer's relation $R = C_p - C_V$ one may express Eq. (5.1.31) in the following form:

$$dQ/T = C_V[d\ln T + (\kappa - 1)d\ln V] = C_V d\ln(TV^{\kappa-1}). \qquad (5.1.33)$$

The specific heat ratio

$$\kappa = \frac{C_p}{C_V} \qquad (5.1.34)$$

is a constant, which by definition is greater than 1 and known as the *adiabatic index*. The meaning of this is clarified when we consider the adiabatic expansion of a gas, According to Eq. (5.1.33) the cooling of a gas due to its expansion must occur in such a way that the following products remain constant:

$$TV^{\kappa-1} = \text{const} \qquad (5.1.35a)$$

$$\text{or} \quad pV^{\kappa} = \text{const}. \qquad (5.1.35b)$$

Formulae (5.1.35) determine the adiabatic line ("adiabatic") which falls more steeply than the isotherm, as is shown in Fig. 5.4. Moving along this curve in an equilibrium process of adiabatic expansion, the system performs work exclusively at the expense of its internal energy and therefore is cooled. In the case of

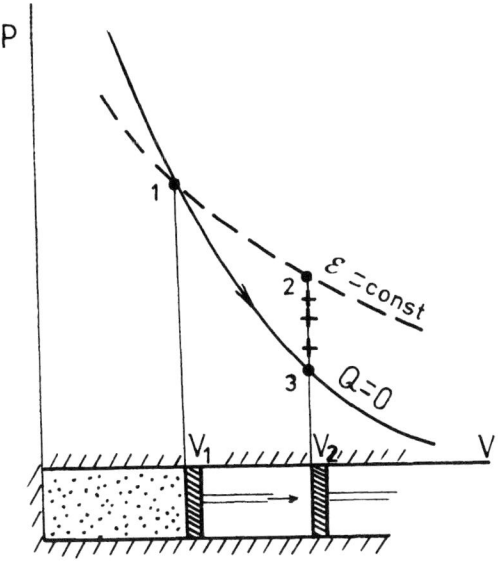

Figure 5.4 Adiabatic expansion from volume V_1 to the volume V_2: quasistatic (from 1 to 3 along solid line) and explosive (nonequilibrium) which may terminate at the states marked with crosses.

adiabatic nonequilibrium expansion of an ideal gas into the vacuum, the final state lies not on the adiabatic line but on the isotherm. Between these two extreme cases corresponding to quasistatic and explosive expansion lie all natural processes which develop at finite speed and bring the system to points along an isochore $V = V_2$, framed between adiabatic line and isotherm. Luckily, expansion at speeds typical for piston engines is much closer to the quasistatic limit than to the explosive one, although it is accompanied by some loss of heat. Neglecting these losses we may analyze the cycles of heat engines in ideal conditions corresponding to the equilibrium change of state of the working medium. It was just this analysis that led to the discovery of the second law of thermodynamics.

5.2 SECOND LAW

Carnot Cycle

In the beginning of the nineteenth century, steam engines began to be widely used in transport and industry. Their construction continued to become more sophisticated, providing increasing economic advantages to the relevant industries. It became clear that the number of ways to convert the energy of burned fuel into work is really unlimited. In principle, all these engines used a furnace to heat the working medium and the atmosphere as a sink where heat waste together with the burnt products and worked steam were ejected. To improve the efficiency one might look to alter the temperature of the heater (or heaters), the working medium or its aggregate state and the method of extracting work, that is, the thermodynamic cycle executed by the engine. The Scot, James Watt, as inventor of one of the best steam engines, clearly understood that the objective to be met was the following: to minimize the amount of coal burned for the required amount of work. More generally stated, one wishes to minimize the amount of heat Q_1 received from a heat source, regardless of the kind of fuel being burned. The ratio of the net work output A to the heat input is the efficiency of a heat engine

$$\eta = \frac{A}{Q_1} \qquad (5.2.1)$$

The question is: what is the actual value of η and in particular what is its maximal achievable value?

A correct answer to this question was first given by the young French engineer Sadi Carnot. Drawing the analogy between caloric and water rotating a turbine, he came to the conclusion that the difference of temperature should play the same role as the difference in the heights of the falling water. The efficiency of heat engines should not depend on the nature of the working body in the same way as the efficiency of a hydro turbine does not depend on the kind of liquid but only on its mass and falling height. Using the model of "the motive power of heat" which he later abandoned, Carnot took into account the energy conservation law in the form (5.1.19), which may be presented as follows:

$$A = Q_1 + Q_0 = Q_1 - |Q_0|, \tag{5.2.2}$$

where Q_0 denotes the heat rejected to the sink (or sinks). Carnot stressed that the presence of heat sinks is as inherent a feature of heat engines as the use of heaters (heat sources). Therefore,

$$\eta = 1 + \frac{Q_0}{Q_1}, \tag{5.2.3}$$

and is always less than 1 (because $Q_0 < 0 < Q_1$). In the case when we have only one heat source of temperature T and only one heat sink at temperature T_0, the efficiency of the heat engine will depend entirely on the difference between these temperatures. Exactly in the same way the efficiency of a turbine depends on the difference in heights of the falling water. Ten years later Clapeyron shaped these speculations into a more rigorous form by refinement of a type of cycle, called the Carnot cycle, in which the use of a single heat sink as well as only a single heat source is possible. Using an ideal gas equation of state Clapeyron showed that such a cycle should consist of two isotherms at temperatures T_1 and T_0 connected by two adiabatics (Fig. 5.5).

Using an ideal gas as an working medium for the heat engine one may deduce from Eq. (5.1.32) the heats Q_1 and Q_0 corresponding to the expansion

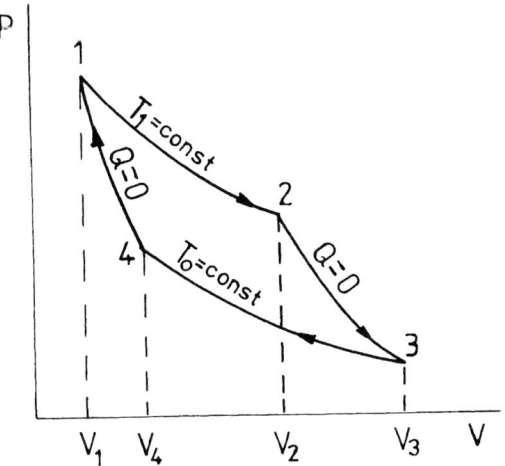

Figure 5.5 Carnot cycle consisting of two isotherms and two adiabatics.

of gas from V_1 to V_2 and to its compression from V_3 to V_4. Substituting these into Eq. (5.2.3) we get:

$$\eta_c = 1 + \frac{T_0 \ln(V_4/V_3)}{T_1 \ln(V_2/V_1)} = 1 - \frac{T_0 \ln(V_3/V_4)}{T_1 \ln(V_2/V_1)}. \qquad (5.2.4)$$

Taking into account that the expansion from V_2 to V_3 occurs along one adiabatic, while that from V_1 to V_4 occurs along the other, these volumes should appear to be related in pairs by equation (5.1.35a):

$$T_1 V_2^{\kappa-1} = T_0 V_3^{\kappa-1} \quad \text{and} \quad T_1 V_1^{\kappa-1} = T_0 V_4^{\kappa-1}.$$

Dividing the first equation by the second we derive the following relation:

$$V_2/V_1 = V_3/V_4. \qquad (5.2.5)$$

Substituting this relation into (5.2.4) we obtain, in accordance with the prognosis of Carnot:

$$\eta_c = 1 - \frac{T_0}{T_1}. \qquad (5.2.6)$$

In other words, the efficiency of the Carnot cycle depends entirely on the difference of temperatures of heat source and sink and increases with the increase of this difference.

The most important achievement of Carnot was that he raised to the level of a principle the impossibility of achieving a higher efficiency of a heat engine regardless of its design. This principle asserts that a perpetuum mobile of the second kind is impossible. As distinct from a perpetuum mobile of the first kind, which would extract work "out of nothing" ($A > Q$) and have efficiency $\eta > 1$, a perpetuum mobile of the second kind does not contradict the law of conservation of energy and has $\eta < 1$. Nevertheless, Carnot's principle claims more than this. The efficiency of a heat engine must satisfy the more rigid demand:

$$\eta \leq \eta_c \qquad (5.2.7)$$

Thus is established the qualitative inequivalency of heat and work. In the conversion of work from one form into another the efficiency may get arbitrarily close to unity in, for example, electrical motors. However, in the conversion of heat into work the upper limit for the efficiency established by formula (5.2.6) appears to be much lower, especially when the difference of temperature between heat source and heat sink is small. The Carnot principle is one of the formulations of the second law of thermodynamics. We will consider this law further from other viewpoints as well, but for now it is more convenient to take it as an initial postulate.

Entropy

The study of the efficiency of heat engines, while aimed at pragmatic goals, certainly contributed to the establishment of thermodynamics as an independent "pure" science with its own system of axioms. As we have seen above the efficiency of the Carnot cycle may be determined either from formula (5.2.3) or from (5.2.6):

$$\eta_c = 1 + Q_0/Q_1 = 1 - T_0/T_1 . \tag{5.2.8}$$

This also implies the following relation:

$$\frac{Q_1}{T_1} + \frac{Q_0}{T_0} = 0, \tag{5.2.9}$$

which Rudolf Clausius had formulated as the theorem of "reduced heats" Q_i/T_i. According to this theorem the sum of reduced heats is equal to 0 for an arbitrary thermodynamic cycle executing an equilibrium process:

$$\sum_i \frac{Q_i}{T_i} = 0 . \tag{5.2.10}$$

The theorem can be easily proved by splitting the cycle, represented by an arbitrary contour on the p–V diagram, into a set of Carnot cycles executed between adjacent adiabatics (Fig. 5.6). The relation (5.2.9) holds for each

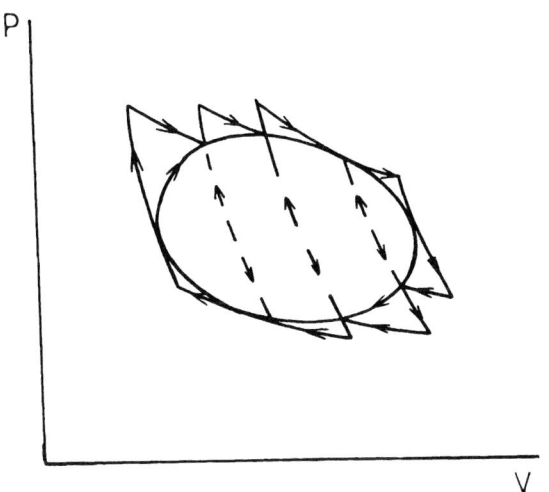

Figure 5.6 An arbitrary cycle, inscribed into polygonal line enveloping a set of Carnot cycles.

Carnot cycle separately, while for the total the equality (5.2.10) is valid. The set of Carnot cycles is carried out by an appropriate number of heat engines in such a manner that each segment of the adiabatic line adjacent to the two cycles is passed twice: in the first pass in the right direction, in the second, in reverse. As this takes place the work performed by one engine is executed on the other and may be excluded from the total balance together with the appropriate segment of the adiabatic line. Due to this, the cycle, the net result of the entire set of the Carnot cycles, is represented as a curved trajectory within which the contour of interest falls. The latter is approximated better the more dense is the mesh of adiabatics. Letting their number tend to infinity, we obtain instead of (5.2.10)

$$\oint \frac{dQ}{T} = 0. \tag{5.2.11}$$

This relation, as well as (5.1.18), is valid in an integration along arbitrary contour. Therefore, the integrand must be an exact differential of some function of state S, which is defined as follows:

$$dS = \frac{dQ}{T}. \tag{5.2.12}$$

When adiabatic expansion is accomplished in an equilibrium way, S remains unchanged, whereas in a nonequilibrium process it changes. This is why Clausius called it "entropy," that is, a measure of "changes". The variation of entropy in a nonequilibrium process is described by Clausius's inequality

$$dS > \frac{dQ}{T} \tag{5.2.13}$$

identical in essence to inequality (5.2.7). We will return later to the proof of this statement.

Substituting expression (5.1.33) into (5.2.12) and integrating we discover that the molar entropy of an ideal gas is given by

$$S = C_V \ln(TV^{\kappa-1}) + \text{const} = C_V \ln(pV^\kappa) + S_0 \tag{5.2.14}$$

As with internal energy, S is defined in thermodynamics up to a constant S_0. Since adiabatics are simultaneously isoentropes, to each of them corresponds a certain value of S. The further an adiabatic line from the origin of coordinates on the p–V diagram, the higher the entropy.

A more distinct parallel may be drawn now between the mechanical variables p and V and the thermal parameters T and S, as well as between work and heat. Actually, in the expressions

$$dQ = TdS \quad \text{and} \quad dA = pdV$$

temperature plays the same role as pressure, while entropy is similar to volume. The latter are extensive factors, while the first are intensive ones. The amount of heat and work done depends on the degree of the change of state of the working medium, characterized by the variation of its entropy and volume accordingly. Both entropy and volume characterize the capacity of a substance for heat and work to be put into it and are proportional, therefore, to the amount of substance. However, neither temperature nor pressure depend on the amount of a substance; they merely characterize the intensity of transfer of heat and work at the same change of state but under different conditions.

The demonstrated symmetry between the mechanical variables p and V and the thermal parameters T and S allows us to consider all processes from the viewpoint of either. In particular, in place of the p–V diagram of Fig. 5.2 one may use a T–S diagram shown in Fig. 5.7. Any state of a system can be equivalently displayed by a point on either of the diagrams, but work is more convenient to calculate as an area on the p–V diagram, while heat is represented by the area under a trajectory of the process on the T–S diagram. The infinitesimal area element is given in the first case by $dpdV$, and in the second by $dTdS$. Any arbitrary cycle can be displayed on either of the diagrams and, though its shape differs, the area inside the cycle remains the measure of the work performed and therefore should remain unchanged. From the mathematical viewpoint this means that

$$|I| = \frac{\partial(T,S)}{\partial(p,V)} = 1 \tag{5.2.15}$$

where I is the Jacobian for the transformation of variables p, V into variables

$$S = C_V \ln(pV^\kappa) + S_0 \quad \text{and} \quad T = pV/R. \tag{5.2.16}$$

Generally speaking, in a such a transformation the area expressed by the new variables is

$$dTdS = |I|dpdV, \tag{5.2.17}$$

but using Eq. (5.2.16) it is easy to verify that identity (5.2.15) is valid:

$$I = \begin{vmatrix} \left(\frac{\partial T}{\partial p}\right)_V & \left(\frac{\partial T}{\partial V}\right)_p \\ \left(\frac{\partial S}{\partial p}\right)_V & \left(\frac{\partial S}{\partial V}\right)_p \end{vmatrix} = \begin{vmatrix} V/R & p/R \\ C_V/p & \kappa C_V/V \end{vmatrix} = \frac{C_p - C_V}{R} = 1.$$

Thereby substituting Eq. (5.2.15) into Eq. (5.2.17), we obtain:

$$dTdS = dpdV. \tag{5.2.18}$$

This property—transformation of the shape of a region without change of its area—is called *calibration invariance*. This is due to the specific form of the $S(p, V)$ dependence obtained in (5.2.14), which in turn is a direct consequence of the second law of thermodynamics in the form of Eq. (5.2.12). Therefore, one may reverse the reasoning: to raise to a principle calibration invariance as given by the relation (5.2.15), then to deduce from it the second law in its conventional formulation. Such an axiomatic derivation of thermodynamics was presented by Rumer and Ryvkin in their book, *Thermodynamics, Statistical Physics and Kinetics*.

Let consider two examples demonstrating the use of the T–S diagram in thermodynamics. It is clearly seen from Fig. 5.7 how the sign and the value of the specific heat might change in various processes. Since

$$C_x = \left(\frac{dQ}{dT}\right)_x = T\left(\frac{dS}{dT}\right)_x \tag{5.2.19}$$

the question is reduced to that of how entropy and temperature change in the appropriate process; the latter is clearly seen from the picture. In an adiabatic process $dS = 0$ and therefore $C_S = 0$ both in heating (compression) and in cooling (expansion). It is evident that at equal dT the change of entropy is higher at $p = $ const and consequently $C_p > C_V > 0$. Finally, in an isothermal process the denominator (5.2.19) becomes 0, while $|C_T| = \infty$. In the first and in the third quadrants the heat capacity is positive because both dS and dT have

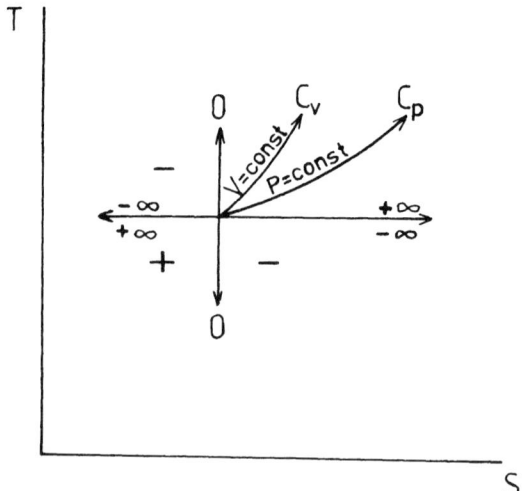

Figure 5.7 The sign and magnitude of specific heat in various processes, shown on a T–S diagram.

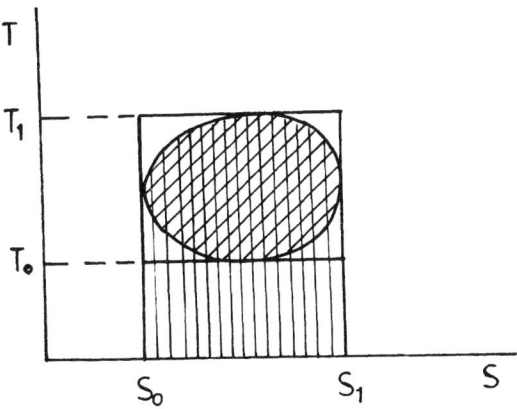

Figure 5.8 Arbitrary reversible cycle, inscribed into a Carnot cycle.

the same sign, while in the second and the third it is negative because dT and dS have opposite signs.

Geometric considerations only applied to the T–S diagram are sufficient to prove the following important statement: the efficiency of an arbitrary reversible cycle η_0 does not exceed the efficiency of the Carnot cycle into which it is inscribed (Fig. 5.8). In other words, the Carnot cycle has the highest efficiency among any reversible heat engines operating in the temperature interval between T_1 (maximal temperature) and T_0 (minimal temperature). In fact,

$$\eta_0 = \frac{A'}{Q'} < \frac{A''}{Q'} \tag{5.2.20}$$

where Q' is obtained by an engine and hence denotes positive heat, equal to the vertically hatched area in Fig. 5.8, while A' is the work performed by the engine, which is equal to the area bounded by the cycle (diagonal hatching). This work will increase up to the value of A'' when the cycle is deformed so as to include the bottom vertically hatched corners of the Carnot cycle, leaving the upper border of the cycle and hence Q' intact. The efficiency of the deformed cycle, given in the right-hand side of the inequality (5.2.20), will increase. Comparing it now with the Carnot cycle, we see that the latter performs more work and attains more heat accordingly. Both amounts are determined by the value of $\Delta = A - A'' = Q - Q'$, measured by the area of the upper nonhatched corners. It is evident that

$$\eta_0 < \frac{A - \Delta}{Q - \Delta} < \frac{A}{Q} = \eta_c . \tag{5.2.21}$$

This inequality increases with the areas of the upper and lower corners, which are the difference between the cycle under consideration and the appropriate Carnot cycle. If this difference is reduced to 0, then the efficiencies of both cycles are equal. Therefore, in the general case inequality (5.2.21) should look as follows:

$$\eta_0 \leq \eta_c. \qquad (5.2.22)$$

The equality is possible only when both cycles align.

We should stress immediately the distinction of this statement from the principle of Carnot as it was given in Eq. (5.2.7). Since it was formulated for all realizable, not only reversible, heat engines, then in comparing two engines working over a Carnot cycle, the inequality should hold when one of these engines is irreversible. Therefore, the second law of thermodynamics unites two independent statements. The first referring to reversible engines, may be expressed in the form (5.2.22), while the other, which is valid for arbitrary engines, claims that their efficiency is not greater than that of a reversible engine working over the same cycle:

$$\eta \leq \eta_0. \qquad (5.2.23)$$

In this formulation, equality refers to the exceptional case when both engines are reversible. Substituting Eq. (5.2.22) into Eq. (5.2.23) we reproduce the general formulation of the Carnot principle as it was given in (5.2.7): *No cycle can be more efficient than the Carnot cycle when operating between the same constant-temperature heat source and heat sink.*

Principles of Thermodynamics

Relation (5.2.22) may be considered as proven, but only in a very particular case. The T–S diagram, which we used, and the entropy itself, deduced from Eq. (5.2.11), were all found using the known properties of an ideal gas. Therefore the statement (5.2.22) as well as the theorem for reduced heats (5.2.10) are so far valid just for engines using ideal gas as an working medium. One might wish to take relation (5.2.23) as a principle, and thus to claim that all reversible engines executing the same cycle have equal efficiency, regardless of their construction and the working substance. In particular, the efficiency of the Carnot cycle (5.2.8), though it was calculated using the specific example of an ideal gas, remains the same for any reversible heat engine and sets an unattainable limit for any irreversible engine executing the same cycle .

Since this principle is postulated, not proved, one must have sufficient background to accept it with an easy heart. Carnot's line of reasoning, based on the analogy of heat engines with hydro turbines may not be convincing to everyone. Therefore, in classical thermodynamics, the provision has been made to offer several equivalent formulations of the same principle, from which everybody may choose a basic one depending on his own taste. Given that, one should take care to prove that all other statements follow immediately from the accepted one.

We restrict ourselves here to the two best-known statements, due to Thomson (Lord Kelvin) and Clausius. The acceptance of either one leads inevitably to the adoption of the rest including the principle of Carnot.

(a) Thomson: *It is impossible to extract heat from a reservoir and convert it wholly into work without causing other changes in the universe.* In other words, conversion of heat energy into work which is accompanied by nothing except the cooling of the heat source is not possible. The process of thermal expansion that leads to Eq. (5.1.15) is not excluded by this principle because the increase in volume of the working medium serves as a compensation. The operation of cyclic thermodynamic engines for which relation (5.1.19) is valid is also not prohibited. In this case the compensation involves the heating of a heat sink, into which the portion of heat not converted into work is ejected. The only processes prohibited are those in which the heat taken from the source is completely transformed into work without any changes either in the engine itself or in the bodies surrounding it. On the other hand, the reverse process of non-compensated transformation of work into heat is not only allowed, but can be witnessed everywhere. This is the routine extraction of heat by means of friction, an electric stove, and so on.

(b) Clausius: *Heat can never, of itself, flow from a lower to a higher temperature.* This had already been understood by Carnot who took as an axiom that heat cannot be made to flow from a cold place to a hot place without the expenditure of work. This statement does not mean that the functioning of a heat pump is impossible. Any heat engine run backwards cools a heat sink and heats a heat source. This is how, for example, any kitchen refrigerator works, ejecting into the room heat extracted from the cooling chambers. However, this happens only due to the work done by the electric current, which is switched on at our will. What is excluded by Clausius is such a process in which the refrigerator chamber becomes cooler of its own accord without any effort from our side—just because heat started to flow outside against the gradient of temperature. This is generally intuitive. Our experience says that houses spontaneously get cooler in winter when we stop heating them, but never get spontaneously warmer due to the transport of heat into the house from outside.

The above formulations of the second law of thermodynamics establish the qualitative inequivalency of heat and work, despite their quantitative equivalence, presented by the first law. One qualitative distinction between spontaneous thermal processes and mechanical ones is in that the former are one-directional. Heat flows from warmer bodies to colder ones and never backwards; diffusion of gases leads to their uniform mixing and never to their separation; heat arises from the work of friction and never turns back into it, and so on.

All these and similar statements are not independent. Each follows from the other and vice versa. Accepting, for instance, the second law in the formulation of Thomson, it is necessary to agree with the formulation given by Clausius. Let us demonstrate this by *reductio ad absurdum*. Let us assume that the formulation of Clausius is false and we have succeeded in some way or another in transferring heat, Q_1, from the heat sink, held at temperature T_0, to the heater, with temperature $T_1 > T_0$. Then, giving the heat, Q_1, to a Carnot engine operating between these two reservoirs, we may obtain from it the work $A = Q_1 - |Q|_0$. Upon the completion of one cycle the engine has returned to the initial state and the heater has released all it has attained, so the heat that was transformed into work was taken only from the heat sink. Thus we have succeeded in the uncompensated transformation into work of the heat $Q_1 - |Q|_0$, which is impossible according to Thomson. Therefore, the initial assumption—the invalidity of the principle of Clausius—is false.

On the other hand, by accepting the principle of Clausius, we must also agree with the principle of Thomson. If the latter is false, then the work obtained without any compensation from the body at temperature T_0 could be converted back into heat by friction and transferred to the body at higher temperature T_1. However, this means that we have managed to realize a process prohibited by the principle of Clausius, which we had taken as an immutable axiom. Consequently, the Thomson principle is true as well.

Let us now convince ourselves that these formulations of the second law are equivalent to the principle of Carnot, which has the mathematical formulation given by Eq. (5.2.7). For this, it is necessary to reproduce the line of reasoning of Clausius about combined heat engines executing the Carnot cycle, that is, using only one heater and only one heat sink (Fig. 5.9). First, we will show that according to the above formulation of the second law no engines can have efficiency η greater than the efficiency η_0 of a reversible heat engine. Let us

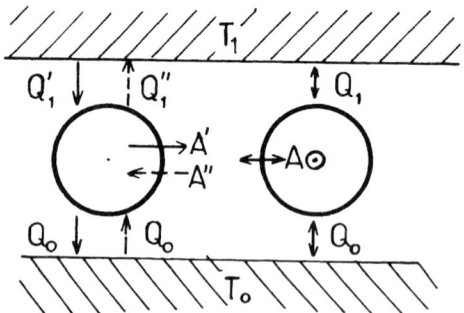

Figure 5.9 Left: heat engine under test in direct (solid arrows) and reverse (dashed arrows) cycles. Right: reversible engine between the same heat reservoirs (hatched).

assume for a moment that it is nevertheless possible to have $\eta > \eta_0 = \eta_c$. This then means that

$$\eta = 1 + \frac{Q'_0}{Q'_1} > 1 + \frac{Q_0}{Q_1} = \eta_c. \tag{5.2.24}$$

The primes denote the heats given or taken from the tested engine, which has by assumption an anomalously high efficiency. To check whether it is actually so high let us connect in parallel with the "better-than-reversible" heat engine a reversible one running backwards (Fig. 5.9). The volume of its working medium may be chosen such that in every cycle the reversible engine will extract from the heat sink exactly the same heat that is transferred to it in the forward cycle by the engine under test: $Q_0 = -Q'_0$. Then, from Eq. (5.2.24) it follows that the reversible engine will eject to the heater a smaller amount of heat $|Q_1| < Q'_1$, and will require for that a smaller amount of work $|A| = |Q_1| - Q_0 < Q'_1 + Q'_0 = Q'_1 - Q_0 = A'$. Thus, if one is to use the portion of work A' to perform work $|A|$, then after both engines have executed one cycle we will arrive at the following. The heat sink has given back all it had obtained $(Q_0 + Q'_0 = 0)$ and both the engines have returned to their initial states, but some amount of work still remains:

$$A' + A = A' - |A| = Q'_1 - |Q_1| = Q'_1 + Q_1 > 0. \tag{5.2.25}$$

The sign of this expression indicates that the only result of this process appears to be a uncompensated production of work out of the heat extracted by the two engines from the heater. The impossibility of such a result according to the second law in the formulation of Thomson proves that the initial assumption, that is the inequality (5.2.24), is invalid.

On the other hand, nothing contradicts the inequality

$$\eta = 1 + \frac{Q'_0}{Q'_1} < 1 + \frac{Q_0}{Q_1} = \eta_c, \tag{5.2.26}$$

that takes place if the efficiency of the engine under test is lower than the reversible one. Executing one cycle together both engines and their common heat sink return to the initial state since $Q_0 = -Q'_0$ as before. At the same time the work produced is negative:

$$A' + A = A' - |A| = Q'_1 - |Q_1| = Q'_1 + Q_1 < 0. \tag{5.2.27}$$

The sign of this result holds for any conversion of work into heat. This is not prohibited by the second law.

At first sight, it seems that starting both of the engines backwards, one may again arrive at a contradiction to the Thomson formulation by the same line of reasoning, although $\eta_c > \eta$. This confusion is immediately dispelled if it is considered that an irreversible engine traversing the cycle in the counterclockwise direction is consuming the work $|A''|$ which is larger than that produced in the direct cycle: $|A''| > A'$. Besides, it ejects to the heater the heat $|Q_1''| > Q_1'$. The second law of thermodynamics is not contradicted if

$$|A''| > A > A' \quad \text{and} \quad |Q_1''| > Q_1 > Q_1'. \qquad (5.2.28)$$

In this case the result of the joint work of both engines in the reverse cycle is again conversion of work into heat:

$$A + A'' = A - |A''| = Q_1 - |Q_1''| = Q_1 + Q_1'' < 0. \qquad (5.2.29)$$

Although uncompensated, this conversion is not prohibited by the second law.

Any real heat engine is irreversible, but the closer it is to the ideal of a reversible engine, the smaller is the difference between the absolute values of works and heats in the direct and reverse cycles: $|A''| \to A'$ and $|Q_1''| \to Q_1'$. The inequalities still apply while taking the limit, so any tested engine, if it is reversible, consumes the same heat and produces the same work as the partner. In other words, all reversible engines regardless of their construction and working medium show the same efficiency in traversing the Carnot cycle:

$$\eta_0 = \eta_c \qquad (5.2.30)$$

When this equality is combined with inequality (5.2.26), then the Carnot principle in the form (5.2.7) is reproduced, where an equality refers to the case of reversible engines only.

The above line of reasoning related to the Carnot cycle may be easily generalized for an arbitrary cycle by again fragmenting it into a set of Carnot cycles (Fig. 5.6). Then the result of such reasoning leads to Eq. (5.2.23). Since all reversible engines have equal efficiency, nothing prevents us from considering as "lawful" an engine using an ideal gas, for which the relation (5.2.22) is proven. Combining this with Eq. (5.2.23) we obtain a hierarchy of inequalities, which confirm and clarify the meaning of the Carnot principle :

$$\eta \leq \eta_0 \leq \eta_c. \qquad (5.2.31)$$

Open Processes

Referring again to the Carnot cycle, we may express Eq. (5.2.7) in the form:

$$1 + \frac{Q_0}{Q_1} = \eta \leq \eta_c = 1 - \frac{T_0}{T_1}.$$

From this we obtain for the sum of reduced heats:

$$\sum_i \frac{Q_i}{T_i} = \frac{Q_1}{T_1} + \frac{Q_0}{T_0} \leq 0. \tag{5.2.32}$$

This is an extension of relation (5.2.9) to irreversible processes, which are the subject of the inequality in (5.2.32). For an arbitrary cycle, repeating the necessary deductions, we obtain instead of (5.2.11):

$$\oint \frac{dQ}{T} \leq 0 = \oint dS,$$

while for open (not cyclic) processes, instead of (5.2.12) one finds

$$dS \geq \frac{dQ}{T}. \tag{5.2.33}$$

This is the most general formulation of the second law of thermodynamics, referring both to reversible processes (equality) and to irreversible ones (inequality).

The second law is said to describe reversible processes quantitatively, while irreversible ones qualitatively only, because the inequality (5.2.33) indicates merely the direction of their development, but not the final result. The most commonly known consequence of the second law in the special case of adiabatic processes ($dQ = 0$) was formulated by Clausius as a principle of "entropy increase in a thermally insulated system." In fact

$$dS \geq 0, \tag{5.2.34}$$

where entropy is unchanged if the adiabatic process is reversible and increases if it is irreversible. An example of the first process is equilibrium adiabatic expansion of a gas, where work is performed during expansion. An alternative example is expansion into a vacuum. As the system is insulated both adiabatically and mechanically work cannot be performed during expansion and the entropy increases. Assuming that in both cases an ideal gas expands to the volume V_2, the final states of the system must be found on the isochore $V = V_2$: in the first case, on its intersection with the adiabatic line; in the second, with the isotherm (5.1.24). In the latter case increase of entropy is maximal. All natural processes occur between these extremes. In expansion with finite speed the work performed by the gas is not equal to 0, but is less than the maximum produced in a reversible process. At the termination of a natural process, the state of the gas falls on the same isochore, but on a point lying between the isothermal and adiabatic lines (Fig. 5.4); the closer it lies to the adiabatic, the slower was the expansion. For all such processes one may write instead of Eq. (5.2.34)

$$S_2 \geq S_1 \tag{5.2.35}$$

Equality occurs in the case of equilibrium expansion, while the inequality increases as the process becomes more nonequilibrium. When the gas is compressed back to the primary volume V_1, it will return to the initial state only if both expansion and compression were equilibrium processes. When a gas has expanded into a vacuum then even equilibrium adiabatic compression to the previous volume will increase the entropy. The entropy will be equal to the entropy of the state of the gas after expansion. The entropy increase will be even greater if the compression is very rapid, that is, also nonequilibrium.

Natural Variables

Combining first and second laws into a single equation we obtain after elimination of dQ from (5.2.33) and (5.1.20):

$$TdS \geq d\mathcal{E} + pdV, \qquad (5.2.36)$$

where equality refers to equilibrium processes. This equality is convenient to express in the form:

$$d\mathcal{E} = TdS - pdV. \qquad (5.2.37)$$

It is important to bear in mind how this equation differs from the very general expression

$$d\mathcal{E} = \left(\frac{\partial \mathcal{E}}{\partial x}\right)_y dx + \left(\frac{\partial \mathcal{E}}{\partial y}\right)_x dy, \qquad (5.2.38)$$

which is just an exact differential of $V(x,y)$ as a function of two arbitrary variables. Any two out of four thermodynamic parameters may be taken as these variables

$$p, V, T, S, \qquad (5.2.39)$$

but only the pair V, S is "natural" for the internal energy $\mathcal{E}$. This is exceptional in that the derivatives of $\mathcal{E}$ with respect to these parameters are thermodynamic parameters themselves. Comparing (5.2.37) to (5.2.38) term by term we obtain:

$$T = \left(\frac{\partial \mathcal{E}}{\partial S}\right)_V, \quad p = -\left(\frac{\partial \mathcal{E}}{\partial V}\right)_S. \qquad (5.2.40)$$

Moreover, according to the theory of functions of several variables, mixed derivatives do not depend on the order of differentiation and therefore

$$\left(\frac{\partial T}{\partial V}\right)_S = \frac{\partial \mathcal{E}}{\partial V \partial S} = \frac{\partial \mathcal{E}}{\partial S \partial V} = -\left(\frac{\partial p}{\partial S}\right)_V. \qquad (5.2.41)$$

Equalities of this kind are known as the Maxwell relations. These are direct corollaries of the first two laws of thermodynamics and demonstrate the substantial reduction of information necessary to describe the system. In particular, relation (5.2.41) shows that it is not necessary to calculate the coefficient $(\partial S/\partial p)_V$, which characterizes the variation of entropy in an isochoric process as pressure increases, because it can be expressed simply via the coefficient of thermal expansion in an adiabatic process $\frac{1}{V}(\partial V/\partial T))_S$.

The essence of the thermodynamic method is in finding answers to posed questions, expressed in terms of a minimal number of easily measured parameters and specific heats. This method can be rigorously formalized, especially when Jacobians are being used for manipulations with derivatives of the type (5.2.41). The latter clearly may be expressed in the form

$$\left(\frac{\partial T}{\partial V}\right)_S = \frac{\partial(T,S)}{\partial(V,S)}, \quad \left(\frac{\partial p}{\partial S}\right)_V = \frac{\partial(p,V)}{\partial(S,V)},$$

while the property of calibration invariance (5.2.15) allows (5.2.41) to be reproduced by means of the following transformation:

$$\frac{\partial(T,S)}{\partial(V,S)} = \frac{\partial(T,S)}{\partial(p,V)}\frac{\partial(p,V)}{\partial(V,S)} = \frac{\partial(p,V)}{\partial(V,S)} = -\frac{\partial(p,V)}{\partial(S,V)}.$$

Here we took into account that interchange of any two variables of the Jacobian means inversion of its sign.

The necessity for such transformations arises constantly in thermodynamics. Let us express, for example, equation (5.2.37) in the form:

$$\left(\frac{\partial \mathcal{E}}{\partial V}\right)_T = T\left(\frac{\partial S}{\partial V}\right)_T - p. \tag{5.2.42}$$

In order to determine the volume dependence of internal energy it is useful to transform the expression

$$\left(\frac{\partial S}{\partial V}\right)_T = \frac{\partial(S,T)}{\partial(V,T)} = \frac{\partial(V,p)}{\partial(V,T)} = \left(\frac{\partial p}{\partial T}\right)_V. \tag{5.2.43}$$

After substitution of this result into equation (5.2.42) it takes the following form, quoted sometimes as the "thermodynamic equation of state":

$$p + \left(\frac{\partial \mathcal{E}}{\partial V}\right)_T = T\left(\frac{\partial p}{\partial T}\right)_V. \tag{5.2.44}$$

This relation allows us to determine $\mathcal{E}(V)$ knowing only the thermal equation of state $p(T,V)$.

Gas of van der Waals

In reality the equations of state are borrowed by thermodynamics from outside, that is, not deduced, but only used. Such is, for example, the equation of van der Waals

$$p + \frac{A}{V^2} = \frac{RT}{V-B}. \qquad (5.2.45)$$

This is an equation of state of a real gas. Its parameters, the molar constants A and B, have a statistical interpretation, discussed in Section 1.9. Using Eq. (5.2.45) in the right-hand side of Eq. (5.2.44) we obtain the following result:

$$p + \left(\frac{\partial \mathcal{E}}{\partial V}\right)_T = \frac{RT}{V-B}. \qquad (5.2.46)$$

This can be easily identified term by term with the equation of van der Waals. Most likely this was the reason that equation (5.2.44) was given the essentially incorrect title of the thermodynamic equation of state. Actually, it is not an equation of state but a general thermodynamic relation. It may be combined with the van der Waals equation to find the following

$$\left(\frac{\partial \mathcal{E}}{\partial V}\right)_T = \frac{A}{V^2}. \qquad (5.2.47)$$

Using this information together with Eq. (5.1.28) we obtain:

$$d\mathcal{E} = \left(\frac{\partial \mathcal{E}}{\partial V}\right)_T dV + \left(\frac{\partial \mathcal{E}}{\partial T}\right)_V dT = \frac{A}{V^2} dV + C_V dT \qquad (5.2.48)$$

Since V and T are not natural variables for $\mathcal{E}$, the coefficients in the exact differential are not thermodynamic parameters as in Eq. (5.2.37). Integrating Eq. (5.2.48) we find:

$$\mathcal{E} = C_V T - \frac{A}{V} + \text{const}. \qquad (5.2.49)$$

From this it is seen that the internal energy of van der Waals gas decreases during compression. The negative term in Eq. (5.2.49) represents the attraction that dominates intermolecular interactions at low temperatures and moderate gas densities.

Even in a gaseous state the presence of this term is really essential despite its smallness. It indicates that the process is not necessarily isothermic when $\mathcal{E}$ is conserved. In particular, for expansion into a vacuum when certainly $\mathcal{E}_1 = \mathcal{E}_2$ according to Eq. (5.1.23), the temperature of a real gas is not constant. Moreover, we are able now to calculate the derivative (5.1.22) and to determine by this whether a real gas gets warm or becomes cool when it expands into a vacuum. In fact,

$$\left(\frac{\partial T}{\partial V}\right)_\mathcal{E} = \frac{\partial(T,\mathcal{E})}{\partial(V,\mathcal{E})} = -\frac{\partial(T,\mathcal{E})}{\partial(T,V)}\frac{\partial(T,V)}{(\partial\mathcal{E},V)} = -\frac{1}{C_V}\left(\frac{\partial \mathcal{E}}{\partial V}\right)_T. \quad (5.2.50)$$

Using Eq. (5.2.44) we reduce this expression to the form:

$$\left(\frac{\partial T}{\partial V}\right)_\mathcal{E} = \left[p - T\left(\frac{\partial p}{\partial T}\right)_V\right]\Big/ C_V. \quad (5.2.51)$$

Now, to estimate the sign of this expression we need only the thermal equation of state $p(T, V)$. Using as an example the van der Waals equation (5.2.47) we obtain:

$$\left(\frac{\partial T}{\partial V}\right)_\mathcal{E} = -A/C_V V^2 < 0. \quad (5.2.52)$$

This means that, unlike an ideal gas, the real gas becomes cool when expanding into a vacuum. The difference between real and ideal gases decreases with the increase of V and disappears completely when $A = 0$.

The thermodynamic method, whose power has been demonstrated here, will be used often in what is to come in situations where we possess complete information. However, let us first consider a simpler way of using the thermodynamic laws that was used in classic works in this field for extending the areas of its application.

5.3 METHOD OF CYCLES

The simplest method we will discuss is based on the fact that all reversible Carnot engines have the same efficiency. This assertion may be expressed by the equality

$$\frac{dA}{Q} = \frac{dT}{T}, \quad (5.3.1)$$

when the temperature of the heater differs from the temperature of the heat sink by an infinitesimal amount dT. Here Q is the heat received from heater in the cycle while dA is work produced. The infinitesimal temperature difference relieves us of the necessity of taking care of adiabatic transitions between isotherms T and $T - dT$. If these transitions are not adiabatic the accompanying losses (or acquisitions) of heat dQ are still infinitesimal. They are negligible in comparison with the heat Q, absorbed (or rejected) isothermally (Fig. 5.10).

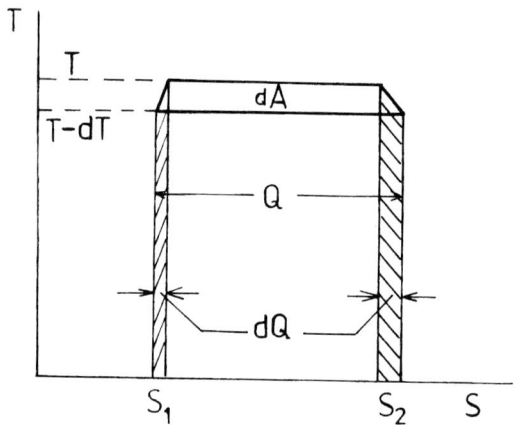

Figure 5.10 Infinitesimal cycle between two close isotherms at T and $T - dT$. The vertical shaded segments are negligible comparable with the areas under isotherms.

Relation (5.3.1) is equivalent to the second law of thermodynamics for reversible cyclic processes. To utilize it one has to express the open process of interest through a specially constructed Carnot cycle and calculate its efficiency. Substituting the result into Eq. (5.3.1), useful information about the process may be obtained.

The Clapeyron–Clausius Relation

Let us refer to a real gas. It may be treated as ideal only at high temperatures and low pressures. The further the gas is from ideal, the closer it is to condensation into a solid or liquid state. The *coexistence line*, which connects the triple point with the critical point on the p–T diagram (Fig. 4.1) is in fact the vapor pressure $p_V(T)$. Only points on this curve represent equilibrium states of a two-phase system: liquid + vapor. In a gravitational field the phases are separated by the horizontal surface of liquid. This boundary can be moved with the help of a plunger placed over the vapor. The volume of vapor may increase by pushing back the plunger and thus performing usable work or, alternatively, one may condense the vapor by pushing the plunger forward. As neither temperature nor pressure is changed in the process, the work is performed at the expense of internal energy, which changes at the phase transition, and evaporation heat, which is conveyed to the system.

Variations of the system volume accompanying these processes may be seen on the p–V diagram of Fig. 4.2. The coexistence line there becomes a two-valued function: at the same pressure of saturated vapor, a homogeneous substance may be either in the liquid state with molar volume V_L or in a gaseous one with molar volume V_V. In the interval $V_L < V < V_V$ there is no single phase, but the liquid with a vapor phase over it. As the volume is varied

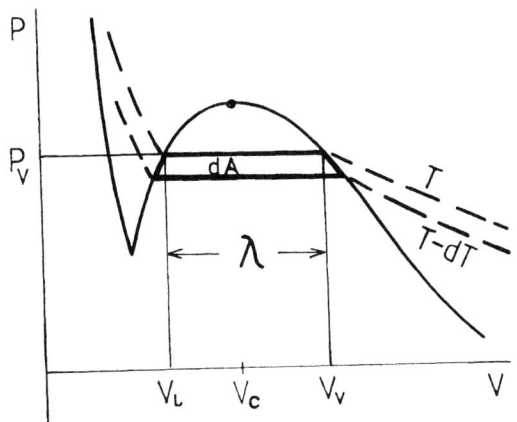

Figure 5.11 "Evaporation–condensation" cycle between two close isotherms (dashed) whose horizontal segments (solid) correspond to liquid–vapor transformation.

the distribution between these two phases changes accordingly. Equally, the existence of a single phase in the corridor separating a regular (crystal) phase from a disordered one (liquid or gas) is impossible. The isotherms, shown on Fig. 5.11, bend, crossing the region which is forbidden for a single-phase existence.

Let us consider a cycle whereby liquid is transformed into vapor by expansion and then returned to a liquid state by compression, but at lower temperature (Fig. 5.11). In the expansion at temperature T from the homogeneous liquid to vapor we gain work $A = p_V(V_V - V_L) > 0$. The subsequent compression at a lower temperature $T - dT$, carried out at a correspondingly lower vapor pressure $p_V - dp_V$, returns most of the work: $A' = (p_V - dp_V)(V_L - V_V) < 0$. The useful output is their difference

$$dA = A + A' = dp_V(V_V - V_L). \qquad (5.3.2)$$

The work performed in moving down and up from one isotherm to another along the side branches of the coexistence line may be neglected because their values are higher orders of the infinitesimal. The same is valid for the appropriate heats. In calculating the efficiency the work (5.3.2) must be used in Eq. (5.3.1) as well as the molar heat of evaporation $Q = \lambda$ consumed by the system undergoing a liquid–vapor phase transition:

$$\frac{dp_V}{\lambda}(V_V - V_L) = \frac{dT}{T}. \qquad (5.3.3)$$

This is just the application of the second law to this cycle. From this result immediately follows the Clapeyron–Clausius relation:

$$\frac{dp_V}{dT} = \frac{\lambda}{T(V_V - V_L)}. \qquad (5.3.4)$$

Recall that the equilibrium vapor pressure is a function of a single variable, temperature. This is associated with the fact that vapor easily penetrates through the boundary of liquid and therefore neither its volume nor amount are conserved. The situation is similar to radiation, which is easily absorbed and radiated by the walls of the vessel confining it (see Chapter 2). As a result, the radiation equation of state is also a function of temperature only.

The Clapeyron–Clausius relation helps one to avoid difficult measurements of the evaporation heat λ. It may be calculated using Eq. (5.3.4) from a given $p_V(T)$ that can be easily determined experimentally with higher accuracy. However, an approximate estimate of the vapor pressure in the vicinity of a triple point can be found, taking into account that the molar volume of vapor at low temperature is much higher than that of liquid, and the vapor can be treated as an ideal gas. With these assumptions one can deduce from Eq. (5.3.4)

$$\frac{dp_V}{dT} \approx \frac{\lambda}{TV_V} \approx \frac{p_V \lambda}{RT^2} \quad \text{at} \quad T - T_T \ll T_c - T_T. \qquad (5.3.5)$$

Assuming also $\lambda \approx$ const we find by integration:

$$p_V \propto \exp(-\lambda/RT) \qquad (5.3.6)$$

Such specific information about the temperature behavior of p_V in the vicinity of the triple point is not entirely thermodynamic because it was deduced from Eq. (5.3.4) using known properties of vapor and liquid at low temperatures. The initial Clapeyron–Clausius equation is more general, and is as incontestable as the laws of thermodynamics themselves. One may judge the validity of any calculations or measurements of λ or $p_V(T)$ by comparison of the determined values with this equation.

Thermoelectric Cycle

Let us now consider a heat engine using as an working medium two straight and uniform current-carrying conductors made out of different solid materials and contacting each other at two junctions: one cold and one hot (Fig. 5.12). Such a system is called a thermocouple and is widely used for the measurement of the temperature of one of the junctions. The measurement of either a hot (T) or a cold temperature (T_0) is based on the Seebeck effect. In the open circuit of the thermocouple, the electromotive force E is monotonously increases with the increase in temperature difference $T - T_0$. The $E(T)$ dependence is, generally speaking, nonlinear:

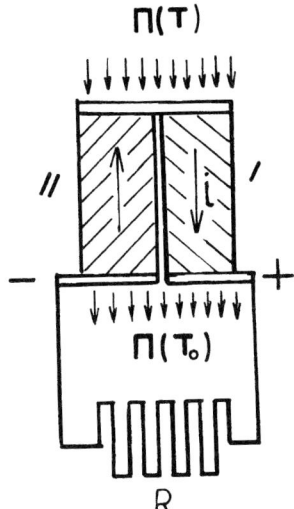

Figure 5.12 Thermocouple with external load R. The upper junction of two different conductors (' and ") has higher temperature T in comparison with the lower junction which is kept at temperature T_0.

$$E = \int_{T_0}^{T} \alpha(T')dT'. \qquad (5.3.7)$$

Therefore, the metric scale of the thermocouple requires calibration, which is easily done if the so-called *thermoelectric power* $\alpha(T)$ is known. It depends exclusively on the temperature and the constituent materials of the thermocouple.

When circuit is closed, the current i circulated around it performs the work

$$A = iE. \qquad (5.3.8)$$

A significant part of the work can be utilized in an external load. The thermocouple is the simplest heat engine. It transforms the heat it receives directly into work, performed by an electrical current. Part of the heat is the Peltier heat, which is absorbed in the hot junction and released in the cold one in proportion to the current i passing through them:

$$Q_\Pi = \pm i \Pi(T). \qquad (5.3.9)$$

Here "$+$" corresponds to the hot junction and "$-$" to the cold one. When the thermocouple works as an electric generator, the current is considered to be positive ($i > 0$). However, it may be also negative since the Peltier effect is reversible. Using an external battery, let current flow through the thermocouple in the reverse direction. Then it becomes a heat pump, absorbing heat at the cold junction and ejecting it at the hot one. The absolute magnitude of heats are the same as in the direct cycle, that is, proportional to a current and the Peltier coefficient $\Pi(T)$. As well as $\alpha(T)$ Peltier's coefficient depends only on temperature and the specifics of the thermocouple.

Applying Eq. (5.3.1) to the thermoelectric cycle we find:

$$\frac{idE}{Q} = \frac{\alpha dT}{\Pi} = \frac{dT}{T}. \qquad (5.3.10)$$

From this immediately follows Kelvin's *second thermoelectric relation*

$$\Pi = \alpha T. \qquad (5.3.11)$$

Such a simple relation between these two thermoelectric effects avoids the need to perform the comparatively complex and inevitably rough heat measurements of Peltier's coefficient. It can be more easily and accurately found from Eq. (5.3.11) with α measured electrically.

However, Thomson (Kelvin) was not satisfied with this result and examined the thermoelectric cycle more scrupulously, using both laws of thermodynamics. His line of reasoning proceeds from the question: what is the shape of the thermoelectric cycle on the T–S diagram (Fig. 5.13)? It was of course possible to assume that it coincided with a Carnot cycle, but nobody knew in advance whether this was in fact true. As in any other case, along with isothermal absorption and ejection of Peltier's heat, heat exchange with the surroundings could occur along the side sections of the cycle connecting the isotherms. This effect was expected by Thomson, who assumed the heat to be proportional to the current and the temperature interval it passes through:

$$dQ = \pm i\tau dT. \qquad (5.3.12)$$

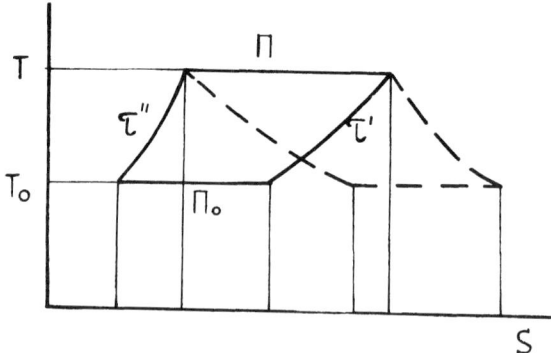

Figure 5.13 Thermoelectric cycles in the T–S plane with τ'', $\tau' > 0$ (solid line) and τ'', $\tau' < 0$ (dashed line). Π and Π_0 are Peltier heats at high and low temperatures.

5.3 METHOD OF CYCLES

The Thomson coefficient τ is by convention positive if heat is received when current flows against the gradient of temperature, and negative in the opposite case. This alternative is reflected in Fig. 5.13 with two possible deviations of the thermoelectric cycle from the Carnot cycle: on the left (at $\tau', \tau'' > 0$) and on the right (at $\tau', \tau'' < 0$). Primes mark different wires of the thermocouple; the positive sign in Eq. (5.3.12) refers to a current flowing against the gradient of temperature, while negative is along the gradient. Therefore the Peltier heat is absorbed in one branch of the circuit and ejected in the other. As a result, the thermocouple receives in total

$$dQ_T = i(\tau'' - \tau')dT$$

when acting as a generator, and ejects the same magnitude of heat when acting as a heat pump. In other words, the Thomson effect, as well as other thermoelectric effects, is reversible.

The most interesting part of this is that the effect was discovered theoretically, by deduction from most general considerations given above. In actual fact the existence of the effect was not strictly predicted, but merely admitted to exist. In order to establish the relation of this newly introduced effect with the others, Thomson applied to the thermoelectric cycle not only the first, but also the second law of thermodynamics, according to which

$$Q_\Pi(T) + Q_\Pi(T_0) + i\int_{T_0}^{T}(\tau'' - \tau')dT = iE(T). \qquad (5.3.13)$$

The second law was used in the form of the theorem of reduced heats (5.2.10):

$$\frac{Q_\Pi(T)}{T} + \frac{Q_\Pi(T_0)}{T_0} + i\int_{T_0}^{T}\frac{\tau'' - \tau'}{T}dT = 0. \qquad (5.3.14)$$

Substituting into Eqs. (5.3.13) and (5.3.14) expressions from (5.3.7) and (5.3.9) we reach the following formulation of the two laws in this particular case:

$$\Pi(T) - \Pi(T_0) + \int_{T_0}^{T}(\tau'' - \tau')dT = \int_{T_0}^{T}\alpha dT, \qquad (5.3.15a)$$

$$\frac{\Pi(T)}{T} - \frac{\Pi(T_0)}{T_0} + \int_{T_0}^{T}\frac{(\tau'' - \tau')dT}{T} = 0. \qquad (5.3.15b)$$

In the limit of an infinitesimal temperature difference, or simply differentiating equations (5.3.15) with respect to the upper limit, we obtain:

$$\frac{d\Pi}{dT} + \tau'' - \tau' = \alpha, \qquad (5.3.16a)$$

$$\frac{d}{dT}\left(\frac{\Pi}{T}\right) + \frac{\tau'' - \tau'}{T} = 0. \qquad (5.3.16b)$$

Differentiating (5.3.16b) and eliminating $d\Pi/dT$ with the help of (5.3.16a) we obtain again the second thermoelectric relation (5.3.11). However, along with the two equations (5.3.16), one can deduce yet another. Direct substitution of (5.3.11) into (5.3.16a) gives the first Kelvin relation:

$$\tau'' - \tau' = -T\frac{d\alpha}{dT}. \qquad (5.3.17)$$

This relation shows that the Thomson effect is absent only if $\alpha = $ const. In this single case the thermoelectric scale does not need calibration because $E = \alpha T$. In reality, experiments demonstrated the nonlinear temperature dependence of $\alpha(T)$ which unambiguously pointed to the existence of the Thomson effect. Such experimental proof of this entirely theoretical prognosis gave evidence that the new science of thermodynamics possessed the force of prediction just as her senior sister, mechanics, did, through which was discovered the existence of planets previously unobserved.

Heat Pumps

To convert a thermocouple into a heat pump it is sufficient to pass a current through it in the opposite direction to that generated in the direct cycle. Then in the cold junction the heat $Q_0 = i\Pi_0$ is absorbed, while in the hot one heat $Q_1 = i\Pi_1$ is ejected. The same construction is shared by all pumps which extract heat from an internal vessel (freezing chamber) and eject it outside, together with the work spent in cooling, which is iE in this particular case.

The efficiency of the heat pump is characterized by the cooling coefficient K, which is the ratio of the heat extracted from the cooling chamber to the work spent in doing this. According to (5.2.28) any real engine takes work $|A''|$ which is higher than the work $|A|$ required by a reversible engine to extract the same amount of heat Q_0. Therefore,

$$K = \frac{Q_0}{A''} \leq \frac{Q_0}{A} = K_0. \qquad (5.3.18)$$

The cooling coefficient K_0 of a reversible heat engine can be easily associated with the efficiency of the same engine in the direct cycle

$$K_0 = \frac{Q_0/Q_1}{A/Q_1} = \frac{1-\eta}{\eta}. \qquad (5.3.19)$$

For a Carnot cycle $\eta_0 = \eta_c = (T_1 - T_0)/T_1$ and, therefore,

$$K_c = \frac{T_0}{T_1 - T_0}. \qquad (5.3.20)$$

It is therefore seen that there is an important difference between the efficiency of the direct and reverse cycles. The maximal efficiency of a heat engine depends on the working temperature difference, but in all cases $0 \leq \eta_c \leq 1$. In contrast there is no upper limit for the cooling coefficient which varies between $0 \leq K_c \leq \infty$. If it is necessary to cool the chamber by a few degrees only, this can be done with very high K_0 and even real K can be made higher than 1. However, the deeper the cooling, the lower is K_0, and of course also $K < K_0$. Semiconductor thermoelectric pumps and thermogenerators appear to have much higher efficiency than metallic ones, but still cannot compete with conventional industrial engines due to irreversible heat transfer and Joule losses in the wires of the thermocouple.

5.4 COOLING OF GASES

The Joule–Thomson Effect

Shortly prior to the discovery of the thermoelectric effect Thomson, together with Joule, performed a modified experiment on the irreversible adiabatic expansion of gas, testing its ideality. The first experiments of such a kind, conducted by Gay-Lussac and Joule, allowed gas to expand from a filled volume into an empty one through a hole in dividing wall. To increase the sensitivity of such experiments, Joule and Thomson decided to perform them in a more gentle way. In their experiment the gas was throttled through the plug separating a thermally insulated vessel into two parts stopped up from both sides by movable plungers (Fig. 5.14). The plug was made of dense cotton that required the gas to be forced through it, and the pressure was greater on the near side than on the far side, where it is usually atmospheric. If in the beginning the plunger on the right made contact with the plug and all the gas was at the left side occupying volume V_1 at pressure p_1, then finally all the gas was found on the right with pressure p_2 and with volume increased to V_2. As both pressures were held constant, during this transition the right plunger performed work $p_2 V_2$, and the left one $p_1 V_1$. This was the only difference of the Joule–Thomson process from the process of Gay-Lussac: in throttling through the plug, the gas performed work

$$A = p_2 V_2 - p_1 V_1, \qquad (5.4.1)$$

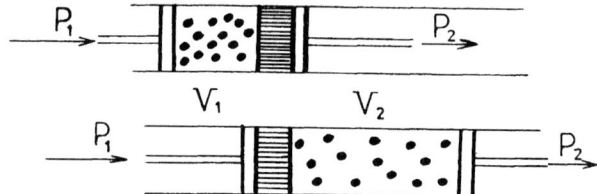

Figure 5.14 Initial (upper) and final (lower) states of the system in the Joule–Thomson experiment of expanding gas in a piston. A plug inside the vessel is shaded.

while on expansion into the vacuum $A = 0$. Although also in this case, the process is adiabatic, we obtain from (5.1.17) and (5.4.1):

$$0 = \mathcal{E}_2 - \mathcal{E}_1 + p_2 V_2 - p_1 V_1. \qquad (5.4.2)$$

Instead of the conservation law for internal energy (5.1.23), peculiar for expansion into a vacuum, equation (5.4.2) generates a conservation law for some new entity H which was called *enthalpy*:

$$H_1 = \mathcal{E}_1 + p_1 V_1 = \mathcal{E}_2 + p_2 V_2. \qquad (5.4.3)$$

How should one pose correctly the question as to whether the temperature of the gas rises or falls during throttling? It must now to be formulated differently than was done in Eq. (5.1.22). Since during expansion the pressure definitely decreases ($dp < 0$), then the temperature will decrease as well ($dT < 0$) if

$$\left(\frac{\partial T}{\partial p}\right)_H > 0, \qquad (5.4.4)$$

where

$$H = \mathcal{E} + pV. \qquad (5.4.5)$$

If condition (5.4.4) is satisfied, then the process could be repeated, conducting it now from right to left and achieving an even greater decrease of temperature rather then returning to its previous value (thus the irreversible character of the process is manifested). Moreover, such a device is suited for the continuous repetition of the process: back–forth, back–forth, and so on, thereby achieving unlimited cooling (and expansion) of the gas. However, one must remember that the derivative in (5.4.4) is a function of T and p. If in the lowering of the temperature the inequality (5.4.4) is inverted, then by the Joule–Thomson effect the gas will heat up and further cooling will be stopped. In order to determine by thermodynamics the limits of gas cooling by throttling, it is necessary first to study the properties of the newly introduced function of state, enthalpy.

Enthalpy

Differentiating (5.4.5) we obtain:

$$dH = d\mathcal{E} + p\,dV + V\,dp \tag{5.4.6}$$

Substituting $d\mathcal{E}$ from Eq. (5.2.37) we find:

$$dH = T\,dS + V\,dp. \tag{5.4.7}$$

From this we conclude that the natural variables for enthalpy are S and p. Knowing $H(S,p)$ it is easy to find by simple differentiation

$$T = \left(\frac{\partial H}{\partial S}\right)_p, \qquad V = \left(\frac{\partial H}{\partial p}\right)_S. \tag{5.4.8}$$

From this follows a Maxwell relation similar to (5.3.41):

$$\left(\frac{\partial T}{\partial p}\right)_S = \frac{\partial(T,S)}{\partial(p,S)} = \frac{\partial(p,V)}{\partial(p,S)} = \left(\frac{\partial V}{\partial S}\right)_p. \tag{5.4.9}$$

If $d\mathcal{E}$ from Eq. (5.2.36) is substituted into Eq. (5.4.6) then relation (5.4.7) is generalized appropriately:

$$dH \leq T\,dS + V\,dP. \tag{5.4.10}$$

This inequality indicates that enthalpy tends to a minimum if the natural process is adiabatic and isobaric.

According to (5.4.5), (5.1.29), and (5.1.30) the enthalpy of ideal gas is

$$H = C_V T + RT = C_p T, \tag{5.4.11}$$

and thus is constant along isotherms just as internal energy. If gas were ideal it would expand isothermally in the Joule–Thomson experiment in exactly the same way as it expands into vacuum. However all real gases behave in various ways in these experiments, due to varying conditions under which they are conducted. For real gases, the lines of constant H and $\mathcal{E}$, do not align either with isotherms, or with each other.

The Sign of the Effect

Let us consider now where the inequality (5.4.4) is actually valid, guaranteeing the lowering temperature by the Joule–Thomson effect. To find the sign of the derivative in (5.4.4) let us represent it in the form:

$$\frac{\partial(T,H)}{\partial(p,H)} = \frac{\partial(T,H)}{\partial(p,S)}\frac{\partial(p,S)}{\partial(p,H)} = \left[\left(\frac{\partial T}{\partial p}\right)_S\left(\frac{\partial H}{\partial S}\right)_p - \left(\frac{\partial T}{\partial S}\right)_p\left(\frac{\partial H}{\partial p}\right)_S\right]\bigg/\left(\frac{\partial H}{\partial S}\right)_p. \tag{5.4.12}$$

Using Eqs. (5.4.8) and (5.4.9) this may be transformed as follows:

$$\left(\frac{\partial T}{\partial p}\right)_H = \left[\left(\frac{\partial V}{\partial S}\right)_p T - \frac{TV}{C_p}\right]\bigg/T = \frac{\partial(V,p)}{\partial(S,p)} - \frac{V}{C_p}. \tag{5.4.13}$$

Here we have used definition (5.2.19) in the form:

$$C_p = T\left(\frac{\partial S}{\partial T}\right)_p, \tag{5.4.14}$$

which we use again to calculate the derivative

$$\frac{\partial(V,p)}{\partial(S,p)} = \frac{\partial(V,p)}{\partial(T,p)}\frac{\partial(T,p)}{\partial(S,p)} = \left(\frac{\partial V}{\partial T}\right)_p \frac{T}{C_p}. \tag{5.4.15}$$

Substituting this into (5.4.13) we finally obtain:

$$\left(\frac{\partial T}{\partial p}\right)_H = \frac{TV}{C_p}\left[\frac{1}{V}\left(\frac{\partial V}{\partial T}\right)_p - \frac{1}{T}\right] > 0. \tag{5.4.16}$$

This condition of cooling is evidently satisfied only when

$$\alpha = \frac{1}{V}\left(\frac{\partial V}{\partial T}\right)_p = -\frac{1}{n}\left(\frac{\partial n}{\partial T}\right)_p > \frac{1}{T}, \tag{5.4.17}$$

where isobaric coefficient of thermal expansion, and $n = N_0/V$ denotes the number density of the gas (N_0—Avogadro's number).

Inversion Curve

Let us calculate the Joule–Thomson coefficient (5.4.16) for real gas, taking the van der Waals Eq. (5.2.45) in the form:

$$p + an^2 = \frac{nkT}{1 - bn}, \qquad (5.4.18)$$

where $a = A/N_0^2$ and $b = b/N_0$. Differentiating at constant p we obtain

$$2an\left(\frac{\partial n}{\partial T}\right)_p = \frac{nk}{1 - bn} + \left(\frac{\partial n}{\partial T}\right)_p \frac{kT}{(1 - bn)^2}.$$

Using the definition of α, we transform this expression into the following:

$$\left[\frac{kT}{(1 - bn)^2} - 2an\right]\alpha = \frac{k}{(1 - bn)}. \qquad (5.4.19)$$

If $1/T$ is substituted for α, Eq. (5.4.19) becomes an inequality identical to (5.4.17):

$$\frac{kT}{(1 - bn)^2} - 2an < \frac{kT}{(1 - bn)}. \qquad (5.4.20)$$

Let us commit to memory that this sign corresponds to cooling of the van der Waals gas in the Joule–Thomson effect.

After a simple transformation of (5.4.20) we obtain:

$$T/2T_B < (1 - bn)^2. \qquad (5.4.21)$$

The *Boyle temperature*,

$$T_B = a/bk = A/BR, \qquad (5.4.22)$$

is an important parameter of the inversion curve, determined by the equality:

$$\sqrt{T/2T_B} = 1 - bn. \qquad (5.4.23)$$

When the left-hand side of (5.4.21) is actually small, then cooling takes place. Otherwise the Joule–Thomson effect results in heating. An exact equality determines the line separating the region of cooling from the region of heating. It appears on the T–n diagram (Fig. 5.15) as a fragment of a parabola with a minimum at the point $n = 1/b$. Cooling is possible only below the inversion curve.

In order to judge whether condensation of gas is possible in the region where the Joule–Thomson effect is cooling, it is convenient to draw the inversion curve on the p–T diagram. For that, let us denote both sides of Eq. (5.4.23) as x and specify the curve by parametric equations:

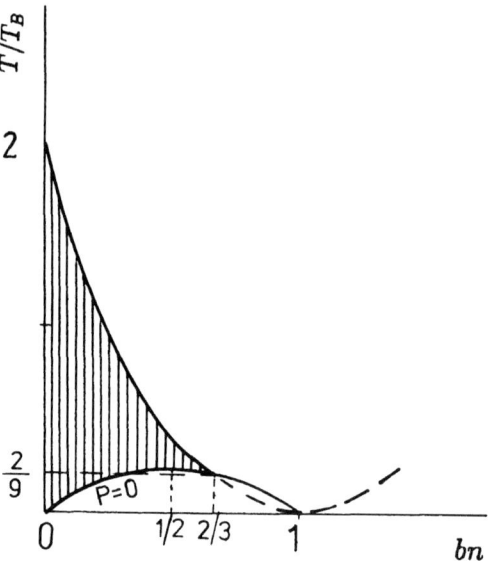

Figure 5.15 Cooling region of the van der Waals gas (hatched). The parabola (partly dashed) is the inversion curve. The inverted parabola (completely solid) corresponds to the state of a substance at zero pressure.

$$T = 2T_B x^2, \qquad n = (1-x)/b. \qquad (5.4.24)$$

Eliminating these variables from (5.4.18) and using (5.4.22), we obtain an equation specifying the inversion curve as $p(x)$:

$$3x^2 - 4x + 1 + \frac{pb}{kT_B} = 0. \qquad (5.4.25)$$

Solutions of this equation have the form:

$$\sqrt{\frac{T}{2T_B}} = x = \frac{2}{3} \pm \sqrt{\frac{4}{9} - \frac{1}{3}\left(1 + \frac{pb}{kT_B}\right)} \qquad (5.4.26)$$

It is seen from this that the inversion curve is a two-valued function $T(p)$, crossing the abscissa at two points:

$$\sqrt{T/2T_B} = 2/3 \pm 1/3 \quad \text{at} \quad p = 0.$$

The corresponding values of reduced temperature T/T_B, are respectively 2/9 and 2, as can be seen in Fig. 5.16. At the peak of the curve where the radical in Eq. (5.4.26) becomes 0, the pressure is:

$$p = kT_B/3b. \qquad (5.4.27)$$

To this extremum corresponds $T/T_B = 8/9$. Since cooling of gas is feasible only below the inversion curve, it is possible only within a limited range of temperatures. At $p = 0$ the upper limit of this interval is the so-called *inversion temperature* T_i, equal to $2T_B$ in the van der Waals theory. For instance, for nitrogen it is 633 K.

Liquefaction is possible when the coexistence curve falls at least partially within the region bounded by the inversion curve as shown on Fig. 5.16. In the course of throttling, the system moves along the curves of equal enthalpy. These have such the property that the derivative (5.4.4) is zero at the intersection with the inversion curve, and accordingly $(\partial p/\partial T)_H$ is infinite. In other words, isoenthalpy lines cross the inversion curve vertically, with positive derivative inside and the negative outside, as shown in Fig. 5.16. For an ideal gas they are vertical lines ($T = $ const), while for a real gas throttling under the inversion curve takes the substance into states with the same enthalpy, but a lower temperature. A portion of the gas can then be compressed isothermally, and this process may be repeated a few times achieving even deeper cooling. Such manipulations may be continued until finally the gas is cooled to such an extent that in isothermal compression the liquid phase is produced (Fig. 5.16).

Such cooling, leading to condensation, begins from room temperature, if it is lower than the inversion temperature T_i of the gas. Otherwise, it is necessary to cool the gas below T_i by some other means: for instance, by putting it into contact with another, already cooled gas. Then one may apply the Joule–Thomson process to obtain even lower temperatures and liquefaction.

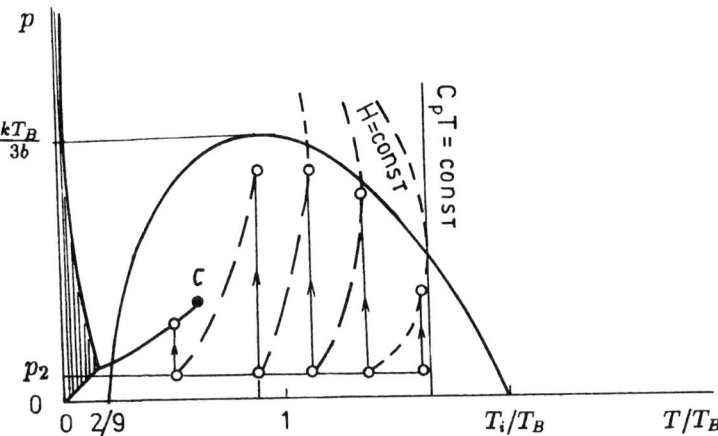

Figure 5.16 Inversion curve and lines of constant enthalpy (dashed) on p–T diagram of state. Vertical lines are isotherms, and the shaded region belongs to the solid state.

Boyle Temperature and Inversion Temperature

As is now clear, the Boyle temperature is the most important characteristic property of a real gas. In particular it determines the inversion temperature $T_i = 2T_B$. However, the definition of T_B given in Eq. (5.4.22) and its relation to T_i are linked to a specific model, associated with the equation of van der Waals, whereas actually T_B and T_i should be defined (and measurable) without referring to any specific equation of state.

Let us note initially that definition (5.4.5) in the case of an ideal gas is reduced to the form:

$$H = \mathcal{E} + RT \tag{5.4.28}$$

For a real gas the situation is different. The relation between enthalpy and internal energy in the general case is the following:

$$H = \mathcal{E} + RTF, \tag{5.4.29}$$

where

$$F = pV/RT \tag{5.4.30}$$

is the compressibility factor. Fig. 1.18 gives an idea of its variation along isotherms with real gas compression. In the state corresponding to the ideal gas (at $p = 0$) F equals 1, but with increase of pressure F either decreases (on low temperature isotherms) or increases (on high temperature ones).

Returning to the van der Waals equation, it is possible to express the compressibility factor as in Eq. (1.9.16):

$$F = 1 + (B - A/RT)/V + (B/V)^2 + (B/V)^3 + \ldots \tag{5.4.31}$$

This equation should be considered as an expansion in density, *virial expansion*, having in general the form:

$$F = 1 + B_0/V + C_0/V^2 + D_0/V^3 \tag{5.4.32}$$

The coefficients $B_0, C_0 \ldots$, known as the second, third, and so on, "virial" coefficients, depend only on T. At moderate pressures the virial expansion describes the behavior of above-critical isotherms much better than the equation of van der Waals. These change continuously, but, generally speaking, not monotonically with respect to p. The low-temperature isotherms $F(p)$ clearly show a minimum at the so-called "Boyle point" p_B (see Fig. 1.18). In order to determine the position of minimum up to the second-order terms in the virial expansion (5.4.32) we require that

$$\left(\frac{\partial F}{\partial p}\right)_T = \left(\frac{\partial F}{\partial V}\right)_T \left(\frac{\partial V}{\partial p}\right)_T = -\frac{1}{V^2}\left[B_0 + \frac{2C_0}{V}\right]\left(\frac{\partial V}{\partial p}\right)_T = 0. \quad (5.4.33)$$

This equality is satisfied when B_0 and C_0 have opposite signs. In case of the van der Waals gas this means that $B_0 < 0 < C_0$. With an increase of temperature the Boyle point moves towards $p = 0$, where the real substance becomes an ideal gas. Equation (5.4.33) is there satisfied at

$$p_B = -\frac{B_0(T)RT}{2C_0(T)}. \quad (5.4.34)$$

The Boyle temperature is defined by the equation $p_B(T_B) = 0$ or

$$B_0(T_B) = 0. \quad (5.4.35)$$

In the van der Waals model

$$B_0 = B - A/RT \quad (5.4.36)$$

Using this expression in equation (5.4.35) it is easy to confirm the definition of T_B given in (5.4.22). At $T < T_B$ and $p = 0$, both B_0 and $(\partial F/\partial p)_T$ are negative. As a result the compressibility factor F initially decreases with gas compression but after passing the minimum begins to grow again (Figs. 1.18 and 1.19). For $T > T_B$ it increases monotonically from the very beginning.

This general property of gases is associated with their ability to cool down by throttling, which was quantified by condition (5.4.17). The latter may be expressed in the form:

$$\frac{T}{V}\left(\frac{\partial V}{\partial T}\right)_p = \frac{1}{F}\left(\frac{\partial FT}{\partial T}\right)_p > 1$$

or

$$\left(\frac{\partial F}{\partial T}\right)_p > 0. \quad (5.4.37)$$

At $p \to 0$ it is possible to use a truncated virial expansion and find the approximate equation for the low-pressure part of the inversion curve:

$$\left(\frac{\partial F}{\partial T}\right)_{p \to 0} = \frac{1}{V}\left[\frac{dB_0}{dT} - \alpha B_0\right] = 0. \quad (5.4.38)$$

From (5.4.17) $\alpha = 1/T$ along the inversion curve, so we obtain from (5.4.38) a general equation for the inversion temperature:

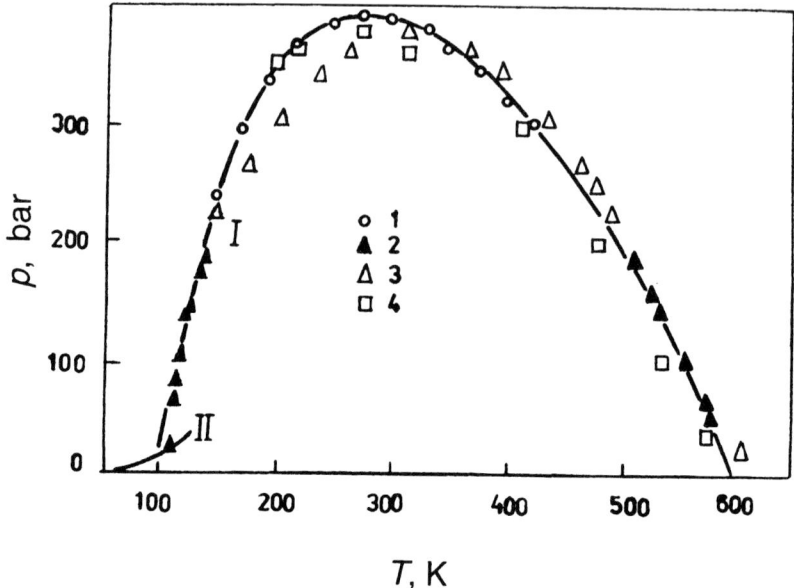

Figure 5.17 (I) Inversion curve for nitrogen with the data of different authors (1–4) and (II) the curve of liquid–gas coexistence, presented in A. A. Vasserman et al, *Russ. J. Phys. Chem.* **38**, 1289 (1964).

$$\left.\frac{dB_0}{dT}\right|_{T_i} = \frac{B_0(T_i)}{T_i}. \tag{5.4.39}$$

Using Eq. (5.4.36) it is easy to verify that in the van der Waals model $T_i = 2T_B$. This simple relationship is certainly a peculiarity of this simple model for a real gas.

In the general case equation (5.4.35) determining T_B and Eq. (5.4.39) defining T_i are independent. Using a measured or calculated second virial coefficient it is possible to extract from these more reliable information about T_B and T_i for any particular gas. This does not detract from the capability of the simple van der Waals model to describe at least qualitatively or semiquantitatively for any p all the important features of real gas behavior in condensation, throttling and expansion into a vacuum. For a real substance only refinement of the shape of the inversion curve is required. In Fig. 5.17 this curve is shown, as an example, together with the coexistence line for nitrogen.

5.5 FREE ENERGY

According to (5.1.16) and (5.2.33) the joint formulation of both laws of thermodynamics is the following:

$$TdS \geq d\mathcal{E} + dA, \qquad (5.5.1)$$

where dA is a work executed by a system not necessarily by expansion. If the work is performed by a system which is thermally stabilized at temperature T, then one may write instead of (5.5.1):

$$T(S_2 - S_1) \geq \mathcal{E}_2 - \mathcal{E}_1 + A. \qquad (5.5.2)$$

So, for any isothermal transition from state 1 into state 2 the work is limited by the following inequality:

$$A \leq (\mathcal{E}_1 - TS_1) - (\mathcal{E}_2 - TS_2) = \mathcal{F}_1 - \mathcal{F}_2 = A_{\max}. \qquad (5.5.3)$$

The maximum available work is consequently the work equal to the reduction of the Helmholtz *free energy*

$$\mathcal{F} = \mathcal{E} - TS. \qquad (5.5.4)$$

It was implied that in an isothermal process not all energy can be converted into work, but only that part of it which is "free," and therefore

$$dA \leq -\mathcal{F} \quad \text{at} \quad T = \text{const}. \qquad (5.5.5)$$

The loss of free energy $d\mathcal{F}$ is the maximum work which can be gained in a reversible isothermal process. In an irreversible process the work is always smaller and can even fall to 0 if the process develops spontaneously. In the latter case

$$\mathcal{F} \leq 0. \qquad (5.5.6)$$

This result may be formulated as a principle of decrease of free energy in a thermally stabilized isolated system. Any process bringing such a system to equilibrium occurs only in the direction of decrease of $\mathcal{F}$, reaching its minimum in the equilibrium state where $d\mathcal{F} = 0$. This implies that the system is characterized by some additional parameters with respect to which free energy is minimized at $T, V = \text{const}$. Below we will consider some of them.

Gibbs–Helmholtz Relation

Free energy is defined via heat parameters in the same way as enthalpy via mechanical ones. Like H and $\mathcal{E}$ it depends only on the state of the system. Differentiating (5.5.4) we find the exact differential of free energy

$$d\mathcal{F} = d\mathcal{E} - TdS - SdT.$$

Using it in (5.5.1), we obtain:

$$d\mathcal{F} \leq -SdT - dA. \tag{5.5.7}$$

In a particular case of an isothermal process this equation reduces to Eq. (5.5.5).

For the reversible processes in which work is performed by expansion we have

$$d\mathcal{F} = -SdT - pdV. \tag{5.5.8}$$

From this follows

$$S = -\left(\frac{\partial \mathcal{F}}{\partial T}\right)_V \tag{5.5.9a}$$

$$p = -\left(\frac{\partial \mathcal{F}}{\partial V}\right)_T \tag{5.5.9b}$$

and the appropriate Maxwell relation:

$$\left(\frac{\partial S}{\partial V}\right)_T = \left(\frac{\partial p}{\partial T}\right)_V. \tag{5.5.10}$$

Formulae (5.5.9) are often used for deducing calorific and thermal equations of state of the system, provided $\mathcal{F}$ can be calculated statistically.

There exists simple relation between internal and free energies, which follows from (5.5.4) and (5.5.9a):

$$\mathcal{E} = \mathcal{F} - T\left(\frac{\partial \mathcal{F}}{\partial T}\right)_V. \tag{5.5.11}$$

This is the Gibbs–Helmholtz equation. It may be expressed in the following form:

$$\mathcal{E} = \left[\frac{\partial(\mathcal{F}/T)}{\partial(1/T)}\right]_V. \tag{5.5.12}$$

For an ideal gas, for example, $\mathcal{E} = C_V T$ and one can easily determine

$$\mathcal{F} = C_V T \int T d\frac{1}{T} = -C_V T [\ln T + \varphi(V)], \tag{5.5.13}$$

up to an arbitrary function of volume $\varphi(V)$.

Thermal Radiation

There exists an ideal gas for which this uncertainty does not occur. Such a gas is the photon gas in a state of thermal equilibrium, or "black body" radiation. One must distinguish between equilibrium and nonequilibrium radiation. Examples of nonequilibrium radiation are the radiation of an electric bulb or a laser and even the thermal radiation of a body removed from equilibrium. Radiation of this type is analogous to a molecular beam emerging through a hole in a vessel filled with gas. A flux of gas or radiation is a nonequilibrium system, but the radiation confined in a "black" (non-transparent) box can be regarded as a thermodynamic system in equilibrium with a "box", that is, the surroundings. The walls of the box must be sufficiently thick (ideally, infinitely thick), and the box itself ought to be thermally stabilized. The radiation confined inside is called thermal or "black" radiation (see Section 2.1).

The most important feature of thermal radiation is that its properties do not depend on the box volume, V, but are determined entirely by temperature. In this respect radiation held in equilibrium with the walls of "black box" is similar to vapor above the surface of a liquid. An increase of the volume occupied by radiation (vapor) leads only to an increase in the number of photons (molecules), while the pressure on the walls and densities of all extensive variables remain unchanged. In other words,

$$N = nV, \qquad \mathcal{E} = uV, \qquad S = sV, \qquad \mathcal{F} = fV, \qquad (5.5.14)$$

where $n(T)$, $u(T)$, $s(T)$, $f(T)$ and $p(T)$ are functions of temperature only.

The pressure of light may be calculated in a very general form using methods of electrodynamics. Regardless of whether the radiation is equilibrium or not its pressure depends exclusively on the energy density in the flux of light:

$$p = \tfrac{1}{3} u. \qquad (5.5.15)$$

This electrodynamic information is sufficient to determine the $u(T)$ dependence of equilibrium radiation by thermodynamic methods.

Let us use for this a relation deduced from Eq. (5.2.44) by substituting into it $\mathcal{E} = u(T)V$:

$$u = T \frac{\partial p}{\partial T} - p. \qquad (5.5.16)$$

From here onward we do not need to use partial derivatives because all quantities are functions of a single variable, temperature. Using (5.5.15) in (5.5.16) we obtain

$$u = \alpha T^4, \qquad (5.5.17)$$

where α must be a universal constant independent of the nature (material) of the black box. If we assume for the moment that this is not true, this implies different densities of radiation in two boxes held at the same temperature. Connecting them with a light guide we get a flux of energy flowing from one box into the another to equalize the difference in radiation densities. This will lead to the heating of one box and cooling of the other in contradiction to the principle of Clausius. In Section 2.2 we have already discussed this problem showed that not only the energy density is a universal function of temperature, but also its distribution over frequencies.

Taking into account Eqs. (5.5.14) let us express the Gibbs–Helmholtz relation (5.5.12) in the form:

$$u = \frac{\partial(f/T)}{\partial(1/T)}. \tag{5.5.18}$$

Using (5.5.17) in (5.5.18) one can easily obtain by integration

$$f = -T \int \frac{u}{T^2} dT = -\frac{1}{3}\alpha T^4 - \text{const } T. \tag{5.5.19}$$

From this follows that entropy density is

$$s = -\frac{df}{dt} = \frac{4}{3}\alpha T^3 + \text{const}, \tag{5.5.20}$$

and the specific heat of equilibrium radiation is

$$c_V = T\frac{ds}{dt} = 4\alpha T^3. \tag{5.5.21}$$

Analogous results are obtained for a vapor if $p_V(T)$ is used instead of (5.5.15).

Two-Phase System

Thermal radiation and vapor are thermodynamically equivalent because these are open systems in which the number of particles is not conserved. Molecules freely cross the boundary between liquid and vapor in both directions and therefore "choose" where they prefer to stay. To reflect this fact it is necessary to generalize the definition of free energy given in (5.5.8). In an open system $\mathcal{F}$ varies not only with temperature and volume, but also with the number of particles N:

$$d\mathcal{F} \leq -SdT - pdV + \mu dN. \tag{5.5.22}$$

5.5 FREE ENERGY

The *chemical potential*, μ, is equal to the to change of free energy with the number of particles, with all other parameters held constant:

$$\mu = \left(\frac{\partial \mathcal{F}}{\partial N}\right)_{T,V}. \tag{5.5.23}$$

With the increase of number of variables comes an increase in the number of Maxwell relations. Besides (5.5.10), obtained at $N = \text{const}$, we get now two more:

$$\left(\frac{\partial \mu}{\partial T}\right)_{N,V} = -\left(\frac{\partial S}{\partial N}\right)_{T,V}; \quad \left(\frac{\partial \mu}{\partial V}\right)_{T,N} = -\left(\frac{\partial p}{\partial N}\right)_{V,T}. \tag{5.5.24}$$

In terms of Jacobians, relation (5.5.10) can be expressed as follows:

$$\frac{\partial(S,T)}{\partial(V,T)} = \frac{\partial(p,V)}{\partial(T,V)} \quad \text{at} \quad N = \text{const},$$

from which immediately follows the property of calibration invariance: $\partial(S,T)/\partial(V,P) = 1$. Similarly from relations (5.5.24) their analogs are obtained:

$$\frac{\partial(S,T)}{\partial(N,\mu)} = -1 \quad \text{at} \quad V = \text{const}, \tag{5.5.25a}$$

$$\frac{\partial(p,V)}{\partial(\mu,N)} = 1 \quad \text{at} \quad T = \text{const}. \tag{5.5.25b}$$

This is true for a vapor and liquid treated independently of one other.

For a two-phase system in a closed volume V at temperature T, the free energy of the united system

$$\mathcal{F} = \mathcal{F}(V_1, V_2; N_1, N_2; T)$$

depends on the distribution of substance between the two phases and the position of the surface separating them. Here, V_1 and V_2 denote the volumes occupied by liquid and vapor respectively and N_1 and N_2 the number of particles contained in them. The total volume and mass are assumed to be constant:

$$V_1 + V_2 = V, \quad \text{and} \quad N_1 + N_2 = N. \tag{5.5.26}$$

Therefore there are actually only two variables, for example, V_1 and N_1, upon which depend the distribution of substance between volumes occupied by liquid and vapor at constant N, V and T.

The free energy of a two-phase system is given by

$$\mathcal{F} = \mathcal{F}_1 + \mathcal{F}_2, \tag{5.5.27}$$

where, according to (5.5.22),

$$d\mathcal{F}_1 = -p_1 dV_1 + \mu_1 dN_1 - S_1 dT \tag{5.5.28a}$$

$$d\mathcal{F}_2 = -p_2 dV_2 + \mu_2 dN_2 - S_2 dT. \tag{5.5.28b}$$

Differentiating (5.5.27) and using Eqs. (5.5.28) and (5.5.26) we obtain

$$d\mathcal{F} = -(p_1 - p_2)dV_1 + (\mu_1 - \mu_2)dN_1 - p_2 dV + \mu_2 dN - Sdt, \tag{5.5.29}$$

where $S = S_1 + S_2$ denotes the total entropy of the system. The total volume, V, and number of particles, N, together with temperature, T, are conserved if the system is closed and thermostated. Then, substituting (5.5.29) into (5.5.6) we obtain

$$d\mathcal{F} = -(p_1 - p_2)dV_1 + (\mu_1 - \mu_2)dN_1 \leq 0. \tag{5.5.30}$$

This expression is 0 only if

$$p_1 = p_2, \tag{5.5.31a}$$

$$\mu_1 = \mu_2. \tag{5.5.31b}$$

These are the equilibrium conditions of the vapor–liquid coexistence: equality of pressures and chemical potentials in both phases.

The first is the condition of hydrostatic equilibrium: the pressure on both sides of liquid surface must be equal if it is to be stationary. If the surface is shifted from the equilibrium position, this means that gas is compressed and liquid is expanded or vice versa. This engenders a the pressure difference $p_1 - p_2$ that tends to return the boundary back to its previous position when the system is released. The resulting displacement ΔV_1 has the same sign as the directing force and therefore $\Delta \mathcal{F}$ is negative, that is, free energy decreases (Fig. 5.18). It decreases until $p_1 - p_2$ is equal to 0.

In order to understand the condition for chemical equilibrium we have to leave the liquid surface where it is but take into account that it is transparent for molecules traveling from one phase to the other. If the chemical potential in one of the phases is higher than in the other, then the number of particles in that phase will start to decrease. The free energy starts to drop and does so until the substance is distributed between the phases in such a way that the chemical potentials become equal as required by condition (5.5.31b).

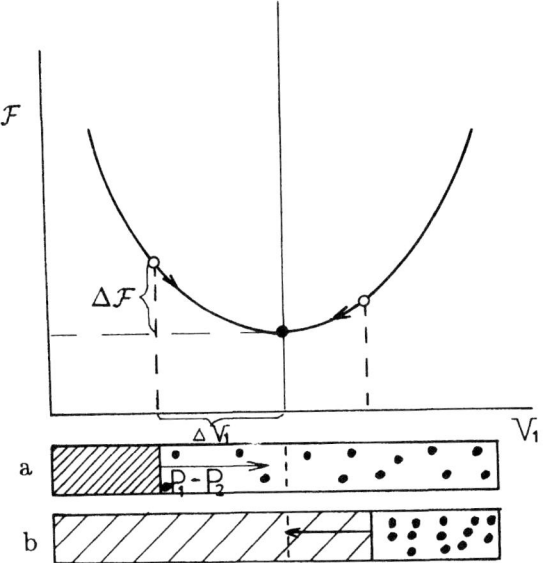

Figure 5.18 Stability of equilibrium towards shifting of the boundary between liquid (hatched) and vapor (points): (a) compression of liquid, (b) compression of vapor.

Surface Tension

If one was to cut a homogeneous liquid, which fills the whole space, into two parts and then remove the substance from one of the semispaces without changing the other, then the liquid–vacuum interface would have a purely mathematical character. If the vapor over such a liquid were also homogeneous up to its surface then the additiveness of free energy of two phases, postulated in (5.5.27), would be reasonable. In reality the interface is not a plane but a transition region with finite thickness and specific properties different from those of homogeneous phases. In fact,

$$\mathcal{F} = \mathcal{F}_1 + \mathcal{F}_2 + \mathcal{F}_\sigma, \qquad (5.5.32)$$

where $\mathcal{F}_\sigma$ denotes the excess of free energy of a real two-phase system as compared to the idealized one separated by an absolutely sharp boundary. In any finite system the excess is proportional to the area of the interface Π that may be varied within some range by changing the shape of the liquid. This requires some effort against the opposing forces of surface tension. If, for instance, the liquid is placed in a frame equipped with a movable plank (Fig. 4.22), then both surfaces of liquid (front and back) act on the plank with force γl per unit of its length l. To reversibly stretch the liquid surface it is necessary to apply a force $2\gamma l$ to the plank. Displacing it by a height dh one performs the work

$$dA = -2\gamma l\,dh = -\gamma\,d\Pi. \tag{5.5.33}$$

Since the work is done to the system it is negative and proportional to the increase in total area of both stretched surfaces $\Pi = 2lh$. Using (5.5.33) and (5.5.5) we find:

$$d\mathcal{F} = \gamma\,d\Pi \quad \text{at} \quad T = \text{const}. \tag{5.5.34}$$

From this, using (5.5.32), it follows that

$$\gamma = \left(\frac{\partial \mathcal{F}}{\partial \Pi}\right)_T = \left(\frac{\partial \mathcal{F}_\sigma}{\partial \Pi}\right)_T = \frac{\mathcal{F}_\sigma}{\Pi}. \tag{5.5.35}$$

The last result is a consequence of the fact that the surface tension γ is a function only of T and nothing else. That is, the interface is yet another system similar to radiation or vapor: it can be enlarged by increasing the surface, while all the other properties remain the same at fixed temperature. γ itself is the specific excess (per unit area) of the free energy of the surface. There is also a similar excess of internal energy

$$u_\sigma = \frac{U_\sigma}{\Pi} \quad \text{where} \quad U_\sigma = U - U_1 - U_2, \tag{5.5.36a}$$

as well as of entropy

$$s_\sigma = \frac{S_\sigma}{\Pi} \quad \text{where} \quad S_\sigma = S - S_1 - S_2. \tag{5.5.36b}$$

In accordance with (5.5.9a)

$$s_\sigma = -\frac{d\gamma}{dT}, \tag{5.5.37}$$

and from the Gibbs–Helmholtz relation (5.5.11) it follows that

$$u_\sigma = \gamma - T\frac{d\gamma}{dT}. \tag{5.5.38}$$

This means that it is sufficient to measure $\gamma(T)$ to obtain full information not only about the free energy of a surface, but also its internal energy and entropy. Besides, it is clear that the *latent heat* of surface formation (per unit area) is

$$\lambda = T\left(\frac{\partial S}{\partial \Pi}\right)_T = Ts_\sigma = -T\frac{d\gamma}{dT}. \tag{5.5.39}$$

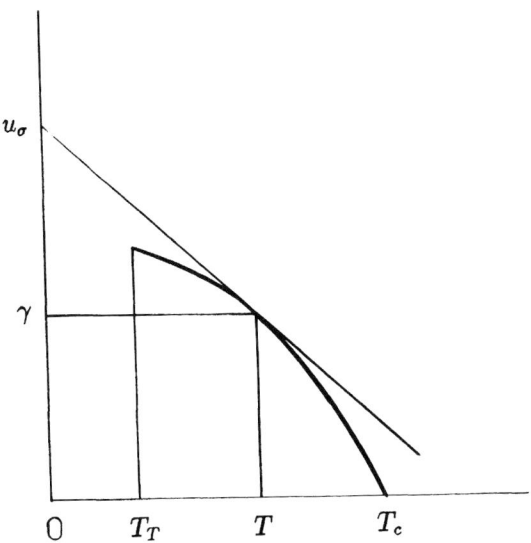

Figure 5.19 Change of surface tension u_σ between triple T_T and critical T_c points (thick line). The construction shows how the surface tension γ and surface energy density u_σ at a given temperature T relate.

As is seen from Fig. 5.19, λ is positive, and together with the work, γ, spent for surface formation it increases the internal energy of a substance transferred from the volume of liquid to the interface.

Taking into account surface effects, formula (5.5.30) has to be refined, because in this case

$$\mathcal{F} = \mathcal{F}(V_1, N_1, \Pi; V, N, T)$$

contains one more parameter. In an equilibrium process

$$d\mathcal{F} = -(p_1 - p_2)dV_1 + (\mu_1 - \mu_2)dN_1 + \gamma d\Pi. \quad (5.5.40)$$

Since deformation of the liquid does not change the number of molecules $dN_1 = 0$, but the other two parameters must vary. For a spherical drop of liquid, $\Pi = 4\pi R^2$ and $V_1 = 4\pi R^3/3$. Using these values in (5.5.40) we obtain the following condition on the minimum of free energy:

$$d\mathcal{F} = -\Delta p\, 4\pi R^2 dR + \gamma\, 8\pi R dR = 0,$$

where Δp denotes the change of pressure between the vapor and liquid, which is

$$\Delta p = \frac{2\gamma}{R}. \tag{5.5.41}$$

This is the well known formula of Laplace, which shows that the pressure inside a drop is higher than above the flat surface of liquid.

5.6 THERMODYNAMIC POTENTIAL

We have already introduced the three energies $U(S, V), H(S, p)$, and $\mathcal{F}(V, T)$. If their natural variables are arranged as the corners of a square and along the sides of the square we place the three energies mentioned (Fig. 5.20), then it becomes clear that there must exist a fourth: the *thermodynamic potential* $\Phi(p, T)$ with natural variables p, T. Its definition is as follows:

$$\Phi = \mathcal{E} - TS + pV = H - TS = \mathcal{F} + pV. \tag{5.6.1}$$

The differential of the latter expression is

$$d\Phi = d\mathcal{F} + pdV + Vdp. \tag{5.6.2}$$

Substituting dF from (5.5.22) in the right-hand side of Eq. (5.6.2) we obtain

$$d\Phi \leq -SdT + Vdp + \mu dN. \tag{5.6.3}$$

From this we derive, as usual,

$$S = \left(\frac{\partial \Phi}{\partial T}\right)_{p,N}; \quad V = \left(\frac{\partial \Phi}{\partial p}\right)_{T,N}; \quad \mu = \left(\frac{\partial \Phi}{\partial N}\right)_{T,p}. \tag{5.6.4}$$

The additional Maxwell relations are:

$$\left(\frac{\partial S}{\partial p}\right)_{T,N} = -\left(\frac{\partial V}{\partial T}\right)_{p,N}; \left(\frac{\partial V}{\partial N}\right)_{p,T} = \left(\frac{\partial \mu}{\partial p}\right)_{N,T}; \left(\frac{\partial S}{\partial N}\right)_{T,p} = -\left(\frac{\partial \mu}{\partial T}\right)_{p,N}. \tag{5.6.5}$$

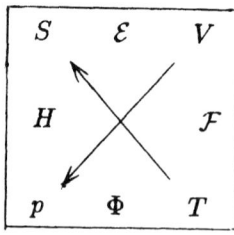

Figure 5.20 Thermodynamic potentials arranged with their pairs of natural variables.

From the last of these equations results another relation complementary to Eqs. (5.5.25):

$$\frac{\partial(S,T)}{\partial(\mu,N)} = 1 \quad \text{at} \quad p = \text{const}. \tag{5.6.6}$$

Using inequality (5.6.3) in the equations (5.6.1) we may generalize inequalities (5.2.36) and (5.4.10)

$$d\mathcal{E} \leq TdS - pdV + \mu dN, \quad \mu = \left(\frac{\partial \mathcal{E}}{\partial N}\right)_{S,V}, \tag{5.6.7}$$

$$dH \leq TdS + Vdp + \mu dN, \quad \mu = \left(\frac{\partial H}{\partial N}\right)_{S,p}. \tag{5.6.8}$$

This adds to the previously earlier deduced relations yet another of the (5.6.6) type:

$$\frac{\partial(p,V)}{\partial(\mu,N)} = 1 \quad \text{at} \quad S = \text{const}. \tag{5.6.9}$$

Chemical Potential

According to the formulae (5.5.23), (5.6.4), (5.6.7), and (5.6.8) several equivalent definitions can be given to the chemical potential:

$$\mu = \left(\frac{\partial \mathcal{E}}{\partial N}\right)_{S,V} = \left(\frac{\partial H}{\partial N}\right)_{S,p} = \left(\frac{\partial \mathcal{F}}{\partial N}\right)_{T,V} = \left(\frac{\partial \Phi}{\partial N}\right)_{T,p}. \tag{5.6.10}$$

The last of these has certain advantages over the others. To appreciate the advantage it is necessary to take into account the fact that all thermodynamic potentials are extensive quantities, additive with respect to the number of molecules contained in the sample:

$$\mathcal{E} = N u_0, \, H = N h_0, \, \mathcal{F} = N f_0, \, \Phi = N \varphi_0, \tag{5.6.11}$$

where the subscript 0 denotes the same quantity, but for a single molecule. The latter in their turn depend on the parameters

$$s = S/N \quad \text{and} \quad v = V/N, \tag{5.6.12}$$

that do not change if the system changes in amount of mass without changing its state. This means that

$$u_0 = u_0(s,v)\,, h_0 = h_0(s,p)\,, \quad f_0 = f_0(T,v)\,, \varphi_0 = \varphi_0(T,p)\,. \qquad (5.6.13)$$

Combining (5.6.11) and (5.6.13) we obtain:

$$\mathcal{E} = N u_0\left(\frac{S}{N},\frac{V}{N}\right), H = N h_0\left(\frac{S}{N},p\right), \mathcal{F} = N f_0\left(T,\frac{V}{N}\right), \Phi = N \varphi_0(T,p)\,.$$
$$(5.6.14)$$

It appears that all the potentials apart from the thermodynamic one become nonlinear functions of N when the number of particles in the system changes at fixed values of the other natural variables. This means that the state of the system changes with a decrease (or increase) of the mass in such a process. For example, adding a substance into a fixed volume held at constant temperature leads to a nonlinear increase in the system's free energy, because the free energy of each molecule f_0 changes with compression.

The only exception is Φ. If, for instance, vapor occupies a volume limited from below by the liquid surface and from above by a movable plunger held under constant external pressure (Fig. 5.21), then the amount of vapor may increase together with the occupied volume when the system is isothermally supplied by the latent vaporization heat. In this process the increase in the thermodynamic potential of the vapor is strictly proportional to the number of particles in it, since $\varphi_0(T,p) = \text{const}$. The same is true for an evaporating

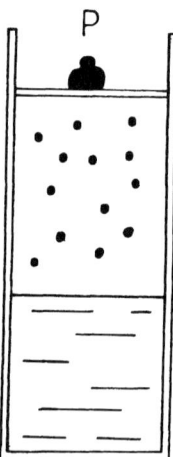

Figure 5.21 Isobaric expansion of a two-phase system under isothermal conditions at the expense of absorption of latent heat of vaporization.

liquid. A mathematical reflection of this peculiarity of Φ is essentially a redefinition of the chemical potential

$$\mu = \left(\frac{\partial \Phi}{\partial N}\right)_{T,p} = \varphi_0(T,p) = \frac{\Phi}{N}. \tag{5.6.15}$$

This has the meaning of the thermodynamic potential per unit molecule.

Phase Equilibrium

The arguments leading to the formulation of conditions on the equilibrium of phases (5.5.31) may be repeated in the case when T and p are fixed, instead of T and V. In this case, it is not the free energy which is minimized but the thermodynamic potential, for which, according to (5.6.3), the following is valid:

$$d\Phi \leq 0 \quad \text{at} \quad T, p, N = \text{const}. \tag{5.6.16}$$

Taking into consideration that in (5.6.2) $dp = 0$ and substituting into this expression $d\mathcal{F}$ from (5.5.29) in which $dN = dT = 0$ we obtain:

$$d\Phi = -(p_1 - p_2)dV_1 - (p_2 - p)dV + (\mu_1 - \mu_2)dN_1 \leq 0. \tag{5.6.17}$$

As a result, the previous conditions for equilibrium are reproduced:

$$p_1 = p_2 = p \tag{5.6.18a}$$

$$\mu_1 = \mu_2 = \mu \tag{5.6.18b}$$

with the only refinement that, if we want to hold the system in the equilibrium state, then not only must the vapor and liquid pressure be equal, but also the external pressure on the system p should be the same. We have encountered this pressure before. It was denoted as p_V—the pressure of the saturated vapor.

It is interesting to consider how both conditions (5.6.18) can be satisfied simultaneously. To understand this it is necessary to consider the relation

$$\mu_0 = f_0 + p_0 v, \tag{5.6.19}$$

which is obtained by term by term division of (5.6.1) by N. Here and further on the subscript 0 denotes the chemical potential and free energy for one molecule and the pressure in a single-phase system (Fig. 5.22). Of course, the gas does not transform continuously into liquid, but rather coexists with it as a vapor. The points representing liquid and vapor equilibrium states on the same isotherm must be arranged such that both the chemical potentials and pressures in contacting phases are simultaneously leveled off. As according to (5.5.9b)

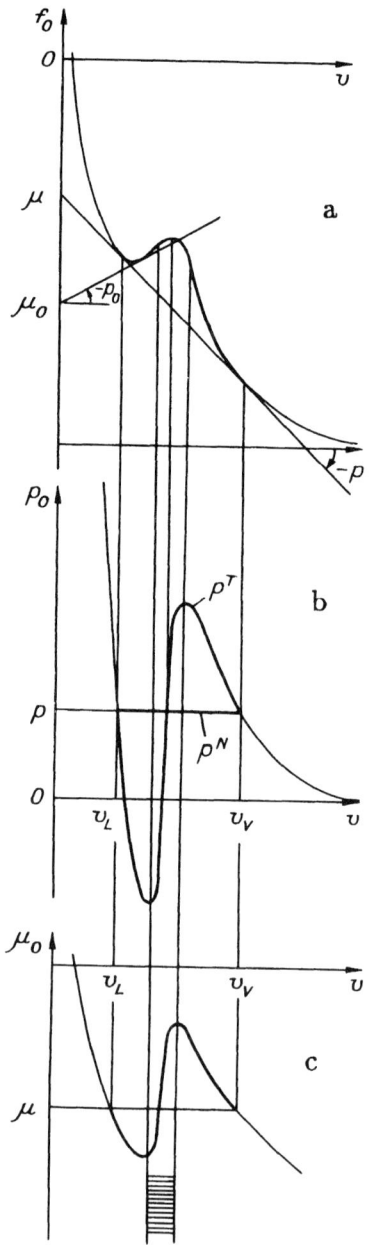

Figure 5.22 Variation of the specific free energy f_0 (a), pressure tensor components p^T and p^N (b), and chemical potential μ_0 (c) of a homogeneous system expanding from liquid to gas phase. Horizontal lines $p = $ const and $\mu = $ const represent the real phase transition.

5.6 THERMODYNAMIC POTENTIAL

$$p_0 = -\left(\frac{\partial f_0}{\partial v}\right)_T, \qquad (5.6.20)$$

it is clear that the pressure is determined by the slope of the tangent to the curve $f_0(V)$ (Fig. 5.22a). Substituting this formula in Eq. (5.6.19) we find that

$$\mu_0 = f_0 - v\left(\frac{\partial f_0}{\partial v}\right)_T \qquad (5.6.21)$$

is the segment intercepted by this tangent on the ordinate axis. Armed with this geometric interpretation one can see directly that the pressures of the two points are equal when the tangents at these points are parallel. This requirement can be satisfied in many ways if one can ignore the equality of chemical potentials. However, since this condition forces the crossing of the tangent lines at a point on the ordinate axis, the tangents must be not only parallel, but coincide. This can be realized in a single way as shown in Fig. 5.22a. Phase equilibrium is possible only between those states of a substance (vapor and liquid) which are represented by tangent points of this straight line with the curve $f_0(V)$. The abscissae of these points indicate the specific volumes of coexisting liquid and vapor. Their ordinates on Fig. 5.22b and 5.22c establish correspondingly the magnitude of the equilibrium vapor pressure $p = p_V$ and chemical potential μ.

The Maxwell relation (1.9.11) we have already discussed in Chapter 1 allows us to make another important prognosis about the magnitude of the equilibrium vapor pressure p_V. To this pressure corresponds the "Maxwell shelf" connecting liquid and vapor states as in Fig. 5.22b. It is possible to obtain certain information about the height of this shelf if the equation of state describing the variation of $p_0(V)$ is known. The equation of van der Waals is an example of such an equation. There exist more complex versions of it which describe real isotherms practically to experimental accuracy (Fig. 1.17). Since these equations are analytic they may be continued into the interface region where $dp_0/dV > 0$, that is, where the existence of a homogeneous substance is impossible. It appears that at a temperature not much lower than critical, where the extrapolation is reliable, the van der Waals loop is reproduced, connecting stable branches of isotherms on which $dp_0/dV < 0$. It is evident that Maxwell shelf lies between the minimum and the maximum of this loop. In fact, from (5.6.18) and (5.6.19) it follows that

$$f_1 + p_V v_1 = f_2 + p_V v_2. \qquad (5.6.22)$$

On the other hand, according to (5.6.20)

$$f_1 - f_2 = -\int_{v_1}^{v_2}\left(\frac{\partial f_0}{\partial v}\right)_T dv = \int_{v_1}^{v_2} p_0(v)\, dv. \qquad (5.6.23)$$

Comparing these expressions we find:

$$p_V(T) = \int_{v_1}^{v_2} p_0(v, T)\, dv / (v_2 - v_1). \tag{5.6.24}$$

The vapor pressure, calculated using this recipe from the empirical equation of state, is in good agreement with that found experimentally.

The equalities (5.6.18) are the conditions for equilibrium coexistence, not only for the liquid–vapor system, but for any other two-phase system, vapor–solid, liquid–solid and also solid–solid where the crystal takes different modifications in different phases. These conditions for any phase pair can be expressed with a single equation

$$\mu_1(T, p) = \mu_2(T, p), \tag{5.6.25}$$

which, as we have already seen, has a single solution defining the coexistence curve $p(T)$. If the curves belonging to two different pairs intersect at some point, then at this point the equilibrium coexistence of three phases becomes possible. Such a *triple point* is shown in Fig. 4.1, where the liquid–vapor coexistence curve intersects the sublimation curve which separates the disordered phases of the substance from the crystal one. Generally these curves do not necessarily cross. This is the case with helium for which the solid, liquid, and gaseous phases cannot be in equilibrium simultaneously (Fig. 5.23). However, two solid phases and a liquid one or three different modifications of the crystal phase can quite often coexist at the intersection points of the curves separating phases. This is the case, for example, with different crystal modifications of ice (I,II,III,V,VI) and water (Fig. 5.24). The only thing which practically never happens is the simultaneous existence of four phases, because the intersection of three different curves at one point is highly improbable.

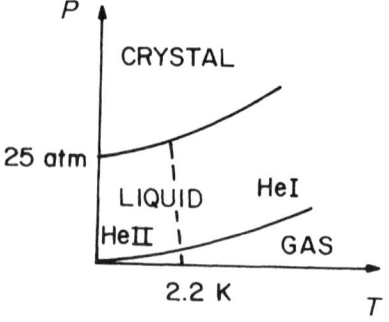

Figure 5.23 Phase diagram of helium at low temperatures.

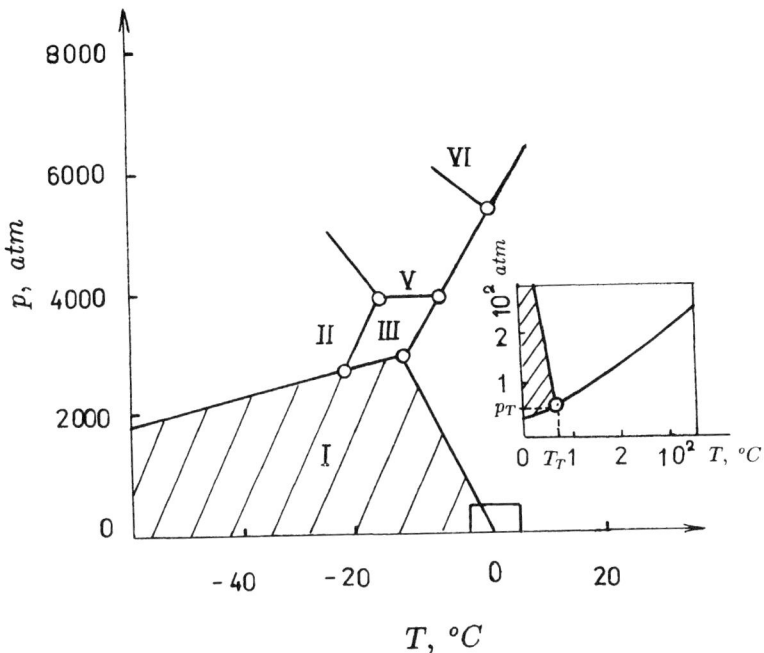

Figure 5.24 Phase diagram of various crystal modifications of ice and water with vapor (on insertion, near the triple point).

Quasithermodynamics

If the system separates into liquid and vapor, the question arises as to the position and structure of the phase boundary. The answer to the first part of the problem has already been found above, using ordinary thermodynamics. The answer to the second part may be obtained only within the framework of so-called *quasithermodynamics* which is an outgrowth of statistical physics.

In reality, the liquid surface is not a geometric boundary of the liquid. Even at low temperatures, when the interface is monomolecular, it is affected by heat fluctuations. In the opposite case, when the temperature is near-critical, the liquid–vapor phase boundary is smooth (multilayered) and one may introduce the local density n at any point within interface. This is a function of the coordinate z perpendicular to the surface. Assuming that the surface crosses the z axis at 0, we take $n(-\infty) = n_L$, and $n(+\infty) = n_V$. Along with $n(z)$, the pressure in the transition region also changes. Due to the spatial inhomogeneity of the medium, the pressure is a tensor $\hat{p}$ involving two components: normal to the surface, p^N, and tangential to it, p^T. The latter is parallel to the surface, and acts upon the plane perpendicular to it. All the other (shearing) tensor components must be equal to zero, otherwise one would observe a nonequilibrium liquid flow. The pressure anisotropy manifesting itself in the difference between p^N and p^T is quite natural. Moreover, this is responsible for the surface tension γ. According to the Laplace definition

$$\gamma = \int_{-\infty}^{+\infty} (p - p^T)\, dz . \tag{5.6.26}$$

It is obvious that the integrand goes to zero deep in the liquid or gas, but is large near $z = 0$ where the tangential pressure is less than the external one, or even negative.

To study the structure and tension of the interface, it is necessary to find the recipe for the calculation of $n(z)$ and $\hat{\mathbf{p}}(z)$ in the transition layer. In the original Hill formulation of quasithermodynamics it was assumed that both the tangential pressure and free energy (per molecule) depend on z only through the local density $n(z)$. The relations between pressure and $\mathcal{F}$ remained the same up to the substitution of p^T for p. Thus the following result was obtained analogous to Eq. (5.6.19):

$$p^T = (\mu - f)n . \tag{5.6.27}$$

This relation was considered as the definition of p^T, while the second component of pressure, p^N, remained uncertain. The voluntary removal of this uncertainty presented considerable problems and results in negative sign of γ in Hill's theory. In fact the actual distribution of density in the transition region may be found only from the conditions of equilibrium identical to (5.6.18):

$$p^N(z) = p , \tag{5.6.28a}$$

$$\mu(z) = \mu . \tag{5.6.28b}$$

To solve this set of equations, it is necessary to unambiguously define the quantities p^N, μ, and f.

This was accomplished not long ago by Yang, Fleming, and Gibbs. The idea that all local characteristics depend on n only was abandoned. Instead, the following definition of the free energy (per unit area) of a two-phase system with plane surface was suggested

$$\mathcal{F} = \int fn\, dz = \int \left(f_0 n + b\frac{\dot{n}^2}{2} \right) dz . \tag{5.6.29}$$

The quantity f_0 which depends solely on n is a local thermodynamic function in the original sense of the word. However, the actual free energy of a unit volume fn differs from $f_0 n$ by an addend quadratic in $\dot{n} = dn/dz$. This is a way to take into account the fact that the interaction between particles at different points of the interface is not local. Inhomogeneity of the mass distribution is allowed for by $\dot{n}$. At near-critical temperatures, quasithermodynamics as based on expression (5.6.29) is, in fact, an expansion for $\mathcal{F}$ in terms of the density gradient where the first nonvanishing component is retained. This expansion was substantiated by the authors, and the parameter b was defined as follows

$$b = \frac{kT}{6} \int r^2 \left[1 - e^{u(r)/kT}\right] g(r, n) d\mathbf{r},$$

where $g(r, n)$ is the *radial distribution function* of the liquid defined by the statement that the number of pairs of molecules which are separated by a distance r is proportional to $g(r) 4\pi r^2 dr$. The value of b depends on n rather than on z, and, as a last resort, may be considered a constant of the theory.

The density distribution $n(z)$ at which the minimum of free energy (5.6.29) is attained is considered as equilibrium. It has to be found from the equation

$$d\mathcal{F} = 0, \qquad (5.6.30)$$

provided that the volume of the system and the number of particles in it remain {\rm const}ant. In fact, only the last restriction

$$\int n(z) \, dz = N,$$

need be taken into account since the volume is fixed constant by the limits of integration over z. In fact, the procedure reduces to finding the unconditional extremum of the functional

$$\mu N - \mathcal{F} = \int (\mu - f) n \, dz = \int p^T \, dz, \qquad (5.6.31)$$

where μ plays the role of the Lagrange multiplier, while p^T is defined in (5.6.27). In view of (5.6.29) it is necessary to optimize the "action"

$$\int p^T dz = \int \left\{ [\mu - f_0(n)] n - \frac{1}{2} b(n) \dot{n}^2 \right\} dz, \qquad (5.6.32)$$

treating it as a function of n and $\dot{n}$. As is well-known from classical mechanics, this leads to the Euler–Lagrange equation

$$\frac{d}{dz} \left(\frac{\partial p^T}{\partial \dot{n}} \right) - \frac{\partial p^T}{\partial n} = 0. \qquad (5.6.33)$$

In view of the particular form of the integrand in (5.6.32), it can be recast as

$$\mu_0(n) - b(n)\ddot{n} - \tfrac{1}{2}b'(n)\dot{n}^2 = \mu, \qquad (5.6.34)$$

where $\mu_0 = (d/dn)[nf_0(n)]$, as in ordinary thermodynamics, and $b' = db/dn$. To elucidate the behavior of the substance density in the transition region, it is sufficient to find $n(z)$ satisfying Eq. (5.6.34). First of all this concretizes the chemical equilibrium condition (5.6.28b). Moreover, it provides automatic fulfillment of hydrostatic equilibrium condition (5.6.28a) as well: multiplication of (5.6.34) by $\dot{n}$ allows one to integrate, yielding the conservation law for normal pressure

$$p^N = [\mu - f_0(n)]n + \tfrac{1}{2}b(n)\dot{n}^2 = p. \qquad (5.6.35)$$

As constant normal pressure is ensured, we can rewrite the Laplace formula as

$$\gamma = \int (p^N - p^T)\, dz. \qquad (5.6.36)$$

Substituting (5.6.35) and (5.6.32) into (5.6.36) gives

$$\gamma = \int b\dot{n}^2 dz. \qquad (5.6.37)$$

This formula for the calculation of the surface tension was first proposed by van der Waals. However, until recently the numerous attempts to use it had led to paradoxes: omitting either μ or p^N, or even the reversal of sign of γ (see A. I. Burshtein, 1979). Due to consistent formulation of a quasithermodynamics, the van der Waals theory has become closed and the calculation of surface tension realizable.

In conclusion it should be noted that the local free energy and tangential pressure are not defined uniquely in quasithermodynamics. For example, they may be introduced by identifying the integrands on both sides of Eq. (5.6.29). In this case

$$f = f_0 + \frac{b(n)}{2n}\dot{n}^2, \qquad (5.6.38)$$

and, according to (5.6.27), tangential pressure is given by

$$p^T = [\mu - f_0(n)]n - \tfrac{1}{2}b(n)\dot{n}^2. \qquad (5.6.39)$$

In the region of essential variation of these quantities the surface properties differ considerably from the homogeneous phase. However, this definition is not unique (Fig. 5.25). In Section 4.3 we have considered the alternative pos-

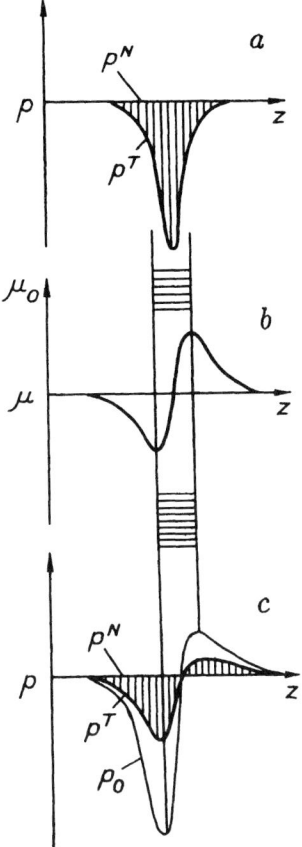

Figure 5.25 The variation of the tangential pressure (hatched) defined in Eq. (5.6.39) (a), in Eq. (4.3.24) (c), and the chemical potential μ (b). The unstable region is sandwiched between two vertical lines.

sibility and calculated actual changes in density and pressure tensor in the transition region. A certain freedom in the choice of f and p^T is due to local description of the interaction which is actually nonlocal. Fortunately, all physically meaningful characteristics of the interfacial region, such as $n(z)$, $p^N(z)$ and γ, do not depend on this choice.

5.7 NERNST'S PRINCIPLE

According to (5.2.14) the entropy of an ideal molecular gas at $V = \text{const}$ falls off as the logarithm of the temperature. Of course, this result has limited generality because at low temperatures a molecular gas ceases to be ideal: it condenses into a liquid or solid, the entropies of which are essentially different.

The heat capacity of the condensed state does not remain constant, as in an ideal gas, but decreases with the decrease of temperature. The same situation occurs with the specific heat of radiation, which is an ideal but quantum gas that may be cooled together with the black body down to $T \simeq 0$. According to (5.5.21) the specific heat falls as the cube of the temperature as does the entropy of radiation (5.5.20). It was found that the specific heat of solids at low temperatures behaves in a very similar manner (see Section 3.3). The tending of the specific heat to 0 at $T \to 0$ is a necessary condition for convergence of the integral

$$\int_0^T C_x \frac{dT}{T} = \int_0^T \frac{dQ}{T} = S(T) - S(0). \tag{5.7.1}$$

If it converges with any method of cooling—isobaric, isochoric, and so on, so that $x = p, V, \ldots$,—then

$$\lim_{T \to 0} S(T, x) = S(0) = \text{const}. \tag{5.7.2}$$

Based on an empirical generalization of extensive studies of the specific heat of the condensed state, Nernst came to the conclusion that this is indeed the case in reality. He stated this assertion as a principle, which is often referred as the "third law of thermodynamics."

Planck refined the formulation of this principle in the following way: the entropy of a body approaching absolute zero tends to the finite limit $S(0)$, independent of the method of cooling (i.e. on the other parameters of state: $p, V, \ldots$). This means the following:

$$S(T, x) = S(0) + A(x) T^n, \quad n > 0 \tag{5.7.3}$$

where $A(x)$ is an arbitrary function, and T is close to 0 (Fig. 5.26). In an isobaric, isochoric or any other process, entropy tends to the same limit $S(0)$. As it is impossible to change $S(0)$ by varying these parameters, $\Delta S(0, x) = 0$. Without much loss of generality one may set $S(0) = 0$. This is actually a statement of Nernst's theorem: *The entropy of any equilibrium system at absolute zero can always be taken equal to zero.* The theorem does not extend to the amorphous solids (glasses, polymers) which remain nonequilibrium when cooled, that is, in a metastable, although very long-lived state.

Postulating (5.7.3) we immediately return to our starting position:

$$C_x = T \left(\frac{\partial S}{\partial T} \right)_x = nA(x) T^n = n[S - S(0)] \to 0. \tag{5.7.4}$$

This does not exhaust the content of the third law. It also implies that the coefficient of thermal expansion tends to 0 at $T \to 0$:

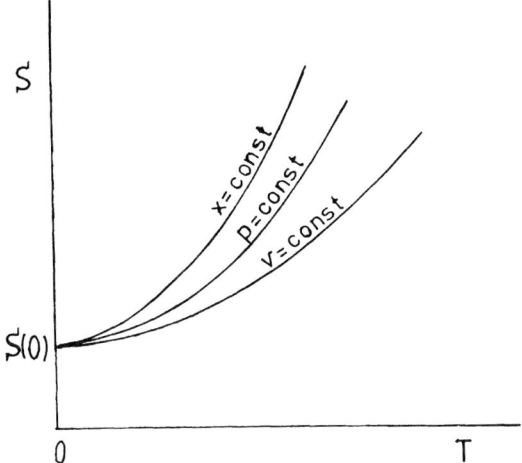

Figure 5.26 Variation of entropy of a system near $T = 0$.

$$\alpha = \frac{1}{V}\left(\frac{\partial V}{\partial T}\right)_p = \frac{1}{V}\frac{\partial(V,p)}{\partial(T,p)} = -\frac{1}{V}\frac{\partial(T,S)}{\partial(T,p)} = -\frac{1}{V}\left(\frac{\partial S}{\partial p}\right)_{T\to 0} \to 0. \quad (5.7.5)$$

Similarly the thermal coefficient of pressure tends to 0 at $T \to 0$:

$$\beta = \frac{1}{p}\left(\frac{\partial p}{\partial T}\right)_V = \frac{1}{p}\frac{\partial(p,V)}{\partial(T,V)} = \frac{1}{p}\frac{\partial(T,S)}{\partial(T,V)} = \frac{1}{p}\left(\frac{\partial S}{\partial V}\right)_{T\to 0} \to 0. \quad (5.7.6)$$

Using this result in Eq. (5.2.44), we see that at $T = 0$

$$p = -\left(\frac{\partial \mathcal{E}}{\partial V}\right)_0, \quad (5.7.7)$$

that is, the pressure is determined entirely by the elastic reaction of the medium.

The Nernst principle establishes that it is impossible to attain absolute zero. This conclusion is easily reached by considering the two isochores on Fig. 5.27, representing the minimal and maximal volumes of the working medium in the Carnot cycle. These isochores are taken as an example only, since one may equally well use any other parameter x, which plays a similar role in the heat pump. Due to (5.7.2) both curves join at one point, approaching which the Carnot cycle becomes so thin that it degenerates into a vertical line. No more heat may be extracted from the cooled body as its temperature approaches 0. Although it is impossible to attain absolute zero, we may approach it arbitrarily close with a sequence of steps, where each succeeding one is smaller and

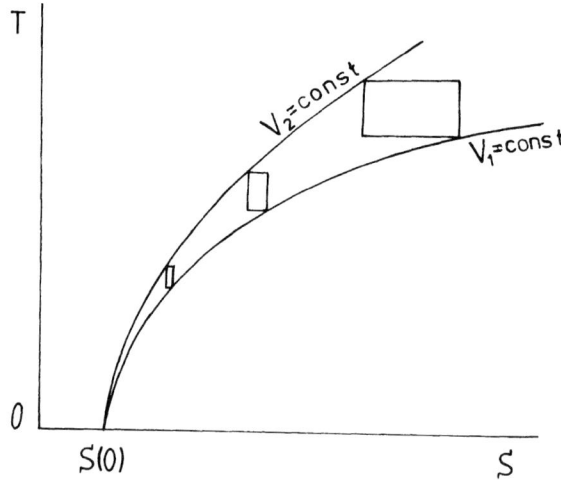

Figure 5.27 The transformation of the Carnot cycle, inscribed between two isochores, with decrease of temperature.

harder to accomplish. The limit of such an infinite sequence of steps is the absolute zero of temperature.

5.8 MIXTURES

One may also consider using thermodynamic analysis the multicomponent systems, such as gas mixtures, and liquid and solid solutions. In some such systems chemical reactions are possible, but even in the absence of such reactions, properties of mixtures are different from those of the pure components constituting them.

Gibbs' Paradox

Upon mixing even ideal gases the entropy of a mixture appears to be different from that achieved in mixing two quantities of the same gas. To prove this, let us express the molar entropy of an ideal gas (5.2.14) in the form:

$$S = N_0 s(T, V) = N_0 \frac{k}{\kappa - 1} \ln(T v^{\kappa - 1}) + S_0. \quad (5.8.1)$$

Here we have taken into account the fact that in accordance with (5.6.12) and (5.6.13) the specific entropy $s(T, v)$ depends exclusively on the specific volume v. Therefore, the generalization to a system containing an arbitrary number of particles N is trivial:

$$S = Ns(T,V) = N\frac{k}{\kappa - 1}\ln\left[T\left(\frac{V}{N}\right)^{\kappa-1}\right] + S_0. \qquad (5.8.2)$$

This formula, referring to any single gas, may be used for the calculation of the entropy of the components in multicomponent systems.

In the simplest case, such a system consists of two containers with different gases held at same temperature. Since the entropy is an extensive variable it is evident that the entropy of system as a whole S' is equal to the sum of the entropies of the two gases (Fig. 5.28a):

$$S' = S'_1 + S'_2, \qquad (5.8.3)$$

where

$$S'_1 = N_1\frac{k}{\kappa - 1}\ln\left[T\left(\frac{V_1}{N_1}\right)^{\kappa-1}\right] + S^0_1, \qquad (5.8.4)$$

$$S'_2 = N_2\frac{k}{\kappa - 1}\ln\left[T\left(\frac{V_2}{N_2}\right)^{\kappa-1}\right] + S^0_2,$$

while N_1 and N_2 denote the number of particles in the vessels of volumes V_1 and V_2 respectively. Removing the barrier between the containers we let the gases mix irreversibly but adiabatically (Fig. 5.28b). The volume occupied by each of them will increase to $V = V_1 + V_2$. If the gases were there alone, then their entropies would be the following:

$$S''_1 = N_1\frac{k}{\kappa - 1}\ln\left[T\left(\frac{V}{N_1}\right)^{\kappa-1}\right] + S^0_1, \qquad (5.8.5)$$

$$S''_2 = N_2\frac{k}{\kappa - 1}\ln\left[T\left(\frac{V}{N_2}\right)^{\kappa-1}\right] + S^0_2.$$

The following question now arises: is it possible to simply add the entropies of two different gases held in the same volume, just as was done in Eq. (5.8.3) when they were spatially separated?

One may obtain a positive answer to this question through the following extremely important thought experiment. Let us assume that the gases, held at the same temperature in two cylinders of equal volume, V, are separated with two partitions as shown on Fig. 5.29a, and are thermally insulated from the surroundings. Let us imagine that one of the cylinders is capable of plunging into the other without friction (Fig. 5.29b). The partitions are semitransparent, that is, transparent only to a gas entering from the outside. Since the gas freely

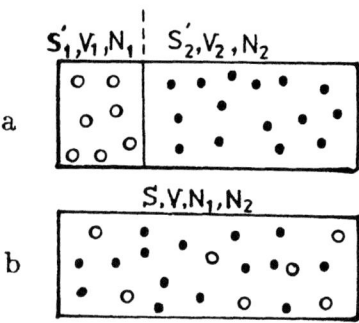

Figure 5.28 Irreversible mixing of different gases by removing the partition inside the container: (a) initial state, (b) final state.

penetrates the cylinders it does not cause pressure on them, and bringing the cylinders closer does not meet any resistance. Consequently, the gases can be joined together in one volume, V (Fig. 5.29c), without spending either work or heat. In other words

$$A = Q = 0$$

and, therefore, $\Delta \mathcal{E} = 0$ in accordance with the first law of thermodynamics. For ideal gases this leads us to the conclusion that the temperature stays unchanged as well as the entropy. It remains the same as in the initial state, $S = S''$, but in this state, where it was obtained, according to (5.8.3), by sum-

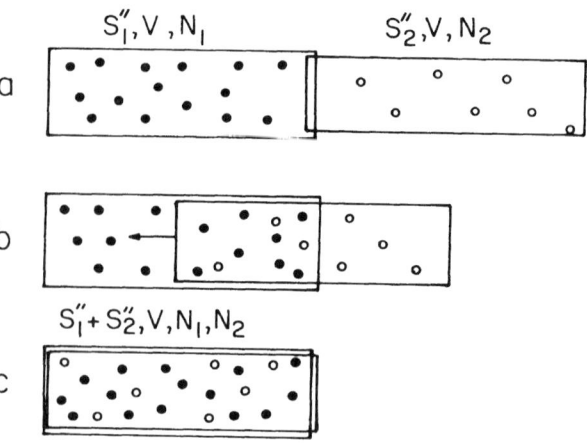

Figure 5.29 Reversible mixing of gases: (a) initial, (b) intermediate, and (c) final states.

ming over the entropies of spatially separated components: $S'' = S''_1 + S''_2$. Consequently, after the irreversible mixing of different gases

$$S = S''_1(T, V) + S''_2(T, V) = S''(T, V). \tag{5.8.6}$$

Although this thought experiment proposed by Gibbs is hardly realizable, it gives the theoretical basis for the calculation of the entropy of a mixture of ideal gases as the sum of the entropies of the components, with the same temperatures and volumes. Similarly one may calculate the internal energy of the final state and the pressure caused by the mixture when one container is completely inserted into another. The double partitions are not transparent to both gases and are subjected to the joint pressure

$$p(T, V) = p_1(T, V) + p_2(T, V). \tag{5.8.7}$$

This is the well-known Dalton law: the pressure of a mixture of ideal gases is equal to the sum of the partial pressures of its components. The number of components may be arbitrary.

Let us find now the total variation of entropy of the system after an irreversible adiabatic mixing of gases by the method shown in Fig. 5.28. In this case it follows from Eqs. (5.8.6) and (5.8.3):

$$\Delta S = S'' - S' = S''_1 + S''_2 - S'_1 - S'_2 = \Delta S_1 + \Delta S_2. \tag{5.8.8}$$

The increase of partial entropies can be easily found from Eqs. (5.8.4) and (5.8.5):

$$\Delta S_1 = S''_1 - S'_1 = N_1 k \ln \frac{V}{V_1} > 0,$$
$$\Delta S_2 = S''_2 - S'_2 = N_2 k \ln \frac{V}{V_2} > 0. \tag{5.8.9}$$

In the simplest case when the mixed gases have equal volumes $V_1 = V_2 = V/2$ one may obtain from Eqs. (5.8.8) and (5.8.9):

$$\Delta S = (N_1 + N_2) k \ln 2 > 0. \tag{5.8.10}$$

As was expected, the spontaneous mixing of gases in an isolated system leads to an increase of the entropy of a system.

The increase of entropy described by Eq. (5.8.10) occurs in the mixing of any gases by removing the barrier separating the cylinder into two parts (Fig. 5.30). Therefore, one might imagine that a similar increase of entropy will occur when the gases in the two parts are of the same kind. However, such a removal of the barrier does not change the specific volume and, consequently, the specific entropy ($s_1 = s_2 = s$), leaving the total entropy of the system unchanged:

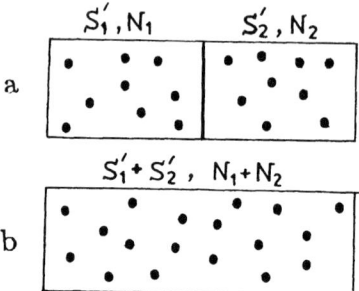

Figure 5.30 Mixing of two portions of the same gas: (a) separated and (b) combined states.

$$S = N_1 s_1 + N_2 s_2 = Ns.$$

The state of the system in this case does not change at all and $\Delta S = 0$. The paradox of Gibbs is the apparent contradiction between this conclusion and the formula (5.8.10). The paradox is resolved if we notice that formula (5.8.10) refers only to *different gases* and is not valid for a single gas. For a single gas the recipe obtained in Eq. (5.8.6) is not correct. Semitransparent partitions for the molecules of the same kind do not exist. They are either nontransparent or transparent in both directions.

This formally correct explanation is still surprising. There is the possibility of gradually reducing the difference between gases by choosing close chemical elements or even isotopes of the same atomic gas. However, the increase in entropy due to their adiabatic and irreversible mixing does not reduce but remains exactly the same. As it was later understood, this difference between the mixing of different gases and the same gas has a fundamental character and is one of the consequences of the principle of identity of particles of the same kind. Such particles are not distinguishable in principle, so that after removal of the barrier it is impossible to say which of them is in which part of the container. This qualitative difference between the mixing of different gases and of a single gas cannot be eliminated smoothly. The quantum consequence of the identity principle is the separation of all particles into two classes (bosons and fermions) which have different statistics. A particular example of Bose-statistics is statistics of photons, described in Section 2.5.

BIBLIOGRAPHY

B. J. Alder and W. G. Hoover "Numerical statistical mechanics." In *Physics of Simple Liquids*. Ed. by H. N. V. Temperley, J. S. Rowlinson, and G. S. Rushbroooke. North-Holland Publishing Company, Amsterdam, 1969, p. 79.

J. A. Barker, "Lattice theories of the liquid state." *The International Encyclopedia of Physical Chemistry and Chemical Physics*. Pergamon Press, Oxford, Vol. 1, 1963, p. 133.

A. Batschinski "Abhudlugen über Zustandsgleichung; Abh. I: Der ortometrische Zustand." *Ann. der Phys.* **19** (1906) 307.

D. Ben-Amotz and D. R. Herschbach "Correlation of Zeno (Z = 1) Line for Supercritical Fluids with Vapor-Liquid Rectilinear Diameters" Israel Journal of Chemistry **30** (1990) 59.

A. I. Burshtein "On the origin of ideal gas curve." *Russ. J. Phys. Chem.* **48** (1974) 1562.

A. I. Burshtein "Simple liquid surface structure and surface tension." In *Advances in Colloid and Interface Science*, Vol. 11, 1979, pp. 315–374. Elsevier Scientific Publishing Company, Amsterdam.

A. I. Burshtein "Internal and thermal pressure in simple liquids." *J. Mol. Liquids* **59**, (1993) 1.

A. I. Burshtein and N. V. Shokhirev "The change in density and pressure tensor at the liquid-vapor interface." *Russ. J. Phys. Chem.* **53** (1979) 1389.

B. S. Carey, L. E. Scriven, H. T. Davis "On gradient theories of fluid interface stress and structure." *J. Chem. Phys.* **69** (1978) 5040.

M. H. Cohen and D. Turnbull "Molecular transport in liquids and glasses." *J. Chem. Phys.*, **31** (1959) 1164.

C. A. Croxton *Liquid State Physics—a Statistical Mechanical Introduction*. Cambridge University Press, 1974.

A. L. Gosman, R. D. McCarty, and J. G. Hust *Thermodynamic Properties of Argon from the Triplet Point to 300 K at Pressures to 1000 Atmospheres*. NSRDS-NBS 27, Washington, DC, 1969.

J. O. Hirschfelder, C. F. Curtiss, and R. B. Bird *Molecular Theory of Gases and Liquids*. John Wiley, New York, 1954.

E. M. Holleran "Linear relation of temperature and density at unit compressibility factor." *J. Chem. Phys.* **47** (1967) 5318.

O. A. Hougen and K. M. Watson *Chemical Process Principles*, Part II. John Wiley, New York, 1947.

C. Kittel *Elementary Statistical Physics*. John Wiley, New York, 1958.

C. Kittel *Introduction to Solid State Physics*. John Wiley, New York, 1986.

N. S. Krylov *Works on Statistical Physics Basis* (in Russian). Academy of Science, Moscow-Leningrad, 1950, p. 207.

R. Kubo *Statistical Mechanics*. North-Holland Publishing Company, Amsterdam, 1965.

R. Kubo *Thermodynamics*. North-Holland Publishing Company, Amsterdam, 1968.

M. C. Kutney, M. T. Reagan, K. A. Smith, J. W. Tester and D. R. Herschbach "The Zeno ($Z = 1$) Behavior of Equations of State: An Interpretation across Scales from Macroscopic to Molecular." J. Phys. Chem B **104** (2000) 9513.

T. A. Litovitz and C. M. Davis "Structural and shear relaxation in liquids." In *Physical Acoustic*, Vol. 2, Part A. Ed. by Warren P. Mason. Academic Press, New York, 1965, p. 281.

J. R. Manning *Diffusion Kinetics for Atoms in Crystals*. Van Nostrand, Princeton, NJ, 1968, p. 257.

V. I. Nedostup, E. P. Gal'kevich, "Method of Ideal Curves in Thermodynamics of Real Gases and Gas Mixtures." J. Eng. Phys. **38** (1980) 424.

V. I. Nedostup, E. P. Gal'kevich, E. S. Kaminski "Thermodynamic Properties of Gases under high Temperatures and Pressures." (in Russian) Kuiv Naukova dumka (1990) pp. 1-196.

S. Ono and S. Kondo *Molecular Theory of Surface Tension in Liquids*. Handbuch der Physik, Vol. X. Springer Verlag, Berlin, 1960.

J. S. Rowlinson "A comparison of the solutions of integral equations for the distribution functions with the properties of model and real systems." In *Physics of Simple Liquids*. Ed. by H. N. V. Temperley, J. S. Rowlinson, and G. S. Rushbrooke. North-Holland Publishing Company, Amsterdam, 1969, p. 59.

Yu. B. Rumer and M. S. Ryvkin *Thermodynamics, Statistical Physics, and Kinetics* (English translation). Mir Publishers, Moscow, 1980.

N. V. Shokhirev and A. I. Burshtein "The change in density and pressure tensor at the liquid–vapor interface." *Progr. Colloid Polym. Sci.* **84** (1991) 285.

V. V. Sychov, A. A. Vasserman, *et al. Thermodynamic Properties of Nitrogen* (in Russian). Publishing House of Standards, Moscow, 1977, p. 352.

A. A. Vasserman "Equation of state for nitrogen." *Russ. J. Phys. Chem.* **38** (1964) 1289.

A. A. Vasserman, Ya. Z. Kazavchinskii, and V. A. Rabinovich *Thermophysical Properties of Air and its Components* (in Russian). Moscow, Nauka, 1966.

H. L. Vörtler "Modified cell theory: equation of state for hard spheres." *Phys. Lett.* **78** (1980) 266.

T. W. Wainwright and B. J. Alder "Molecular dynamics computations for the hard sphere system." *Nuovo Cimento* **9**, Suppl. 1 (1958) 116.

A. I. M. Yang, P. D. Fleming III, and J. H. Gibbs "Molecular theory of surface tension." *J. Chem. Phys.* **64** (1976) 3732.

V. N. Zharkov and V. A. Kalinin *Equations of State for Solids at High Pressures and Temperatures*. Plenum Publishing, New York, 1971.

INDEX

Absorption coefficient, 124, 126, 128
Absorption spectrum, 129, 147
Acoustic vibrations, 157, 163–165, 173
Adiabatic expansion, 264, 267, 268, 272, 281, 293
Adiabatic index, 267
Amorphous state, 175, 199, 245, 324
Arhenius factor, 250
Attraction, 11, 33, 35–38, 42, 44, 46, 47, 61, 113, 151, 179, 202, 204, 216, 219, 230, 232, 236, 284
Avogadro number, 10

Barometric formula, 19, 23, 24, 36
Binodal, 39, 40
Body-centered cubic lattice, 154
Bohr postulate, 78, 79
Boltzmann constant, 10, 24
Boltzmann distribution, 22, 24, 25, 187
Boyle point, 300, 301
Boyle's temperature, 44, 45, 297, 300, 301
Brownian particles, 23, 24, 105, 251, 253–256

Calibration invariance, 274, 283
Caloric, 258, 259, 261, 268
Caloric equation, 266, 300
Calorimeter, 259
Canonical ensemble, 49, 219
Canonical variables, 48, 49, 69, 73
Carnot cycle, 268, 269, 271, 272, 275, 276, 279, 280, 286, 290, 291, 293, 325
Carnot principle, 270, 276, 277, 279, 280
Characteristic temperature, 21, 72, 73, 75, 79, 81, 82
Chemical potential, 307, 308, 313, 315, 316
Clapeyron–Clausius relation, 288
Clausius principle, 277, 279, 281, 306
Coexistence curve, 46, 202, 286, 287, 299, 302, 318

Collective entropy, 64, 66, 210, 211, 218, 219, 221, 228
Compressibility factor, 41, 42, 45, 62, 66, 188, 219, 228, 300, 301
Cooling coefficient, 292, 293
Correlation function, 202, 254
Corresponding states, 41, 42
Critical point, 40, 41, 46, 201, 202, 215, 216, 230, 234, 286
Critical temperature, 39, 46, 316, 321
Crystallization, 66, 218, 219, 230
Current, 118, 196, 197, 277, 288–292
Current carriers, 119–121, 123

Dalton law, 329
Debye approximation, 167, 168
Debye law, 170, 173
Debye temperature, 169–172, 179, 180, 182, 188, 199, 210, 225
Density distribution, 23, 321
Density of states, 160, 161, 166
Detailed balance principle, 102, 125, 139, 145, 146, 191
Diatomic molecule, 27, 67, 71, 82, 105, 162
Dieterici equation, 37, 41
Diffusion, 109, 110, 113, 189, 192, 193, 195, 246, 252, 253
Diffusion coefficient, 110, 195, 248
Dislocation, 173, 191, 193
Dulong–Petit law, 171, 173, 182, 188, 210

Einstein coefficients, 142, 144, 145
Electric conduction, 11, 118–120, 123, 196, 246
Electric polarization, 31, 32, 87–90
Electric susceptibility, 32
Electron temperature, 121–123
Emission spectrum, 129, 147
Emissive power, 126–128
Empirical equation of state, 39, 230, 318

333

INDEX

Energy quantization, 75, 78
Enthalpy, 294, 295, 299, 300, 303
Entropy, 51–55, 64, 92, 199, 248, 249, 271–274, 276, 281–283, 306, 308, 310, 323, 324, 326–330
Equation of state, 5, 9, 11, 37, 39, 55, 60, 151, 153, 174, 182, 200, 214, 216, 222, 225, 227, 228, 248, 284, 285, 288, 300, 316, 318
Equilibrium radiation, 130, 135, 140, 146, 147, 305, 306
Equipartition law, 70–74, 131, 142, 171, 255
Evaporation, 66, 149, 201, 218, 259, 286, 287
Excluded volume, 33, 34

Face-centered cubic lattice, 65, 153, 154, 203, 224
First law of thermodynamics, 257, 261–263, 265, 266, 277, 302
First thermoelectric relation, 292
Fluctuations, 2, 36, 52, 190, 214, 250, 253, 319
Force characteristic, 178, 214, 215
Force correlation time, 254
Fourier law, 107, 114, 259
Free energy, 53–55, 59, 64, 176, 233, 236, 240, 241, 302, 303, 306–311, 314, 315, 320
Free path length, 97, 98, 111, 112, 115, 123, 174, 243, 245
Free path time, 95, 97, 98, 100, 120, 242, 250
Free volume theory, 62, 65, 188, 202, 203, 205, 208–211, 221, 223
Free volume theory of viscosity, 248, 251
Frenkel defects, 190, 192, 193

Generalized coordinates, 26–28, 30
Generalized momentum, 28
Generalized velocities, 28
Gibbs canonical distribution, 25, 27, 29, 32, 50, 51, 54–56, 71, 74, 77
Gibbs microcanonical distribution, 48, 49, 55, 56, 91
Gibbs' paradox, 326, 330
Gibbs-Helmholtz relation, 303, 304, 306, 310
Grüneizen constant, 210
Gravitation, 4, 19–21, 24, 72, 135, 138, 139, 168, 286

Harmonic approximation, 68, 69, 176, 177, 181, 187, 188, 205–207, 209, 227
Harmonic vibrations, 69, 74, 157, 165, 173, 213
Heat of evaporation, 288
Heat capacity, 66–68, 72–74, 82, 85, 257, 258, 274, 324

Heat conduction, 106, 115, 173, 174, 181, 245, 246
Heat expansion, 183, 185–187, 207, 210, 215, 216, 250, 263, 264
Hill theory, 320

Ideal crystals, 150, 151
Ideal gas equation of state, 1, 5, 9, 32, 51, 269
Ideal molecular gas, 11, 42, 52, 55, 59, 64, 67, 92, 175, 260, 261, 264–266, 268, 269, 272, 276, 280, 281, 285, 295, 299–301, 326
Identity paticles principle, 56, 330
Induced radiation, 142, 145, 147
Interface, 39, 233, 234, 309–311, 316, 319
Internal energy, 51, 54, 261–264, 266, 267, 272, 282–284, 286, 294, 295, 300, 310, 311, 329
Internal pressure, 38, 174, 175, 177, 179, 182, 183, 203–206, 215, 216, 221, 223–225, 227, 230, 233, 234
Inverse lattice, 161
Inversion curve, 296–299, 301, 302
Inversion temperature, 299, 301
Ionic crystals, 151, 152, 169, 192, 193
Isothermal compression, 266, 299
Isothermal expansion, 266

Joule-Thomson coefficient, 296
Joule-Thomson effect, 293, 294, 296, 297

Kirchhoff's theorem, 126, 129
Knot point, 66, 188, 205, 226
Kolmogorov-Feller equation, 101, 103

Lambert's laws, 124, 126, 259
Langevin equation, 253–255
Langevin function, 31, 32
Laplace formula, 322
Latent heat of melting, 221, 258, 259
Latent heat of surface formation, 310
Latent vaporization heat, 314
Lattice heat capacity, 160, 168–171, 173
Lattice period, 152, 154, 159, 161, 179, 193, 196, 207, 209
Lattice specific heat, 174
Libration, 165, 200
Local equilibrium, 105, 110, 123

Magnetic moment, 85, 86, 88
Magnetic polarization, 90
Magnetic susceptibility, 88
Maxwell distribution, 12, 13, 17, 18, 22, 95, 134
Maxwell relations, 283, 295, 304, 307, 312, 316

Maxwell shelf, 39, 316
Maxwell's law, 109
Maxwell-Boltzmann distribution, 25, 27
Mayer equation, 260, 267, 300
Mean collision rate, 94
Melting temperature, 185, 193, 219, 221, 225, 245, 258
Metals, 151, 152, 169, 185, 190, 193
Microcanonical ensemble, 49
Mie-Grüneizen equation, 182, 183, 227
Mobility, 119-122, 246, 248, 253, 256
Molecular beam, 96, 97, 305
Molecular crystals, 152, 154, 164, 165, 169
Molecular dynamics, 219
Molecular jet, 17
Molecular vibrations, 164
Monomolecular interface, 234-236, 241, 319
Monte Carlo simulation, 219
Motional pressure, 64, 174, 175, 179-183, 186, 206, 208, 209, 213-219, 221, 228, 233-235
Multiplication theorem, 7, 50

Natural variables, 282, 284, 295, 312, 314
Nearest-neighbor approximation, 176, 204
Nernst principle, 323-325
Normal pressure, 233, 236, 238, 241, 319, 322

Ohm's law, 119
Optical vibrations, 162, 173
Orthometric curve, 45, 46, 221-225

Paramagnetic polarization, 86-89
Peltier coefficient, 290
Peltier effect, 289
Peltier heat, 289-291
Percus-Yevick equation, 211
Phase space, 26, 49, 73, 75, 78, 89, 91, 102, 135
Phase transition, 38, 39, 201, 210, 211, 219, 258, 261, 286, 287
Phonon gas, 1, 11, 165-168, 172, 174, 180, 181
Photon gas, 11, 124, 133-137, 139-142, 166, 168, 171, 305
Photon gas density, 135
Photon gas pressure, 140
Photon model, 133, 135, 142
Planck constant, 74, 78, 132
Planck law, 132, 136, 145
Point masses, 1, 2, 11, 26, 34, 91, 93, 103
Poissonian statistics, 99, 100, 244
Pseudoideal state, 45, 46

Quantization, 74, 76, 78, 79, 81, 86, 89, 132, 133, 135, 160, 200

Quasi-harmonic approximation, 179, 205
Quasi-thermodynamics, 234, 238, 319-322

Radial distribution function, 321
Radiation equation of state, 141
Radiation flux, 132, 134, 139, 142, 305, 306
Radiation heat capacity, 130
Radiation specific heat, 324
Rarefied gas, 42, 58
Rate of collisions, 92-95, 100
Rayleigh law, 131
Real crystals, 167, 168, 173, 189
Real gas, 11, 32, 35, 38, 39, 41, 45, 51, 57, 64, 175, 179, 181, 204, 222, 227, 230, 284-286, 295, 296, 299, 300, 302
Rectangular approximation, 208-210, 213, 217
Relaxation rate, 103, 254
Relaxation time, 91, 92, 213, 254
Repulsion, 33, 34, 37, 38, 42, 44, 46, 47, 61, 113, 151, 202, 204, 209, 230, 236
Root-mean-square velocity, 9, 122, 172, 211

Sackur-Tetrode formula, 56
Saturated vapor pressure, 39, 218, 239, 286-288, 315, 316, 318
Schottky defects, 192, 193
Second law of thermodynamics, 128, 268, 270, 274, 276, 277, 279-282, 286, 287, 291, 302
Seebeck effect, 288
Simple cubic lattice, 153, 177, 204
Simple liquids, 203, 204, 226, 230, 237
6-12 Lennard-Jones potential, 61, 66, 154, 203, 204, 208
Smooth interface, 237, 319, 320
Sound velocity, 160
Spatial anisotropy, 153, 167, 170, 177, 198, 233
Specific heat, 67, 68, 71, 72, 82, 84, 89, 258-260, 265, 267, 274, 283, 306, 324
Spectral density of radiation, 127, 130, 145, 146, 148
Speed distribution, 18
Spinodal, 40, 41, 215
Spontaneous radiation, 143, 145, 147
Standing waves, 160, 161
Statistical independence, 8
Statistical sum, 54, 219
Statistical weight, 14, 15, 26, 27, 30, 53, 167
Stefan-Boltzmann law, 132
Stirring hypothesis, 91, 92, 105, 116
Summation theorem, 13, 14

Supercritical region, 201, 243
Surface tension, 199, 232–235, 237, 239, 309, 310, 320, 322

Tangential pressure, 233–235, 240, 241, 319, 320, 322
Thermal conductivity, 114, 115, 118, 259
Thermal pressure, 38, 64
Thermodiffusion, 113
Thermodynamic equation of state, 283, 284
Thermodynamic potential, 312–315
Thermoelectric cycle, 288, 290, 291
Thomson coefficient, 291
Thomson effect, 290–293
Thomson principle, 277, 279
Translational energy, 1, 80
Traveling waves, 161
Triple point, 201, 230, 234, 286, 288, 318

Ultrararefied gas, 92, 99, 115–117
Ultraviolet catastrophe, 74, 130, 132

Vacancy, 173, 190–193, 195, 196, 244, 246–248
Vacancy diffusion coefficient, 195, 248
Valence crystals, 151, 152, 154
van der Waals equation, 38, 39, 41, 42, 45, 61, 202, 205, 223, 229, 285, 300, 316
van der Waals interface theory, 239, 322
van der Waals loop, 38, 39, 215, 219, 241, 316
van der Waals theory, 39, 44, 45, 61, 222, 299, 302
Velocity correlation time, 242
Velocity of sound, 165, 166, 169, 174
Velocity relaxation, 100, 103
Virial coefficients, 44, 60, 61, 201, 229, 300, 302
Virial expansion, 42, 61, 62, 66, 201, 205, 222, 300
Viscosity, 106, 108, 109, 199, 245, 246, 248, 250, 256

Young modulus, 178